VLSI AND COMPUTER ARCHITECTURE

VLSI AND COMPUTER ARCHITECTURE

KENZO WATANABE
EDITOR

Nova Science Publishers, Inc.
New York

For permission to use material from this book please contact us:
Telephone 631-231-7269; Fax 631-231-8175
Web Site: http://www.novapublishers.com

LIBRARY OF CONGRESS CATALOGING-IN-PUBLICATION DATA

VLSI and computer architecture / editor, Kenzo Watanabe.
p. cm.
Includes index.
ISBN 978-1-60692-075-6 (hardcover)
1. Integrated circuits--Very large scale integration--Design and construction. 2. Computer architecture. 3. Wireless communication systems--Equipment and supplies--Design and construction. 4. Microcontrollers--Design and construction. I. Watanabe, Kenzo.
TK7874.75.V56 2009
621.39'5--dc22

2009000973

Published by Nova Science Publishers, Inc. ✦ New York

CONTENTS

PREFACE

Very-large-scale integration (VLSI) is the process of creating integrated circuits by combining thousands of transistor-based circuits into a single chip. The first semiconductor chips held one transistor each. Subsequent advances added more and more transistors, and as a consequence more individual functions or systems were integrated over time. The first integrated circuits held only a few devices, perhaps as many as ten diodes, transistors, resistors and capacitors, making it possible to fabricate one or more logic gates on a single device. Now known retrospectively as "small-scale integration" (SSI), improvements in technique led to devices with hundreds of logic gates, known as large-scale integration (LSI), i.e. systems with at least a thousand logic gates. Current technology has moved far past this mark and today's microprocessors have many millions of gates and hundreds of millions of individual transistors.

As of early 2008, billion-transistor processors are commercially available, an example of which is Intel's Montecito Itanium chip. This is expected to become more commonplace as semiconductor fabrication moves from the current generation of 65 nm processes to the next 45 nm generations. Another notable example is Nvidia's 280 series GPU. This microprocessor is unique in the fact that its 1.4 Billion transistor count, capable of a teraflop of performance, is almost entirely dedicated to logic (Itanium's transistor count is largely due to the 24MB L3 cache).

At one time, there was an effort to name and calibrate various levels of large-scale integration above VLSI. Terms like Ultra-large-scale Integration (ULSI) were used. But the huge number of gates and transistors available on common devices has rendered such fine distinctions moot. Terms suggesting greater than VLSI levels of integration are no longer in widespread use. Even VLSI is now somewhat quaint, given the common assumption that all microprocessors are VLSI or better.

This new book focuses on VLSI and computer architectures.

Chapter 1 - WiMAX is an ideal solution for providing high data rate communications where traditional landlines are either unavailable or too costly to be installed. It is set to become the most popular way to meet escalating business demand for rapid Internet connection and integrated data, voice, and video services. In this book chapter, the authors consider a number of alternative techniques to achieve high data rate and high quality of service requirements in these systems, including orthogonal frequency division multiplexing (OFDM), single-carrier frequency-domain equalization (SC-FDE), turbo equalization as well as multiple-input multiple-output (MIMO) techniques. In particular, time-domain turbo equalization algorithms suitable for combating intersymbol interference (ISI) in broadband WiMAX type of systems with long symbol dispersion are proposed and compared with the frequency-domain OFDM and SC-FDE schemes. The comparative results show that both frequency and time-domain solutions enable the systems to operate reliably ata high data rate, and the proposed turbo equalizers outperform the frequency-domain solutions with only a reasonable increase in the complexity, which is mainly incurred by the iterative decoding process. Various design considerations and system configuration issues, such as the choice of channel coding schemes, code rate, impact of different decoding algorithms, the data block length and channel delay spread on the receiver complexity are also discussed, and the results presented in this chapter are of practical interests for the development and design of high performance broadband Fixed WiMAX systems.

Chapter 2 - The feature size of integrated circuits has been aggressively reduced in the pursuit of improved speed, power, silicon area and cost characteristics. Semiconductor technologies with feature sizes of several tens of nanometers are currently in development. As per, International Technology Roadmap for Semiconductors (ITRS), the future nanometer scale circuits will contain more than a billion transistors and operate at clock speeds well over 10GHz. Distributing robust and reliable power and ground lines; clock; data and address; and other control signals through interconnects in such a high-speed, high-complexity environment, is a challenging task.

The performance of a high-speed chip is highly dependent on the interconnects, which connect different macro cells within a VLSI/ULSI chip. With ever-growing length of interconnects and clock frequency on a chip, the effects of interconnects cannot be restricted to *RC* models. The importance of on-chip inductance is continuously increasing with faster on-chip rise times, wider wires, and the introduction of new materials for low resistance interconnects. It has become well accepted that interconnect delay dominates gate delay in current deep sub micrometer VLSI circuits. With the continuous scaling of technology and increased die area, this behavior is expected to continue.

To avoid prohibitively large latencies, designers scale down global wire dimensions more slowly than the transistors dimensions and this causes a rapid growth in gap between transistors and interconnects densities on a chip. Thereby, as technology advances to Giga-Scale Integration (GSI), global interconnect resource becomes more and more valuable and it is essential to use global interconnects optimally.

Propagation delay in global interconnects have become a core research problem. Therefore, a lot of work is being carried out to address these problems. Various models have been suggested in literature to analyze interconnects. The present chapter reviews in detail the works carried out in various research aspects and associated problems of VLSI interconnects. A case study is undertaken to understand the methodology to analytically calculate the output waveform and propagation delay. For this case study the effect of short circuit current on the propagation delay is also considered. Furthermore, this chapter also discusses various state-of-the-art delay minimization techniques.

Chapter 3 - Several research institutions have worked together in the development of a 55 Kg low Earth orbit (LEO) microsatellite, aiming the development of university technology toward the space field. As it is known, LEO operation implies a dephased orbital dynamics among satellites and our planet, achieving time limited communications either for downloading high bandwidth scientific telemetry or for uploading command and/or missions for space vehicles. By this reason, among others, long-life space computer architectures constitute an important research field oriented to preserve satellite operations, satellite autonomy and communications among space vehicles and their control Earth stations. Under this scenario, few years of research efforts were dedicated towards the development of a reconfigurable space qualified long-life computer server (SQLLCS) that integrates cold spare redundancies in single points of failure from the architecture to improve hardware reliability. The computer architecture aims for the extension of satellite life even in the presence of important SQLLCS failures. This chapter describes the SQLLCS hardware architecture and underlines the features integrated in the hardware that allow the computer to withstand the harsh space environment. Moreover, it explains some software operations for both the space and ground segments. In addition, the chapter outlines the hardware and software tools specially created for SQLLCS validation purposes, as well as results of a reliability study to predict the server behaviour based on the exponential failure law, the military standard MIL-HDBK217f notice 2 and MatLab software.

Chapter 4 - This chapter is a review of the research of the authors from the last decade and focuses on the mathematical analysis of the Schrödinger model for nano-scale semiconductor devices. They discuss *transparent boundary conditions* (TBCs) for the time--dependent Schrödinger equation on a two dimensional domain.

First the authors derive the two dimensional discrete TBCs in conjunction with a conservative Crank--Nicolson--type finite difference scheme and a compact nine--point scheme. For this difference, equations the authors derive *discrete transparent boundary conditions* (DTBCs) in order to get highly accurate solutions for open boundary problems. The presented discrete boundary--valued problem is unconditionally stable and completely reflection--free at the boundary.

Then, since the DTBCs for the Schrödinger equation include a convolution w.r.t. time with a weakly decaying kernel, the authors construct *approximate* DTBCs with a kernel having the form of a finite sum of exponentials, which can be efficiently evaluated by recursion.

In several numerical tests the authors illustrate the perfect absorption of outgoing waves independent of their impact angle at the boundary, the stability, and efficiency of the proposed method. Finally, they apply inhomogeneous DTBCs to the transient simulation of quantum waveguides with a prescribed electron inflow.

Chapter 5 - Designing multi-modal perceptual systems is a non-trivial endeavor that requires interdisciplinary effort, solid architectural infrastructure, and efficient development tools. The authors introduce an architecture for building intelligent services in smart rooms i.e., environments equipped with a collection of sensors ranging from low-level fire detectors to complex perceptual components processing signals from an array of cameras and microphones. The architecture presents a unified approach towards designing, implementing and testing multi-modal perceptual systems. The authors describe the reference architecture they co-authored for such applications, and present a set of tools they have developed to facilitate the collaborative building of such context-aware applications and services. The authors also show the benefits of the tools for application authoring in the context of the presented architecture. Specifically, they introduce a modeling tool that captures the use cases in a form of virtual animated 3D scenarios, facilitates the controlled triggering of sensor events at proper time intervals as reactions to scenario activities, and provides means to aggregate such events into higher-level information i.e., the situation model, utilized by the contextual services. Their tool supports direct replacement of virtualized sensors by real sensors and perceptual components once they become available in the development cycle, without the necessity to change the API contracts. The authors have used the architectural principles and the tooling successfully in two large international projects dealing with multi-modal perception. In addition, the reference architecture has passed a process of formal and functional evaluation.

Chapter 6 - This paper describes a mixed-signal VLSI design utilizing the controlled conductance properties of MOSFETs, for tunable filters, low power amplifiers, and analog-mixed signal processing. The programmable linear conductance is realized by the pair of MOSFETs in the triode region, and it produces the voltage-controlled tunable function or the analog computing element. It demonstrates the application areas of neural synaptic functions, electronic neurons with biologically plausible characteristics, and tunable analogue complex filters for RF communication. The programmable conductance by MOSFETs is analyzed by the experimentation of prepared MOSFETs and the SPICE simulation.

The amplifier based on the proposed MOSFET conductance is employed to enforce the tunable filters for RF applications. It has the tuning range of +/- 50% and the bandwidth over 1 GHz is exhibited. For applications to analog computing, the design of an analog multiplier and biologically plausible neuromorphic devices are illustrated by utilizing MOSFET's programmable conductance.

Chapter 7 - This chapter gives a hardware-efficient algorithm and a VLSI architecture for finding a path for a mobile robot on an environment image. The algorithm constructs a distance map to identify the collision-free region for a given robot and then finds a path on the region. The path obtained from a start to a goal is the shortest path in terms of the number of steps. The time-critical part of the algorithm is mapped on to a two-dimensional cellular array VLSI architecture that consists of a locally interconnected array of identical processing elements. Due to this local interconnection and regular structure, the architecture can be operated at a high speed and is easily scalable. The design has been implemented on the XCV8000 device of Xilinx. The maximum frequency of operation obtained is 246 MHz. This leads to computing a collision-free path on images of size 100 x100 in less than 41 μs. The hardware is capable of processing images at a video rate for real-time navigation in a dynamic environment.

Chapter 8 - Mobile robots operating autonomously are valuable in a number of situations. These could involve environments/tasks that are either difficult to reach by humans or those that are dangerous for humans. The latter may include detection of land mines, maintenance of hazardous material etc. Considerable research has been done on sensing, planning and control of robots with different models and systems for each of them. Sensors ranging from sonar to highly sophisticated ones (such as laser range finders) have been studied. A variety of control strategies have also been investigated.

Recently, there has been an increasing emphasis on saving energy during various tasks performed by a mobile robot. A significant source of energy consumption is the embedded processing aboard the robot. Traditional mobile robots typically include a laptop as the embedded computer besides sensors, motors and microcontroller(s) for low-level controls. One approach to reducing energy consumption is by examining architectural alternatives for the efficient implementation of algorithms for mobile robots. Algorithm-centric VLSI architectures for mobile robots with a view to run-time configurability, system integration and design productivity in terms of area, time and energy-efficiency seem appropriate. This review paper examines the state of the art in VLSI architectures and algorithms for embedded computing on mobile robots and suggests directions for further research.

Chapter 9 - This chapter presents a very unique work at the MAC layer that spans from analytical and simulative investigations, to the architectural design and implementation of an enhanced MAC protocol and the consequent experiments in real scenarios. This work was spiked by some seminal papers, which demonstrated the shortcomings of using the 802.11 technology for distributed mobile environments like multi-hop ad hoc networks. Thus, the authors decided to redesign the MAC architecture and to realize a prototype of a new MAC protocol integrating features and mechanisms more suitable for ad hoc communications. This work was very successful as, on one side, the authors designed a very flexible architecture that overcomes some of the drawbacks of traditional 802.11 solutions and, on the other side, they concretely implemented and tested an enhanced 802.11 backoff algorithm, showing the alignment of the analytical-simulative work with the implementation-experimental one.

In: VLSI and Computer Architecture
Editor: Kenzo Watanable

ISBN 978-1-60692-075-6

Chapter 1

DESIGN CONSIDERATIONS AND ALGORITHMS FOR BROADBAND FIXED WIMAX SYSTEMS

Pei Xiao[1]*, ***Ioannis Chatzigeorgiou***[2]†, ***Miguel R.D. Rodrigues***[3]‡, ***Rolando Carrasco***[4]§ ***and Ian J. Wassell***[2]¶

[1]The Institute of Electronics, Communications and Information Technology (ECIT)
Queen's University Belfast, Queen's Road, BT3 9DT, United Kingdom
[2]Computer Laboratory, University of Cambridge
15 J.J. Thomson Avenue, Cambridge, CB3 0FD, United Kingdom
[3]Instituto de Telecommunicações
Department of Computer Science, University of Porto
Rua Campo Alegre 1021/1055, Porto 4169-007, Portugal
[4]School of Electrical, Electronic and Computer Engineering
University of Newcastle, Merz Court, NE1 7RU, United Kingdom

Abstract

WiMAX is an ideal solution for providing high data rate communications where traditional landlines are either unavailable or too costly to be installed. It is set to become the most popular way to meet escalating business demand for rapid Internet connection and integrated data, voice, and video services. In this book chapter, we consider a number of alternative techniques to achieve high data rate and high quality of service requirements in these systems, including orthogonal frequency division multiplexing (OFDM), single-carrier frequency-domain equalization (SC-FDE), turbo equalization as well as multiple-input multiple-output (MIMO) techniques. In particular, time-domain turbo equalization algorithms suitable for combating intersymbol interference (ISI) in broadband WiMAX type of systems with long symbol dispersion are proposed and compared with the frequency-domain OFDM and SC-FDE schemes. The comparative results show that both frequency and time-domain solutions enable

*E-mail address: pei.xiao@ecit.qub.ac.uk. Tel: +44 28 9097 1762; Fax: +44 28 9097 1702
†E-mail address: ic231@cam.ac.uk. Tel: +44 1223 767 031; Fax: +44 1223 767 009
‡E-mail address: mrodrigues@dcc.fc.up.pt. Tel: +351 220 402 941; Fax: +351 220 402 950
§E-mail address: r.carrasco@ncl.ac.uk. Tel: +44 191 222 7332; Fax: +44 191 222 8180
¶E-mail address: ijw24@cam.ac.uk. Tel: +44 1223 767 031; Fax: +44 1223 767 009

the systems to operate reliably at a high data rate, and the proposed turbo equalizers outperform the frequency-domain solutions with only a reasonable increase in the complexity, which is mainly incurred by the iterative decoding process. Various design considerations and system configuration issues, such as the choice of channel coding schemes, code rate, impact of different decoding algorithms, the data block length and channel delay spread on the receiver complexity are also discussed, and the results presented in this chapter are of practical interests for the development and design of high performance broadband Fixed WiMAX systems.

Keywords: WiMAX, IEEE 802.16, OFDM, SC-FDE, MIMO, turbo equalization.

1. Introduction

WiMAX (the Worldwide Interoperability for Microwave Access) is a standards-based technology enabling the delivery of last mile wireless broadband access as an alternative to cable and digital subscriber line (DSL). This wireless alternative enables operators in a competitive environment to roll-out broadband wireless access (BWA) services in a rapid and cost effective manner [1]. The standardization activities have been performed under the auspices of the IEEE 802.16 working group, divided between 802.16d, i.e. Fixed WiMAX, and 802.16e, i.e. Mobile WiMAX [2]. The WiMAX Forum was formed in June 2001 to promote conformance and interoperability of the IEEE 802.16 standard. In this work, we focus on the former case, and discuss design considerations and algorithms suitable for broadband Fixed WiMAX applications. Both orthogonal frequency division multiplexing (OFDM) and single-carrier solutions have been adopted in IEEE 802.16 standard as two alternatives for Fixed WiMAX systems operating at 2-11 GHz bands [3].

One of the limiting factors in outdoor wireless transmission is the multipath channel between the transmitter and the receiver giving rise to intersymbol interference (ISI), which degrades the system performance and limits the maximum achievable data rate. The conventional approach to combating ISI is time-domain equalization [4]; Douillard *et al.* proposed turbo equalization [5], which is based on the idea of turbo decoding [6]. Turbo equalization performs equalization and channel decoding jointly in an iterative manner, which significantly improves the performance over separate equalization and decoding. In its original form, turbo equalization employs a maximum a posteriori probability (MAP) based approach for both equalization and decoding [5]. For channels with large delay spreads and for large constellation sizes, it suffers from prohibitive computational complexity due to an increasing number of trellis states. To tackle this problem, the filter-based approach has been proposed, e.g., see [7, 8, 9, 10]. The equalizer is typically implemented by a linear transversal finite impulse response (FIR) filter, the coefficients of which are adjusted to minimize the mean-square error. It was shown that the performance of this approach is similar to the MAP-based receiver, while providing a significant reduction in the computational complexity.

A Fixed WiMAX system should be designed to provide broadband wireless access with wire-line quality. The high requirement for quality arises because it has to compete with cable modems and DSL approaches which operate over static and non-fading channels and hence are able to provide very good quality. In order to be competitive, a Fixed WiMAX system must offer similar data rates to their wire-line counterparts. A fundamental chal-

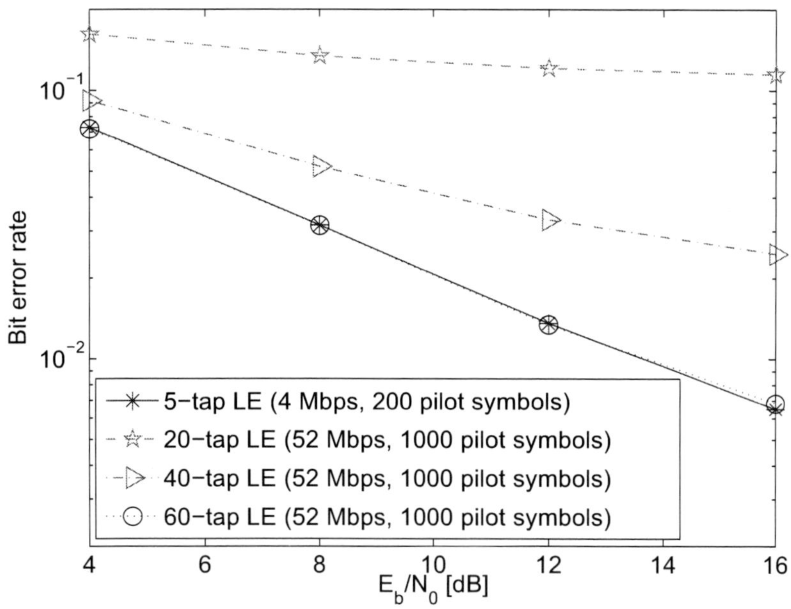

Figure 1. Performance of LE for quadrature phase shift keying (QPSK) modulated uncoded Fixed WiMAX type system with different data rate.

lenge in transmitting high data rates over radio channels is to overcome the channel dispersion which grows linearly with the data rate. The main obstacle with the existing time-domain equalization or the turbo equalization approach mentioned previously is that its complexity grows linearly (sometimes quadratically) with the channel dispersion, and it is therefore infeasible for high-speed wireless data links. To elaborate on this point, we show a simple example in Fig. 1, which compares the performance of linear equalizer (LE) for a 4×10^6 bits/sec (4 Mbps) and a 52 Mbps Fixed WiMAX type system without channel coding. The channel delay spread is 1.0μs, which corresponds to 3 symbols of dispersion in the former case and 26 symbols of dispersion in the latter case. The equalizer coefficients are designed under the minimum mean square error (MMSE) criterion, which requires training with pilot symbols. As indicated by the figure, the length of LE has to be increased to 60 taps for the high data rate system in order to achieve the same performance as the low speed system with a 5-tap LE. Apparently, the complexity of the equalizer increases linearly with data rate. In addition to higher complexity, the high data rate system also requires a longer training sequence (the number of pilot symbols is increased from 200 to 1000 in this example), which imposes overhead and decreases system spectrum efficiency. If the channel dispersion is further increased to, e.g., 60 symbols, time-domain equalizers with over 100 taps would be needed for sufficient ISI reduction. Similar behaviors are also observed for decision feedback equalizers [4]. Equalizers with such complexity are simply unaffordable with today's technology. It is therefore generally believed that only frequency-domain solutions, such as OFDM [11], and single-carrier frequency-domain equalization (SC-FDE) [12, 13] are applicable for high data rate broadband wireless transmission. However, in this book chapter, we show that time-domain equalization can also cope with ISI channels with long symbol dispersion if properly designed, and we introduce a low complexity turbo equalization scheme for reliable transmission in high data rate Fixed WiMAX type systems. Unlike the traditional equalizers or turbo equalizers, the complexity of the proposed scheme is comparable to that of OFDM and SC-FDE, the added complexity mostly comes from multi-stage channel decoding due to the nature of the iterative process. The remainder of the chapter is organized as follows: the frequency-domain OFDM and SC-FDE solutions are described in Section **??** The time-domain turbo equalization schemes for broadband Fixed WiMAX type systems are proposed in Section 2., and the performance of different schemes are quantitatively compared in Section 2.3. Finally, conclusions are drawn in Section 3. based on the comparative results.

Throughout this chapter, $(\cdot)^{\mathcal{H}}$ denotes matrix/vector conjugate transpose, $(\cdot)^{\top}$ denotes matrix/vector transpose, $(\cdot)^{*}$ denotes matrix/vector and complex scalar conjugate, and $\mathrm{E}[\cdot]$ denotes the expectation operation. Lower case bold letters, e.g., $\mathbf{x}$, denote vectors; upper case bold letters, e.g., $\mathbf{X}$, denote matrices. The operators $|\cdot|$, $\angle\cdot$, $\mathrm{Re}\{\cdot\}$ and $\mathrm{Im}\{\cdot\}$ denote the magnitude, angle, real part and imaginary part of a complex number, respectively.

1.1. Introduction

In this section, we consider two frequency-domain techniques which are potential candidates in Fixed WiMAX systems. More specifically, in Section 1.2., we study the principles of OFDM, while in Section 1.3., we present an alternative technique, known as SC-FDE. We consider systems both with and without antenna diversity; in the former case, diversity is exploited by means of space-time block codes (STBC). Finally, in Section 1.4., we compare the performance of OFDM-based systems to that of systems using SC-FDE, and discuss the benefits of each technique in Fixed WiMAX environments.

For STBC, we mainly consider the use of the Alamouti's space-time block code [14], which has been proposed in several wireless standards due to its many attractive features such as achieving full spatial diversity at full transmission rate without requiring channel state information at the transmitter, facilitating maximum likelihood decoding of the STBC with simple linear processing.

1.2. Orthogonal Frequency Division Multiplexing (OFDM)

The first frequency domain solution we consider is space-time coded and OFDM based Fixed WiMAX type systems [16] as depicted in Fig. 2, where N_T and N_R represent the number of transmit and receive antennas, respectively. We consider both single-input single-output (SISO) configurations, where $N_T = N_R = 1$, as well as multiple-input multiple-output (MIMO) arrangements, where $N_T, N_R \geq 1$. At the transmitter, the information sequence $\mathbf{b}$ is encoded generating a sequence $\mathbf{u}$ of code bits, which is then interleaved to produce a sequence $\mathbf{v}$. The mapper converts groups of 2^M code bits into one of M complex symbols from a unit power signal constellation. The modulated symbols are then grouped into blocks of size N each, where N is the number of sub-carriers used in OFDM.

In single transmit antenna systems ($N_T = 1$), the space-time processing unit does not further process the modulated symbols; instead, each block of N symbols is directly sent

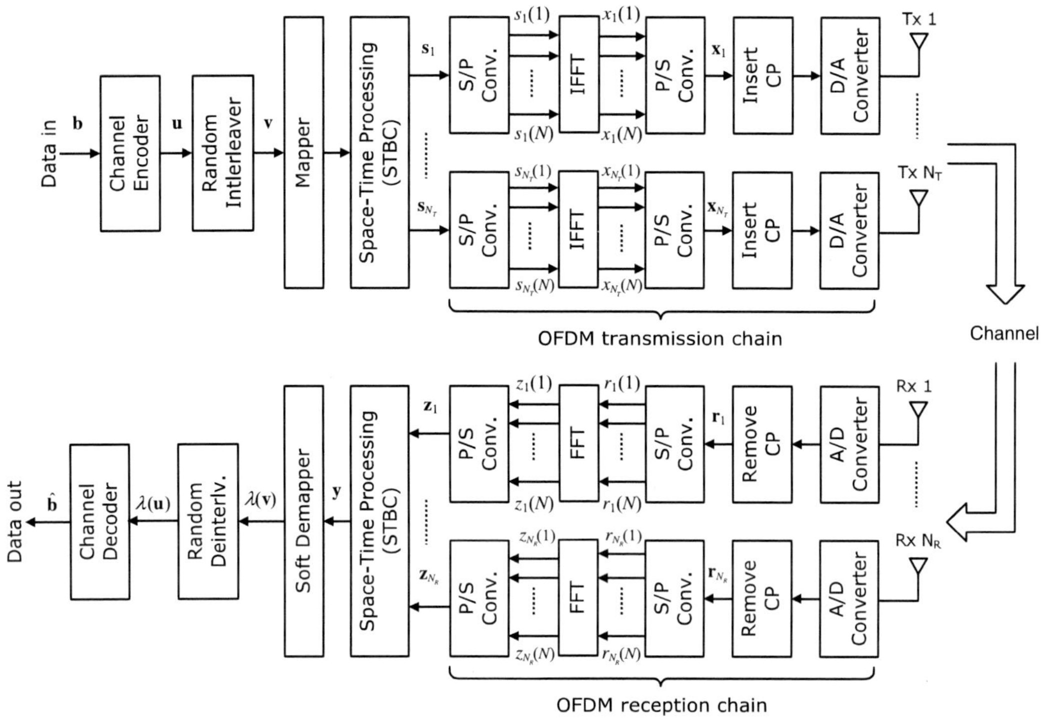

Figure 2. Block diagram of an OFDM based Fixed WiMAX type system.

to an OFDM chain. However, in multiple transmit antenna systems $(N_T > 1)$, a space-time block code is implemented according to a generator matrix $\mathbf{G}$ given in [14, 15]. At each signaling interval, the space-time processing unit generates N_T sequences of length N symbols each, denoted as $\mathbf{s}_i = \{s_i(1), s_i(2), \ldots, s_i(N)\}$ for $i = 1, \ldots, N_T$.

For example, let us consider the system with $N_T = 2$, whose transmission block diagram is shown in Fig. 3. Unlike the original Alamouti transmission scheme [14], which transmits two symbols from two antennas at a time, the OFDM-STBC encoder groups the symbols into two symbol blocks $\mathbf{s}_1$ and $\mathbf{s}_2$ at each antenna, each containing N symbols. The generator matrix $\mathbf{G}$ has the form

$$\mathbf{G} = \begin{bmatrix} \mathbf{s}_1 & \mathbf{s}_2 \\ -\mathbf{s}_2^* & \mathbf{s}_1^* \end{bmatrix} \quad (1)$$

where $\mathbf{s}_i^*$ denotes the conjugate version of $\mathbf{s}_i$. Consequently, sequences $\mathbf{s}_1$ and $\mathbf{s}_2$ are processed by separate OFDM chains in the first signaling interval, followed by sequences $-\mathbf{s}_2^*$ and $\mathbf{s}_1^*$ in the second signaling interval. The transmitted sequence at each signaling interval undergoes a serial-to-parallel (S/P) conversion and the output block of N parallel symbols is modulated using an inverse fast Fourier transform (IFFT). Note that both the FFT and the IFFT operations can be expressed as $N \times N$ matrices denoted as $\mathbf{Q}$ and $\mathbf{Q}^{-1}$, respectively.

bn[][]b_n rn[][]$\mathbf{r}$ vn[][]$\mathbf{w}$ un[][]u_n dn[][]s_n un$_p$[][]v_n Pi[][]$\prod$ ISI[][]MIMO TR-STBC[][]STBC Encoder[][]Encoder Channel[][]Channel QPSK[][]M-ary SP[][]SP IFFT[][]IFFT Modu.[][]Modu. Conv.[][]Conv. sn0 -sn1*[][]$\mathbf{s}_1$ $-\bar{\mathbf{s}}_2$ sn1 sn0*[][]$\mathbf{s}_2$ $\bar{\mathbf{s}}_1$

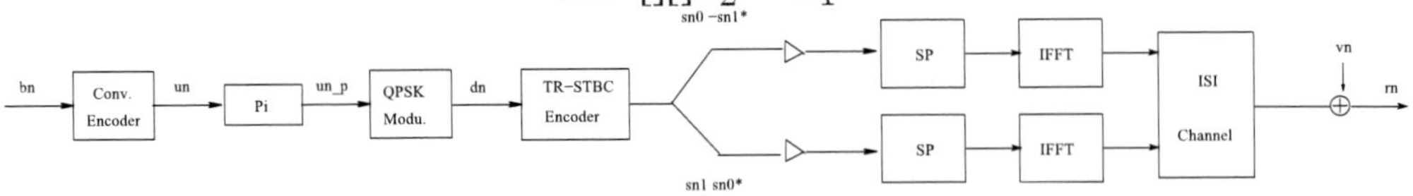

Figure 3. Diagram for the STBC coded OFDM system with 2 transmit antennas.

The (k,l)-th entry of the FFT matrix $\mathbf{Q}$ is

$$\mathbf{Q}_{k,l} = \frac{1}{\sqrt{N}} e^{-j(2\pi/N)kl}; \quad k,l = 0,\ldots,N-1.$$

Note that $\mathbf{Q}$ is an unitary matrix, therefore $\mathbf{Q}^{-1} = \mathbf{Q}^{\mathcal{H}}$, and $\mathbf{Q}\mathbf{Q}^{\mathcal{H}} = \mathbf{Q}^{\mathcal{H}}\mathbf{Q} = \mathbf{I}$, where $\mathbf{I}$ denotes the identity matrix.

The output of the IFFT is a block of symbols $\mathbf{x}_i = \{x_i(1), x_i(2), \ldots, x_i(N)\}$ given by $\mathbf{x}_i = \mathbf{Q}^{\mathcal{H}}\mathbf{s}_i$. A parallel-to-serial (P/S) converter multiplexes the N parallel symbols and a cyclic prefix (CP) is inserted with duration longer than the impulse response of the channel to combat intersymbol and intercarrier interference. Finally, the OFDM signal is digital-to-analogue (D/A) converted and transmitted over the channel.

At the receiver, N_R streams are processed by an equal number of OFDM reception chains. In particular, the cyclic prefix of each OFDM signal is removed at each signaling interval. The received signal $\mathbf{r}_j$ of length N symbols at the j-th OFDM chain can be expressed as a linear combination of the transmitted signals $\mathbf{x}_1, \ldots, \mathbf{x}_{N_T}$ as follows

$$\mathbf{r}_j = \begin{bmatrix} \mathbf{H}_{j,1} & \ldots & \mathbf{H}_{j,N_T} \end{bmatrix} \begin{bmatrix} \mathbf{x}_1 \\ \vdots \\ \mathbf{x}_{N_T} \end{bmatrix} + \mathbf{w}_j,$$

or, equivalently,

$$\mathbf{r}_j = \mathbf{H}_j \mathbf{x} + \mathbf{w}_j, \tag{2}$$

where $\mathbf{H}_j = [\mathbf{H}_{j,1} \ \ldots \ \mathbf{H}_{j,N_T}]$ and $\mathbf{x} = [\mathbf{x}_1 \ \ldots \ \mathbf{x}_{N_T}]^{\top}$. Here, $\mathbf{H}_{j,i}$ is a $N \times N$ circulant matrix that contains the channel impulse response from transmit antenna i to receive antenna j, whilst $\mathbf{w}_j$ is a column vector of length N whose elements are uncorrelated circularly symmetric complex Gaussian random variables with mean zero and variance N_0; for brevity, we use the notation $w_j \sim \mathcal{CN}(0, N_0)$, where w_j is an element of $\mathbf{w}_j$.

Circulant matrices, such as $\mathbf{H}_{j,i}$, hold the important property of being diagonalizable by the Fourier transformation matrix under the condition that the channel is time-invariant over an OFDM symbol, i.e.,

$$\mathbf{H}_{j,i} = \mathbf{Q}^{\mathcal{H}} \mathbf{\Xi}_{j,i} \mathbf{Q}, \tag{3}$$

where $\Xi_{j,i}$ is a $N \times N$ diagonal matrix whose elements $\xi_n^{(j,i)}$, $n=1,\ldots,N$, are the eigenvalues of $\mathbf{H}_{j,i}$. Expression (3) can be easily extended to

$$\mathbf{H}_j = \mathbf{Q}^{\mathcal{H}}\Xi_j\mathbf{Q}, \tag{4}$$

where $\Xi_j = [\Xi_{j,1} \ \ldots \ \Xi_{j,N_T}]$. Performing FFT operation on $\mathbf{r}_j$, we obtain

$$\begin{aligned} \mathbf{z}_j &= \mathbf{Q}\mathbf{r}_j = \mathbf{Q}\left[\mathbf{H}_j\mathbf{x} + \mathbf{w}_j\right] \\ &= \mathbf{Q}\left[\left(\mathbf{Q}^{\mathcal{H}}\Xi_j\mathbf{Q}\right)\mathbf{Q}^{\mathcal{H}}\mathbf{s} + \mathbf{w}_j\right] \\ &= \Xi_j\mathbf{s} + \mathbf{v}_j, \end{aligned} \tag{5}$$

where $\mathbf{s} = \left[\mathbf{s}_1 \ \ldots \ \mathbf{s}_{N_T}\right]^{\top}$. The above equation holds since $\mathbf{Q}$ is an unitary matrix and $\mathbf{x} = \mathbf{Q}^{\mathcal{H}}\mathbf{s}$, $\mathbf{v}_j = \mathbf{Q}\mathbf{w}_j$.

At each signaling interval, the outputs of the N_R OFDM chains are multiplexed to form a sequence of vectors $\mathbf{z} = \{\mathbf{z}_1, \ldots, \mathbf{z}_{N_R}\}$, which is input to a space-time decoder. In SISO systems, the space-time decoder does not process the input vectors, hence the output sequence $\mathbf{y}$ is identical to the input sequence $\mathbf{z}$. In MIMO systems, however, the space-time decoder combines a set of input sequences, namely $\mathcal{S}(\mathbf{z})$, that have been received in successive signaling intervals to generate a set of output sequences $\mathcal{S}(\mathbf{y})$. The size of both sets depends on the adopted MIMO configuration. For example, let us consider the case when $N_T = 2$ and $N_R = 1$. In this case, $\mathcal{S}(\mathbf{z}) = \{\mathbf{z}_1, \mathbf{z}_2\}$, where $\mathbf{z}_1$ and $\mathbf{z}_2$ are the outputs of the OFDM chain at the end of the first and second signaling interval, respectively. Performing conjugate operation on $\mathbf{z}_2$, and assuming that the channels remain static over two consecutive blocks, we obtain

$$\underbrace{\begin{bmatrix}\mathbf{z}_1 \\ \mathbf{z}_2^*\end{bmatrix}}_{\mathbf{z}} = \underbrace{\begin{bmatrix}\Xi_{1,1} & \Xi_{1,2} \\ \Xi_{1,2}^{\mathcal{H}} & -\Xi_{1,1}^{\mathcal{H}}\end{bmatrix}}_{\Xi}\underbrace{\begin{bmatrix}\mathbf{s}_1 \\ \mathbf{s}_2\end{bmatrix}}_{\mathbf{s}} + \underbrace{\begin{bmatrix}\mathbf{v}_1 \\ \mathbf{v}_2^*\end{bmatrix}}_{\mathbf{v}}. \tag{6}$$

Given perfect channel state information i.e., Ξ is assumed to be known at the receiver, $\mathbf{s}_1$ and $\mathbf{s}_2$ are decoupled at the output of the space-time decoder by multiplying both sides of (6) by $\Xi^{\mathcal{H}}$, resulting in two sequences, $\mathbf{y}_1$ and $\mathbf{y}_2$, in series

$$\begin{bmatrix}\mathbf{y}_1 \\ \mathbf{y}_2\end{bmatrix} = \begin{bmatrix}\Xi_{1,1}^{\mathcal{H}} & \Xi_{1,2} \\ \Xi_{1,2}^{\mathcal{H}} & -\Xi_{1,1}\end{bmatrix}\begin{bmatrix}\mathbf{z}_1 \\ \mathbf{z}_2^*\end{bmatrix} = \begin{bmatrix}\mathbf{\Lambda} & \mathbf{0} \\ \mathbf{0} & \mathbf{\Lambda}\end{bmatrix}\begin{bmatrix}\mathbf{s}_1 \\ \mathbf{s}_2\end{bmatrix} + \mathbf{\Upsilon}, \tag{7}$$

where $\mathbf{\Lambda} = |\Xi_{1,1}|^2 + |\Xi_{1,2}|^2$ is an $N \times N$ diagonal matrix whose (n,n)-th element $\lambda_n = |\xi_n^{(1,1)}|^2 + |\xi_n^{(1,2)}|^2$ is equal to the sum of the squared n-th DFT coefficients of first and second channel frequency response. The filtered noise sequence $\mathbf{\Upsilon} = \Xi^{\mathcal{H}}\mathbf{v}$ is still white since the autocorrelation matrix of $\mathbf{\Upsilon}$ is a diagonal matrix

$$\mathbf{R}_{\Upsilon\Upsilon} = \mathrm{E}[\mathbf{\Upsilon}\mathbf{\Upsilon}^{\mathcal{H}}] = \mathrm{E}[\Xi^{\mathcal{H}}\mathbf{v}\mathbf{v}^{\mathcal{H}}\Xi] = \Xi^{\mathcal{H}}\,\mathrm{E}[\mathbf{v}\mathbf{v}^{\mathcal{H}}]\Xi = N_0\,\mathrm{diag}(\mathbf{\Lambda}, \mathbf{\Lambda}). \tag{8}$$

Consequently, a set of $2N$ parallel channels have been created at the output of the space-time decoder as shown in Fig. 4. Each element of the signal vector s can thus be decoded independently. Let us denote

$$y_n = \lambda_n s_n + \Upsilon_n, \tag{9}$$

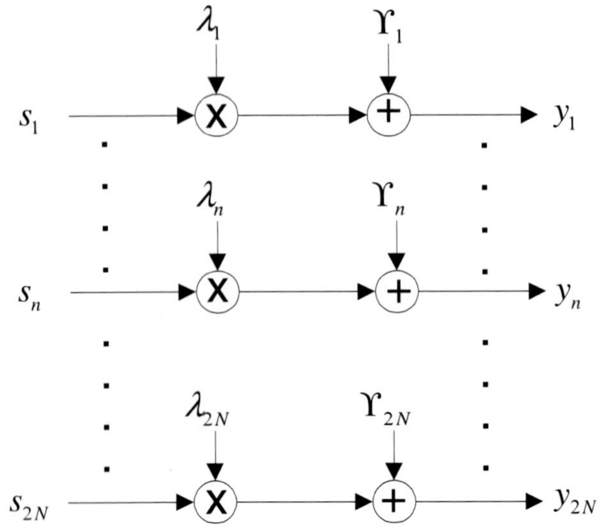

Figure 4. The equivalent parallel flat fading channels for the space-time coded MIMO (2 Tx, 1 Rx) OFDM system.

where y_n, s_n, Υ_n are the n-th element of the vectors $\mathbf{y}$, s, $\boldsymbol{\Upsilon}$, respectively. According to (8), $\Upsilon_n \sim \mathcal{CN}(0, \lambda_n N_0)$.

A single-antenna OFDM system (i.e., when $N_T = N_R = 1$) also results in equivalent parallel channels similar to the one illustrated in Fig. 4. In this case, each element of the received vector $\mathbf{y}$ is formed as

$$y_n = \xi_n s_n + \eta_n, \quad n = 1, \ldots, N \tag{10}$$

where ξ_n is the n-th coefficient of channel frequency response, and the noise term $\eta_n \sim \mathcal{CN}(0, N_0)$.

The generated set of sequences $\mathcal{S}(\mathbf{y})$ ($\mathcal{S}(\mathbf{y}) = \{\mathbf{y}_1, \mathbf{y}_2\}$ for the 2 transmit antennas MIMO system and $\mathcal{S}(\mathbf{y}) = \mathbf{y}$ for the SISO system) is then input to a soft demapper to derive log-likelihood ratio (LLR) values in order to facilitate soft-input channel decoding. In what follows, we derive LLR mapping schemes for both non-iterative and iterative OFDM based Fixed WiMAX type systems.

1.2.1. Non-iterative OFDM Scheme

The soft demapper computes the LLR value for each interleaved code bit v_n^m, denoted as $\lambda(v_n^m; O)$; here, v_n^m corresponds to the m-th bit of the n-th modulated symbol, where $m = 1, \ldots, M$ (e.g., $M = 2$ for QPSK) and $n = 1, \ldots, N$. We use the notation $\lambda(\cdot; I)$ and $\lambda(\cdot; O)$ throughout this chapter to represent soft input and soft output LLR values, respectively. Computation of $\lambda(v_n^m; O)$ requires knowledge of $\mathcal{S}(\mathbf{y})$, or simply y_n due to the equivalent parallel channels as explained previously. In particular, the LLR value of v_n^m

is given by

$$\lambda(v_n^m;O) = \ln\frac{P(v_n^m=+1|\mathcal{S}(\mathbf{y}))}{P(v_n^m=-1|\mathcal{S}(\mathbf{y}))} = \ln\frac{f(\mathcal{S}(\mathbf{y})|v_n^m=+1)P(v_n^m=+1)}{f(\mathcal{S}(\mathbf{y})|v_n^m=-1)P(v_n^m=-1)}$$
$$= \ln\frac{f(\mathcal{S}(\mathbf{y})|v_n^m=+1)}{f(\mathcal{S}(\mathbf{y})|v_n^m=-1)} = \ln\frac{f(y_n|v_n^m=+1)}{f(y_n|v_n^m=-1)}. \tag{11}$$

Equation (11) holds since for non-iterative schemes, no a priori knowledge is available at the receiver, the code bits are assumed to be equiprobable, i.e, $P(v_n^m=+1)=P(v_n^m=-1)=0.5$. The soft-input channel decoder uses the de-interleaved LLRs, denoted as $\lambda(v_n^m;I)$, to generate estimates of the information bits.

In order to give insight into the operation of the soft demapper, we provide analytical expressions for two example cases, in which binary phase shift keying (BPSK) and QPSK modulations are used. In the former case, we drop the index m from v_n^m and use v_n instead, since each BPSK symbol carries one bit. The bit to symbol mapping rule is simply $s_n = \sqrt{E_b}v_n$, where E_b refers to the average transmitted bit energy.

For the SISO system expressed by (10), let us denote $\xi_n = |\xi_n|\exp(j\angle\xi_n)$, the received signal y_n after phase correction becomes

$$z_n = \mathrm{Re}\{y_n\cdot\exp(-j\angle\xi_n)\} = |\xi_n|s_n + \eta_I,$$

where $\eta_I = \mathrm{Re}\{\exp(-j\angle\xi_n)\eta_n\} \sim \mathcal{N}(0, N_0/2)$. The conditional probability density function (PDF) of z_n given the transmit symbol s_n can be expressed as

$$f(z_n|s_n) = \frac{1}{\pi N_0/2}\exp\left(-\frac{(z_n-|\xi_n|s_n)^2}{N_0/2}\right), \tag{12}$$

Substituting (12) into (11) yields

$$\lambda(v_n;O) = \ln\frac{f(\mathcal{S}(\mathbf{y})|v_n^m=+1)}{f(\mathcal{S}(\mathbf{y})|v_n^m=-1)} = \ln\frac{f(y_n|s_n=+\sqrt{E_b})}{f(y_n|s_n=-\sqrt{E_b})} = \ln\frac{f(z_n|s_n=+\sqrt{E_b})}{f(z_n|s_n=-\sqrt{E_b})}$$
$$= \frac{(z_n+|\xi_n|\sqrt{E_b})^2-(z_n-|\xi_n|\sqrt{E_b})^2}{N_0/2}$$
$$= 8\frac{E_b}{N_0}|\xi_n|z_n = 8\frac{E_b}{N_0}|\xi_n|\,\mathrm{Re}\{y_n\cdot\exp(-j\angle\xi_n)\}. \tag{13}$$

For the space-time coded MIMO system, λ_n in (9) is real valued, no phase correction is necessary since it has been done as a result of spatial diversity combining at the space-time decoding stage, the conditional PDF of $z_n = \mathrm{Re}\{y_n\}$ given the transmit symbol s_n is

$$f(z_n|s_n) = \frac{1}{\pi\lambda_n N_0/2}\exp\left(-\frac{(z_n-\lambda_n s_n)^2}{\lambda_n N_0/2}\right), \tag{14}$$

The LLR value can thus be derived as

$$\lambda(v_n;O) = \ln\frac{f(z_n|s_n=+\sqrt{E_b})}{f(z_n|s_n=-\sqrt{E_b})} = \frac{(z_n+\lambda_n\sqrt{E_b})^2-(z_n-\lambda_n\sqrt{E_b})^2}{\lambda_n N_0/2}$$
$$= \frac{8E_b\lambda_n z_n}{\lambda_n N_0} = 8\frac{E_b}{N_0}\mathrm{Re}\{y_n\}.$$

x[][]$\psi_I(t)$ y[][]$\psi_Q(t)$ s0[][]s_4 s1[][]s_2 s2[][]s_1 s3[][]s_3 R0[][]

$R1$[][]$R2$[][]$R3$[][](0, 0)[][]

(1, 1)(0, 1)[][](1, 0)(1, 0)[][](0, 1)(1, 1)[][](0, 0)(+1, +1)[][]($\sqrt{E_b} + \sqrt{E_b}j$) (+1, -1)[][]($\sqrt{E_b} - \sqrt{E_b}j$) (-1, +1)[][]($-\sqrt{E_b} + \sqrt{E_b}j$) (-1, -1)[][]($-\sqrt{E_b} - \sqrt{E_b}j$)

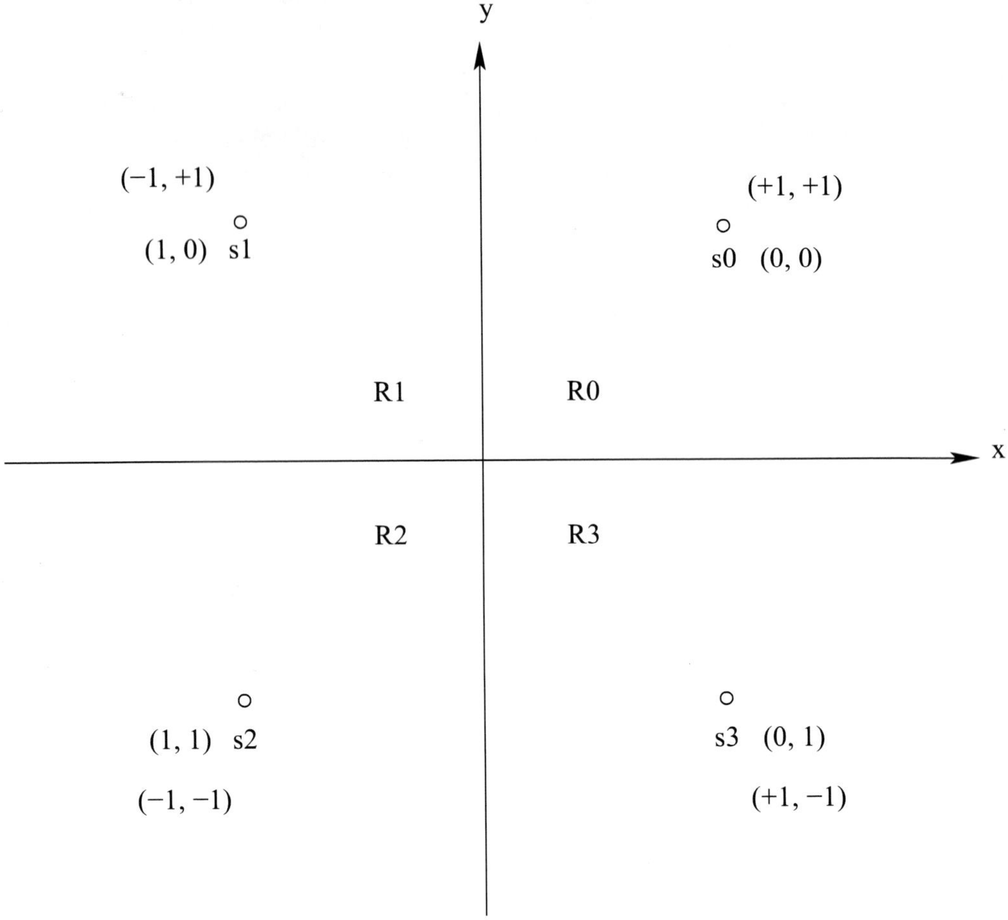

Figure 5. QPSK constellation, bit-to-symbol Gray mapping.

In the case of QPSK modulation, each block of two coded and interleaved bits v_n^0, v_n^1 is mapped into one of the four QPSK symbols $s_n = x_n^0 + jx_n^1$, where $x_n^0 = \pm\sqrt{E_b}$, $x_n^1 = \pm\sqrt{E_b}$ denote the real and imaginary parts of s_n, respectively. For the SISO system, the conditional PDF of y_n (expressed by (10)) given s_n is

$$f(y_n|s_n) = \frac{1}{\pi N_0} \exp(-|y_n - \xi_n s_n|^2/N_0), \tag{15}$$

based on which the LLR values can be derived. For the QPSK constellation with Gray mapping shown in Fig. 5, the LLR value of v_n^0 can be derived as

$$\lambda(v_n^0; O) = \ln \frac{f(y_n|s_3) + f(y_n|s_4)}{f(y_n|s_1) + f(y_n|s_2)} \approx \ln \frac{\exp\left(-|y_n - \xi_n s_+|^2/N_0\right)}{\exp\left(-|y_n - \xi_n s_-|^2/N_0\right)} \tag{16}$$

$$= \frac{1}{N_0}\left\{|y_n - \xi_n s_-|^2 - |y_n - \xi_n s_+|^2\right\} = \frac{2}{N_0}\,\mathrm{Re}\left\{(\xi_n s_+)^* y_n - (\xi_n s_-)^* y_n\right\}, \tag{17}$$

where s_+ denotes the QPSK symbol corresponding to $\max\{f(z_i|s_3), f(z_i|s_4)\}$, and s_- denotes the QPSK symbol corresponding to $\max\{f(y_n|s_1), f(y_n|s_2)\}$ since the real part of the symbols s_3, s_4 corresponds to 1, and the real part of the symbols s_1, s_2 corresponds to -1 as shown in Fig. 5. The dual maxima rule is used in (16) utilizing the fact that one term usually dominates each sum. Similarly,

$$\lambda(v_n^1; O) = \ln \frac{f(y_n|s_2) + f(y_n|s_4)}{f(y_n|s_1) + f(y_n|s_3)} = \frac{2}{N_0}\,\mathrm{Re}\left\{(\xi_n s_+)^* y_n - (\xi_n s_-)^* y_n\right\},$$

where s_+ denotes the QPSK symbol corresponding to $\max\{f(y_n|s_2), f(y_n|s_4)\}$, and s_- denotes the QPSK symbol corresponding to $\max\{f(y_n|s_1), f(y_n|s_3)\}$ since the imaginary part of the symbols s_2, s_4 corresponds to 1, and the imaginary part of the symbols s_1, s_3 corresponds to -1 as shown in Fig. 5. The LLR values for MIMO systems with QPSK modulation can be derived in a similar way.

The LLR values $\lambda(v_n^m; O)$ computed by the soft demapper are first de-interleaved to $\lambda(u_n^m; I)$ and then fed to a soft-input channel decoder, which generates estimates of the transmitted information bits.

1.2.2. Iterative OFDM Scheme

Fig. 6 depicts an iterative demapping and decoding scheme for OFDM systems. The demapper takes y_n and a priori information $\lambda(v_n^m; I)$ as inputs and computes its extrinsic information $\lambda(v_n^m; O)$, which is subsequently de-interleaved to $\lambda(u_n^m; I)$. The interleaver and deinterleaver are denoted as Π and Π^{-1}, respectively in the figures throughout the chapter. With a priori input $\lambda(u_n^m; I)$, a soft-input, soft-output (SiSo) channel decoder computes log-likelihood ratio (LLR) $\lambda(u_n^m; O)$ for the coded bits and $\lambda(b_n; O)$ for the information bits. The latter is used at the final iteration to make a hard decision on the transmitted information bits; whereas the former is interleaved and fed back to the demapper as a priori information. Several SiSo algorithms can be applied to compute the channel decoder output. For the purpose of this study, we consider the use of the Log-MAP, Max-Log-MAP and soft-output Viterbi algorithm (SOVA) [17].

LogMAP[][]Log-MAP LDC[][]LLR Decoder[][]Demapper Decoding[][]Decoder

Pi[][]$\prod$ Pip[][]$\prod^{-1}$ LcpI[][]$\lambda(v_n^m; I)$ LcpO[][]$\lambda(v_n^m; O)$ LcI[][]$\lambda(u_n^m; I)$

LcO[][]$\lambda(u_n^m; O)$ LbO[][]$\lambda(b_n; O)$ r[][]y_n -[][]$-$

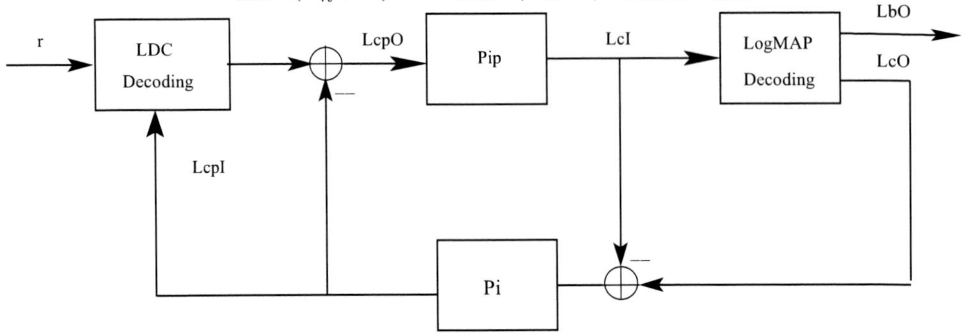

Figure 6. Structure of the iterative demapping and decoding for OFDM systems.

Here, we use the MIMO space-time coded OFDM system with QPSK anti-gray mapping (shown in Fig. 7) as an example to see how the iterative OFDM scheme is designed. Recall that the received signal at the output of the space-time decoder can be expressed as $y_n = \lambda_n s_n + \Upsilon_n$, where s_n conveys two bits v_n^0, v_n^1. The conditional PDF of y_n is given by

$$f(y_n|s_n) = \frac{1}{\pi\lambda_n N_0} \exp\left(\frac{-|y_n - \lambda_n s_n|^2}{\lambda_n N_0}\right), \tag{18}$$

The LLR value of the bit v_n^0 conditioned on y_n can be calculated as [19]

$$\lambda(v_n^0|y_n) = \ln \frac{P(v_n^0 = 1|y_n)}{P(v_n^0 = 0|y_n)} = \ln \frac{P(v_n^0 = 1, v_n^1 = 0|y_n) + P(v_n^0 = 1, v_n^1 = 1|y_n)}{P(v_n^0 = 0, v_n^1 = 0|y_n) + P(v_n^0 = 0, v_n^1 = 1|y_n)}.$$

Since v_n^0 and v_n^1 are approximately independent due to the interleaver between the encoder and mapper, $P(v_n^0, v_n^1) = P(v_n^0)P(v_n^1)$. Using Bayes' rule, we obtain

$$\lambda(v_n^0|y_n) = \lambda(v_n^0; I) + \ln \frac{P(y_n|v_n^0 = 1, v_n^1 = 0) + P(y_n|v_n^0 = 1, v_n^1 = 1)e^{\lambda(v_n^1;I)}}{P(y_n|v_n^0 = 0, v_n^1 = 0) + P(y_n|v_n^0 = 0, v_n^1 = 1)e^{\lambda(v_n^1;I)}}, \tag{19}$$

where the priori value $\lambda(v_n^0; I)$ is defined as

$$\lambda(v_n^0; I) = \ln \frac{P(v_n^0 = 1)}{P(v_n^0 = 0)}.$$

The extrinsic information at the output of the demapper is derived by subtracting $\lambda(v_n^0; I)$ from $\lambda(v_n^0|y_n)$, resulting in

$$\lambda(v_n^0; O) = \ln \frac{P(y_n|v_n^0 = 1, v_n^1 = 0) + P(y_n|v_n^0 = 1, v_n^1 = 1)e^{\lambda(v_n^1;I)}}{P(y_n|v_n^0 = 0, v_n^1 = 0) + P(y_n|v_n^0 = 0, v_n^1 = 1)e^{\lambda(v_n^1;I)}}. \tag{20}$$

x[][]$\psi_I(t)$ y[][]$\psi_Q(t)$ s0[][]s_4 s1[][]s_2 s2[][]s_1 s3[][]s_3 R0[][]R1[][] R2[][] R3[][] (0, 0)[][](0, 0) (0, 1)[][](0, 1) (1, 0)[][](1, 1) (1, 1)[][](1, 0) (+1, +1)[][]$(\sqrt{E_b} + \sqrt{E_b}j)$ (+1, -1)[][]$(\sqrt{E_b} - \sqrt{E_b}j)$ (-1, +1)[][]$(-\sqrt{E_b} + \sqrt{E_b}j)$ (-1, -1)[][]$(-\sqrt{E_b} - \sqrt{E_b}j)$

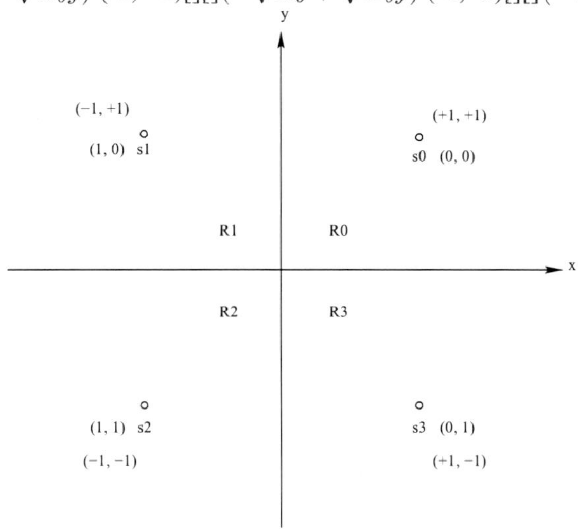

Figure 7. QPSK constellation, anti-Gray mapping.

Substituting (15) into (20) yields

$$\lambda(v_n^0; O) = \max{}^*\left\{-\frac{|y_n - \lambda_n s_{10}|^2}{\lambda_n N_0}, -\frac{|y_n - \lambda_n s_{11}|^2}{\lambda_n N_0} + \lambda(v_n^1; I)\right\} - \max{}^*\left\{-\frac{|y_n - \lambda_n s_{00}|^2}{\lambda_n N_0}, -\frac{|y_n - \lambda_n s_{01}|^2}{\lambda_n N_0} + \lambda(v_n^1; I)\right\},$$

where the function $\max^*\{\cdot\}$ is defined as $\max^*\{x, y\} = \ln(e^x + e^y) = \max\{x, y\} + \ln(1 + e^{-|x-y|})$, i.e., the max operation compensated with a correction term $\ln(1 + e^{-|x-y|})$.

Similarly,

$$\lambda(v_n^1; O) \approx \max{}^*\left\{-\frac{|y_n - \lambda_n s_{01}|^2}{\lambda_n N_0}, -\frac{|y_n - \lambda_n s_{11}|^2}{\lambda_n N_0} + \lambda(v_n^0; I)\right\} - \max{}^*\left\{-\frac{|y_n - \lambda_n s_{00}|^2}{\lambda_n N_0}, -\frac{|y_n - \lambda_n s_{10}|^2}{\lambda_n N_0} + \lambda(v_n^0; I)\right\}.$$

At the first iteration, no priori information is available, so the initial values of $\lambda(v_n^0; I)$ and $\lambda(v_n^1; I)$ are set to 0.

1.3. SC-FDE

Another solution to achieve high data rate transmission with low complexity is SC-FDE [12, 13, 20, 21]. It shares some common elements with OFDM, and potentially offers a similar

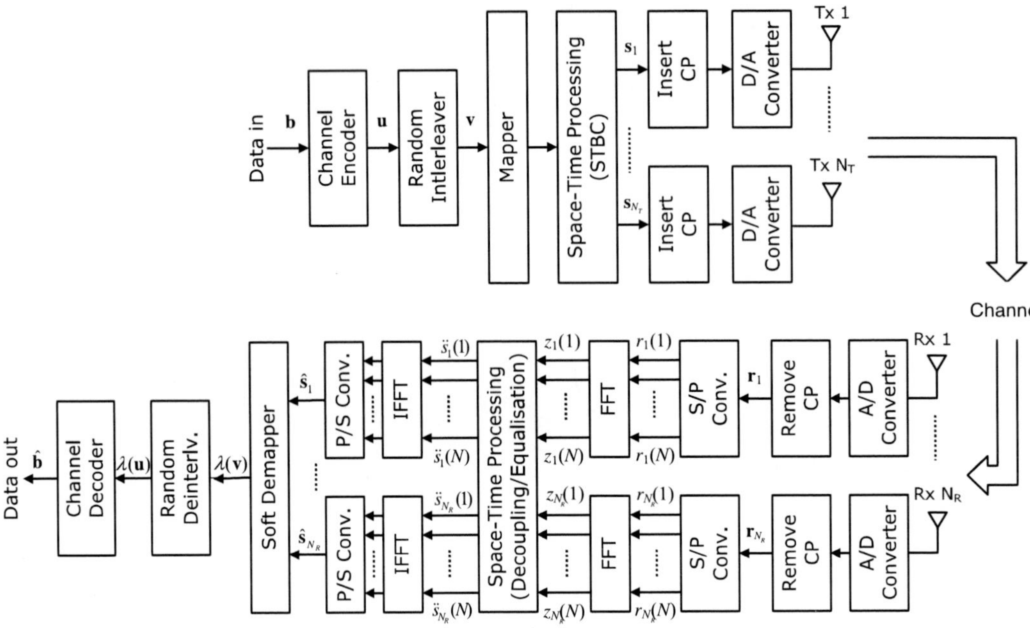

Figure 8. Block diagram of SC-FDE based Fixed WiMAX type system.

performance and complexity. Frequency-domain equalization in a single carrier system is simply the frequency-domain analog of what is done by a conventional linear time-domain equalizer. For channels with severe delay spread, it is simpler than corresponding time domain-equalization due to the use of the computationally efficient fast Fourier transform. Compared to OFDM, SC-FDE requires less power amplifier back-off due to reduced peak-to-average power ratio (PAPR), and is also less sensitive to carrier frequency offset and phase noise [13]. It is shown in [13] that without channel coding, the performance of SC-FDE is superior to OFDM.

Fig. 8 depicts the STBC coded and SC-FDE based Fixed WiMAX type system using multiple transmit and multiple receive antennas. It differs from the OFDM system shown in Fig. 2 in that the transmitter's inverse FFT block has been moved to the receiver to convert frequency-domain equalized signals back into time-domain symbols. The signal processing complexity of these two systems are essentially the same for equal FFT block length.

We use the previous notation in the transmission chain. Thus, sequences $\mathbf{b}$, $\mathbf{u}$ and $\mathbf{v}$ correspond to the information bits, code bits and interleaved code bits, respectively. A mapper converts the binary sequence of interleaved code bits into a sequence of M-ary symbols. As in the case of OFDM-based systems, operations are performed on a block by block basis. The block length should be equal to the number of FFT points, N. At each signaling interval, the space-time block code uses successive blocks of modulated symbols to generate N_T sequences, denoted as $\mathbf{s}_i$ for $i = 1, \ldots, N_T$, of length N symbols each. A cyclic prefix is created, as in OFDM, and appended to the front of $\mathbf{s}_i$. Finally, the space-time coded signal is digital-to-analogue converted and sent to the transmit antenna i.

At the receiver, the N_R signals are analogue-to-digital converted and their cyclic prefix is removed. The relationship between the received signal $\mathbf{r}_j$ at the j-th reception chain and the transmitted signals $\mathbf{s}_1, \ldots, \mathbf{s}_{N_T}$ can be expressed as

$$\mathbf{r}_j = \mathbf{H}_j \mathbf{s} + \mathbf{w}_j, \tag{21}$$

where $\mathbf{H}_j = [\mathbf{H}_{j,1} \ \ldots \ \mathbf{H}_{j,N_T}]$ and $\mathbf{s} = [\mathbf{s}_1 \ \ldots \ \mathbf{s}_{N_T}]^\top$. As previously mentioned, the channel matrix $\mathbf{H}_{j,i}$ can be written as the product of the FFT and IFFT matrices $\mathbf{Q}$ and $\mathbf{Q}^{\mathcal{H}}$, respectively, and a diagonal matrix $\Xi_{j,i}$ such that (3) and (4) are satisfied. Note that the elements of $\mathbf{w}_j$ are modeled as zero-mean circularly symmetric complex Gaussian random variables with autocorrelation matrix $\mathbf{R}_{\mathbf{ww}} = \mathrm{E}[\mathbf{ww}^{\mathcal{H}}] = N_0\mathbf{I}$.

At each signaling interval, the FFT block of the j-th chain generates a sequence $\mathbf{z}_j = \mathbf{Qr}_j$, which is input to the space-time processing and equalization (STPE) unit. If we consider SISO systems and drop indices i and j, this unit equalizes the input sequence using a matrix $\mathbf{F}^{\mathrm{eq}}$, which can be derived from the diagonal matrix Ξ. The output sequence of the FFT block $\mathbf{z}$ is given by

$$\begin{aligned}\mathbf{z} = \mathbf{Qr} &= \mathbf{Q}(\mathbf{Q}^{\mathcal{H}}\Xi\mathbf{Qs} + \mathbf{w}) \\ &= \Xi\ddot{\mathbf{s}} + \ddot{\mathbf{w}},\end{aligned} \tag{22}$$

where $\ddot{\mathbf{s}} = \mathbf{Qs}$, $\ddot{\mathbf{w}} = \mathbf{Qw}$, are the frequency-domain symbol and noise vectors, respectively, and Ξ is a diagonal matrix whose (n, n)-th diagonal element ξ_n is the n-th coefficient of channel frequency response. The filtered noise sequence $\ddot{\mathbf{w}} = \mathbf{Qw}$ is still white since the autocorrelation matrix of $\ddot{\mathbf{w}}$ is

$$\mathbf{R}_{\ddot{\mathbf{w}}\ddot{\mathbf{w}}} = \mathrm{E}[\ddot{\mathbf{w}}\ddot{\mathbf{w}}^{\mathcal{H}}] = \mathrm{E}[\mathbf{Q}^{\mathcal{H}}\mathbf{ww}^{\mathcal{H}}\mathbf{Q}] = \mathbf{Q}^{\mathcal{H}}\,\mathrm{E}[\mathbf{ww}^{\mathcal{H}}]\mathbf{Q} = N_0\mathbf{I}. \tag{23}$$

The above equation holds since $\mathbf{Q}$ is an unitary matrix. Therefore, the SC-FDE system before IFFT block can also be viewed as a set of N parallel channels similar to what is shown in Fig. 4 for the OFDM system. Let us denote

$$z_n = \xi_n \ddot{s}_n + \ddot{w}_n, \tag{24}$$

where $z_n, \ddot{s}_n, \ddot{w}_n$ are the n-th element of the vectors $\mathbf{z}$, $\ddot{\mathbf{s}}$, $\ddot{\mathbf{w}}$, respectively. According to (23), $\ddot{w}_n \sim \mathcal{CN}(0, N_0)$. A 1-tap ZF eqaulizer $f_n^{\mathrm{zf}} = \xi_n^{-1}$ or minimum mean square error (MMSE) equalizer f_n^{mmse} is then applied to z_n. The latter is designed to minimize the mean square error (MSE)

$$f_n^{\mathrm{mmse}} = \arg\min_{f_n} |f_n^* z_n - \ddot{s}_n|^2. \tag{25}$$

Denoting $\sigma_s^2 = \mathrm{E}[\ddot{s}_n^* \ddot{s}_n]$ as the average symbol energy, the solution to (25) is given by $f_n^{\mathrm{mmse}} = R_n^{-1} P_n$, where

$$R_n = \mathrm{E}[z_n z_n^*] = \sigma_s^2 |\xi_n|^2 + N_0; \quad P_n = \mathrm{E}[\ddot{s}_n^* z_n] = \sigma_s^2 \xi_n.$$

Therefore,

$$f_n^{\mathrm{mmse}} = R_n^{-1} P_n = \frac{\sigma_s^2 \xi_n}{\sigma_s^2 |\xi_n|^2 + N_0} = \frac{\xi_n}{|\xi_n|^2 + N_0/\sigma_s^2}. \tag{26}$$

An estimate of the frequency-domain symbol $\ddot{s}_n$ can be obtained as

$$\hat{\ddot{s}}_n = \begin{cases} f_n^{\text{mmse}*} z_n = \frac{\xi_n^*}{|\xi_n|^2 + N_0/\sigma_s^2} z_n & \text{for MMSE;} \\ f_n^{\text{zf}} z_n = \xi_n^{-1} z_n & \text{for ZF.} \end{cases}$$

Zero forcing equalizers, although simple to implement, may significantly amplify noise and hence deteriorate the system performance. If the noise power spectral density is known at the receiver, e.g., by transmitting a known signal over the channel, MMSE equalizers provide superior performance.

The output of the SC-FDE, $\hat{\ddot{\mathbf{s}}} = [\hat{\ddot{s}}(1) \ \ldots \ \hat{\ddot{s}}(N)]^\top$, is then transformed back to the time-domain by IFFT operation, i.e.,

$$\hat{\mathbf{s}} = \mathbf{Q}^{\mathcal{H}} \hat{\ddot{\mathbf{s}}} = [\hat{s}(1) \ \ldots \ \hat{s}(N)]^\top.$$

Based on (22), the above process can be summarized in a vector/matrix format as

$$\hat{\mathbf{s}} = \mathbf{Q}^{\mathcal{H}} \hat{\ddot{\mathbf{s}}} = \mathbf{Q}^{\mathcal{H}} \mathbf{F}^{\text{eq}} \mathbf{z} = \mathbf{Q}^{\mathcal{H}} \mathbf{F}^{\text{eq}} (\mathbf{\Xi} \mathbf{Q} \mathbf{s} + \ddot{\mathbf{w}}), \tag{27}$$

where $\mathbf{z}$ is given by (22), and

$$\mathbf{F}^{\text{eq}} = \begin{cases} \mathbf{\Xi}^{\mathcal{H}} (\mathbf{\Xi} \mathbf{\Xi}^{\mathcal{H}} + \frac{N_0}{\sigma_s^2} \mathbf{I})^{-1} & \text{for MMSE;} \\ \mathbf{\Xi}^{\mathcal{H}} (\mathbf{\Xi} \mathbf{\Xi}^{\mathcal{H}})^{-1} & \text{for ZF.} \end{cases}$$

A close examination at (27) reveals that no diagonalized struture, as shown in (5) and (7) for the OFDM systems, can be formed for the SC-FDE systems. Therefore, there do not exist equivalent parallel channels after IFFT block, and each element of the signal vector $\mathbf{s}$ cannot be decoded separately. The optimum decoding of each symbol entails the processing of the whole vector $\hat{\mathbf{s}}$ [22], which leads to much higher computational complexity compared to the OFDM approach presented previously. In order to reduce the complexity, we introduce an approximation method here, and the effectiveness of this simplified approach will be verified in Section 1.4. by computer simulations. Since $\hat{\ddot{s}}_n$ is an estimate of frequency-domain symbol $\ddot{s}_n$ (this is certainly the case for ZF equalizer, and is also obvious for MMSE equalizer as indicated by Equation (25)), $\hat{s}_n$ also approximates the time-domain symbol s_n. Consequently, the whole SC-FDE system can be approximated as a set of N parallel AWGN channel as shown in Fig. 4, however, with the channel coefficient $\lambda_n = 1, \ n = 1, \ldots, N$. The LLR values derived in (13) and (17) are still applicable for SC-FDE with the value of ξ_n set to unity.

The two symbol sequences $\mathbf{s}_1$ and $\mathbf{s}_2$ are processed by separate chains in the first signalling interval, followed by sequences $\mathbf{s}_1'$ and $\mathbf{s}_2'$ in the second signalling interval. In SC-FDE MIMO systems, the SC-FDE-STBC encoding is slightly different from the OFDM-STBC encoding shown by (1). The two sequences transmitted in the second signalling interval are defined as

$$\mathbf{s}_1'(n) = -\mathbf{s}_2^*((-n)_N); \qquad \mathbf{s}_2'(n) = \mathbf{s}_1^*((-n)_N),$$

where $n = 0, 1, \ldots, N-1$, and $(\cdot)_N$ denotes modulo-N operation.

Depending on the antenna configuration, two or more sequences $\mathbf{z}_j$ obtained at successive signaling intervals, are first combined and then equalized. We use the notation $\mathcal{S}(\mathbf{z})$ and $\mathcal{S}(\mathbf{y})$ to represent the sets of sequences which are input to and output from the STPE unit, respectively. For example, if $N_T=2$ and $N_R=1$, the input set will be $\mathcal{S}(\mathbf{z})=\{\mathbf{z}_1,\mathbf{z}_2\}$, where $\mathbf{z}_1$ and $\mathbf{z}_2$ are sequences generated by the FFT block at successive signaling intervals. The output set of the STPE unit will be $\mathcal{S}(\mathbf{y})=\{\mathbf{y}_1,\mathbf{y}_2\}$. The relationship between the members of the two sets is as follows

$$\begin{bmatrix}\mathbf{y}_1\\ \mathbf{y}_2\end{bmatrix}=\begin{bmatrix}\mathbf{F}^{\text{eq}} & 0\\ 0 & \mathbf{F}^{\text{eq}}\end{bmatrix}\cdot\begin{bmatrix}\Xi_{1,1}^{\mathcal{H}} & \Xi_{1,2}\\ \Xi_{1,2}^{\mathcal{H}} & -\Xi_{1,1}\end{bmatrix}\cdot\begin{bmatrix}\mathbf{z}_1\\ \bar{\mathbf{z}}_2\end{bmatrix}, \tag{28}$$

where

$$\mathbf{F}^{\text{eq}}=\mathbf{\Lambda}^{\mathcal{H}}(\mathbf{\Lambda}\mathbf{\Lambda}^{\mathcal{H}})^{-1}, \tag{29}$$

or [21]

$$\mathbf{F}^{\text{eq}}=\mathbf{\Lambda}^{\mathcal{H}}\left(\mathbf{\Lambda}\mathbf{\Lambda}^{\mathcal{H}}+\frac{N_0}{\sigma_s^2}\mathbf{I}\right)^{-1}, \tag{30}$$

depending on whether ZF equalization (29) or MMSE equalization (30) is used, respectively. In the above equations, matrix $\mathbf{\Lambda}$ is defined as $\mathbf{\Lambda}=|\Xi_{1,1}|^2+|\Xi_{1,2}|^2$.

The N_R output sequences of the STPE unit, $\mathbf{y}_1,\ldots,\mathbf{y}_{N_R}$, are converted back to the time-domain by means of an IFFT operation to produce an estimate of the transmitted symbols. The generated time-domain sequence is then passed to a soft demapper to produce LLR values for the coded bits v_n^m, which can be derived in a similar fashion as shown in the previous section.

The last stage of a system using SC-FDE involves de-interleaving the LLR values and feeding them to a soft-input channel decoder which produces estimates of the source bits.

1.4. Quantitative Results

In this section, we compare the error rate performance of systems using either OFDM or SC-FDE with BPSK modulation[1]. In our simulations the FFT in both OFDM and SC-FDE uses 256 sub-carriers, the signal duration per transmit chain is $12.8\mu s$, while the respective length of the cyclic prefix is $3.2\mu s$. The source information bits are encoded by a rate-1/3 systematic turbo code with generator polynomials $(1, 5/7, 5/7)$ in octal form. The channel decoder implements the exact log-MAP decoding algorithm [28] and obtains estimates of the source information bits after eight iterations, unless otherwise stated. Note that the size of the turbo code interleaver, i.e., the length of the input information sequence is 1000 bits. The Fixed WiMAX channel has been measured and six Stanford University Interim (SUI) models have been specified for particular scenarios [23, 24]. All of them are simulated using 3 taps, having either Ricean or Rayleigh amplitude distributions. In this chapter, we have adopted the SUI-3 model, which corresponds to average British suburban conditions. The line-of-sight (LOS) component is relatively small and the channel is slowly fading as well as mildly frequency selective. We assume that the channel is essentially constant during the transmission of a frame of data. More specifically, the SUI-3 channel model includes three

[1]The results with QPSK modulation are found to be similar to those with BPSK modulation and thus omitted here.

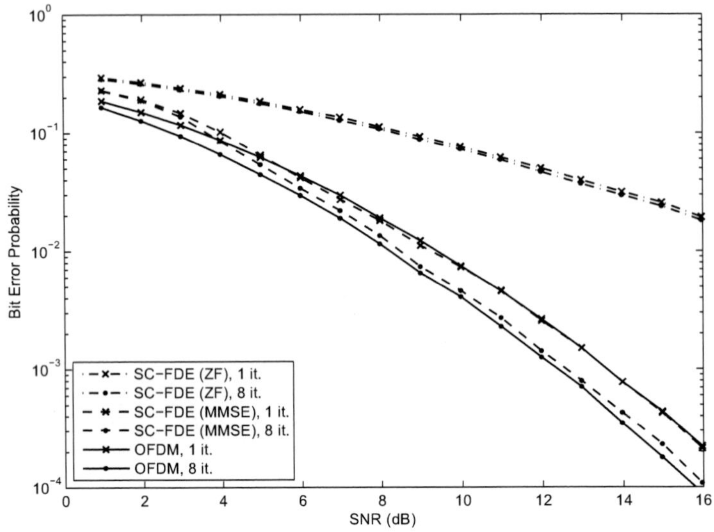

Figure 9. BER comparison of SISO rate-1/3 turbo-coded systems employing OFDM/SC-FDE.

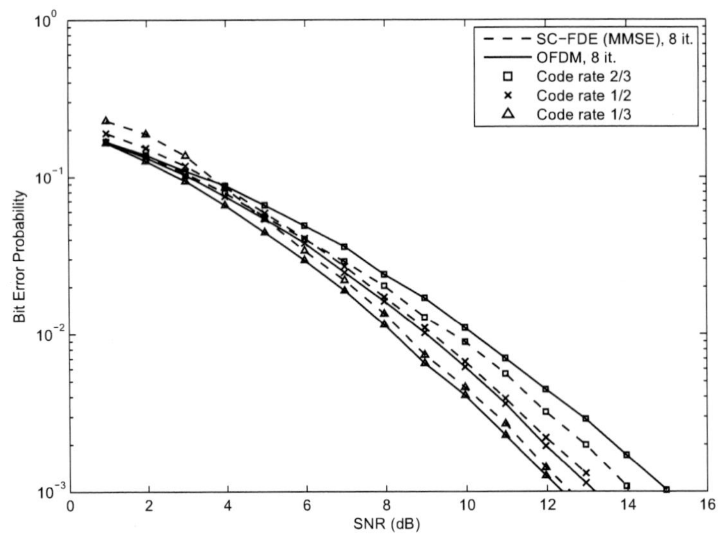

Figure 10. Performance comparison for various code rates between turbo-coded OFDM and turbo-coded SC-FDE systems.

fading taps at delays $0\mu s$, $0.5\mu s$ and $1.0\mu s$, with relative powers 0 dB, -5 dB and -10 dB, and K-factors 1, 0 and 0, respectively. The rms delay spread is 0.264 microseconds. The SUI-3 channel model also specifies an antenna correlation coefficient of 0.4 for the case when multiple antennas are employed.

Fig. 9 depicts the bit error rate (BER) of various SISO turbo-coded systems. As expected, the SC-FDE system using MMSE equalization performs significantly better than that using ZF equalization. On the other hand, OFDM provides a performance advantage at low SNR values compared to SC-FDE. In the high SNR region and for an increasing number of decoding iterations, turbo-coded OFDM achieves a small coding gain of 0.2 dB at a BER of 10^{-4}. However, the rate of the channel code directly affects the relative performance of OFDM and SC-FDE systems. As shown in Fig. 10, increasing the code rate causes a more significant performance degradation to the OFDM system compared to that of the SC-FDE system. Observe that, for a code rate of 2/3, SC-FDE eventually outperforms OFDM since a system using high rate code is close to an uncoded system for which SC-FDE outperforms OFDM [13]. In our simulations, the turbo codes of various rates have been obtained by periodically eliminating parity check bits from the output of a systematic rate-1/3 turbo code, a technique commonly known as *puncturing* [25].

The impact of multiple transmit and multiple receive antennas on the performance of turbo-coded systems is illustrated in Fig. 11 and Fig. 12. In both cases, the iterative decoder implements suboptimal decoding algorithms. Fig. 11 depicts the performance of systems using the max log-MAP algorithm, whilst Fig. 12 presents the BER of systems using the SOVA decoding algorithm. Suboptimal algorithms are often used in practice owing to their reduced computational complexity compared to that of the optimal exact log-MAP algorithm, at the expense of a performance loss. We observe that, with suboptimal decoding algorithms, both SC-FDE and OFDM based systems achieve similar performance for the same number of transmit and receive antennas. The BER performance in the AWGN channel has also been included for reference; note that in AWGN the performance of an SC-FDE system is identical to that of an OFDM system since frequency selectivity is absent, hence processing in the frequency domain does not offer an advantage.

We now focus on systems using OFDM and investigate the impact of the decoding algorithm as well as the number of decoding iterations on the overall system performance. Fig. 13 depicts the coding gain between one and eight iterations of the exact log-MAP decoding algorithm. Observe that the coding gain increases from 1.2 dB to 1.4 dB at a BER of 10^{-3} when the number of antennas is doubled at both ends. Fig. 14 compares the performance of systems using either the exact log-MAP algorithm or the suboptimal max log-MAP algorithm. As previously mentioned, implementation of the max log-MAP algorithm results in a less computational intensive decoding process at the expense of about 0.5 dB loss in gain at a BER of 10^{-3}.

We conclude that SC-FDE, compared to OFDM, is more appropriate for systems that either do not use channel coding or use simple, very high rate codes. Furthermore, the peak-to-average power ratio of the transmitted signal in SC-FDE systems is considerably less than that in OFDM systems, leading to a reduced power back off and, thus, allowing less costly power amplifiers to be used. However, when our objective is to offer a high reliability link between two users, then coded OFDM is the best candidate amongst the two frequency-domain solutions. Note that OFDM and SC-FDE can actually co-exist in a duplex communication system, if OFDM is implemented in the downlink direction whilst SC-FDE is used in the uplink direction. In that case, complexity will be concentrated at the base station, while users will only need to perform FFT operations when receiving OFDM signals.

Fig. 15 shows the performance of the iterative OFDM algorithm described in Section 1.2.2. for the 2Tx-2Rx system with QPSK anti-Gray mapping at different numbers of iterations and compares its performance with the non-iterative system with Gray-mapping.

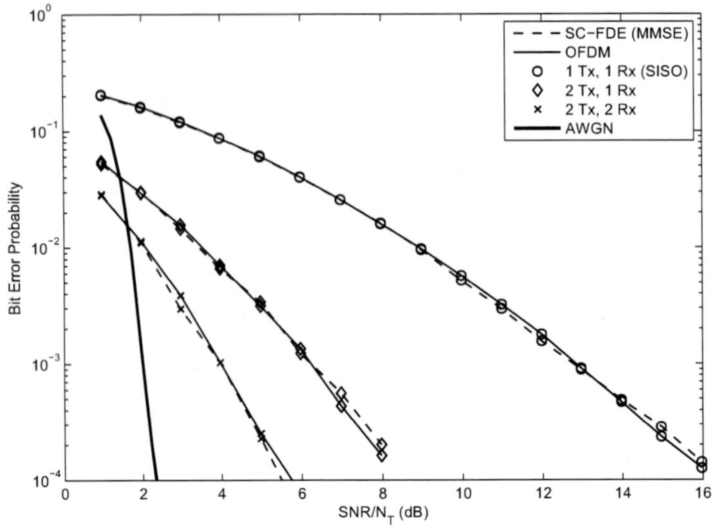

Figure 11. Performance comparison of MIMO rate-1/3 turbo-coded systems employing OFDM/SC-FDE. The max log-MAP decoding algorithm is used (8 iterations).

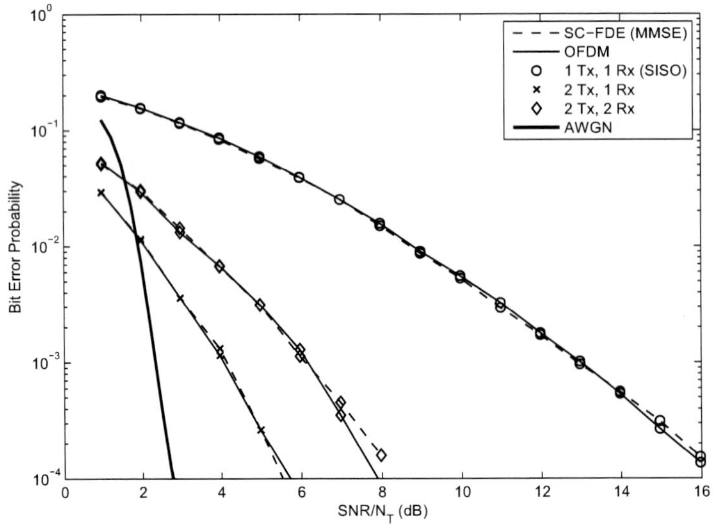

Figure 12. Performance comparison of MIMO rate-1/3 turbo-coded systems employing OFDM/SC-FDE. The soft-output Viterbi algorithm is used (8 iterations).

We use a convolutional code with code rate $R_c = 1/2$, constraint length $L_c = 3$, generator ploynomial $(5, 7)$ in octal form, and the Max-Log-MAP algorithm for channel decoding. Clearly, for the iterative OFDM system, there is a significant performance improvement by applying iterative process if we compare the topmost curve representing the first-iteration between demapping and decoding (the non-iterative case) with the bottom curve representing the performance of iterative process upon convergence. It takes 4 iterations for the algorithm to reach steady state, and further iterations do not yield noticeable performance

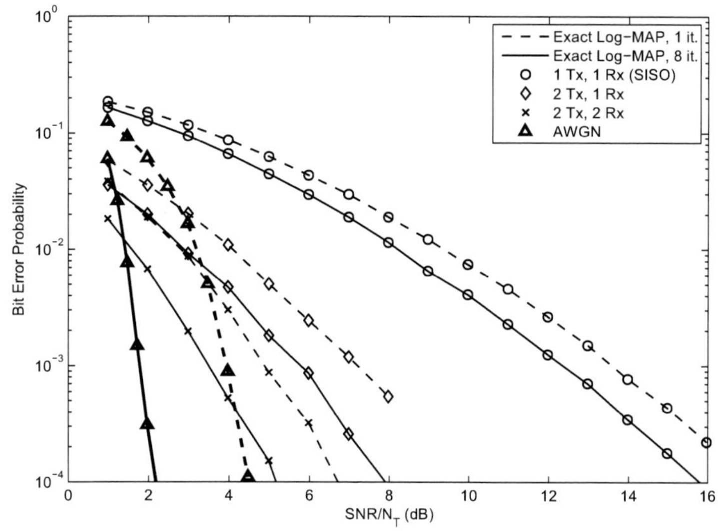

Figure 13. Performance of various OFDM-based turbo-coded systems using the exact log-MAP decoding algorithm after 1 or 8 iterations.

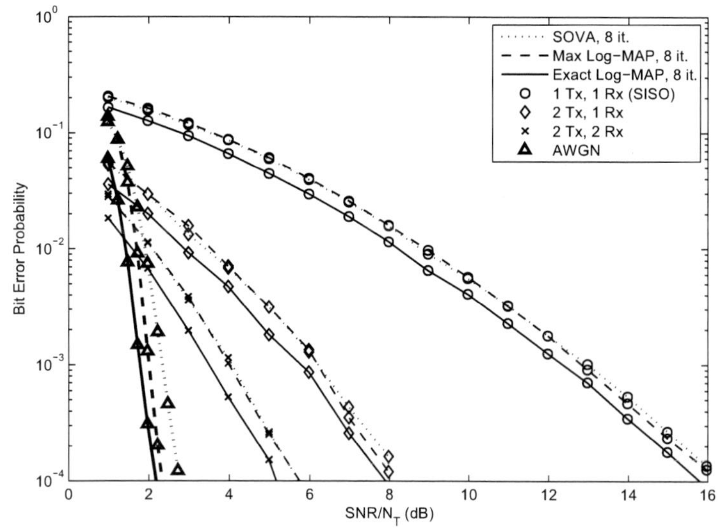

Figure 14. Performance of various OFDM-based turbo-coded systems using either the exact log-MAP or the max log-MAP decoding algorithm after 8 iterations.

improvement. The most significant gain is obtained at the second iteration. Note that no gain can be obtained by performing the iterative process for the systems with Gray mapping, in which the bits are mapped to I and Q channels independently [19]. The iterative scheme with anti-Gray mapping outperforms the non-iterative scheme with Gray mapping at the 2nd iteration when $E_b/N_0 > 5$ dB. A performance gain up to 0.8 dB is observed at target BER$= 10^{-4}$ by applying anti-Gray mapping and turbo processing.

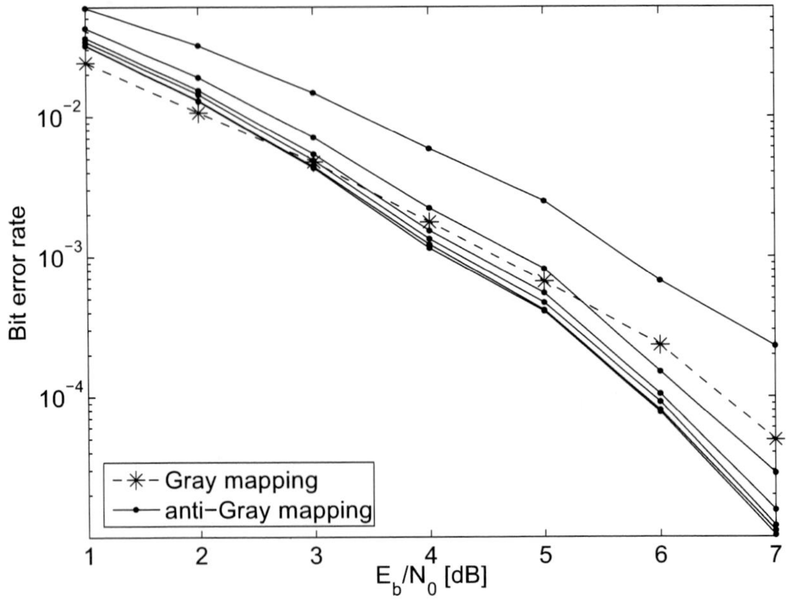

Figure 15. Performance of the iterative OFDM scheme for 2Tx-2Rx (MIMO) system. For the system with anti-Gray mapping, the top curve represents the first iteration of demapping and channel decoding, and the bottom curve represents the 6th iteration of demapping and decoding.

2. Time-Domain Solutions

We introduce time-domain turbo equalization schemes for broadband transmission in SISO and MIMO Fixed WiMAX type systems in Section 2.1. and **??**, respectively. A complexity comparison between time and frequency-domain solutions will also be presented in Section 2.1.3..

2.1. Time-Domain Turbo Equalization for SISO System

The block diagram of the transmitter in the SISO Fixed WiMAX type system being studied is shown in Fig. 16. The information sequence $\{b_n\}$ is encoded into code bits $\{u_n\}$, which

Pi[][]Π Conv.[][]Channel Encoder[][]Encoder ISI[][]SUI-3 QPSK[][]QPSK Mod.[][]Modul. Channel[][]Channel rn[][]r_n bn[][]b_n Sn[][]s_n un[][]u_n un$_p$[][]v$_n$ v[][]w_n h[][]**h**

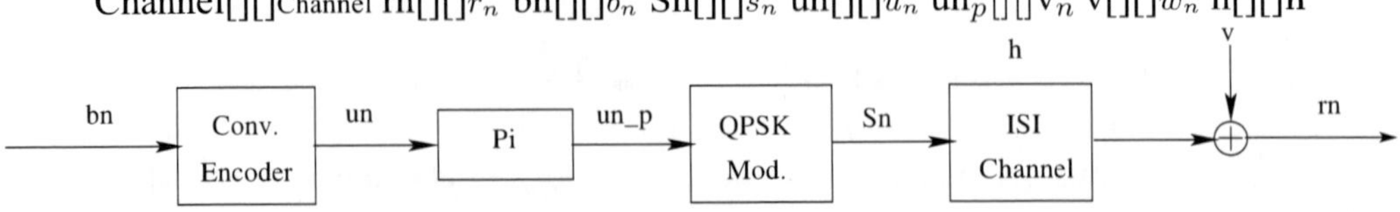

Figure 16. Block diagram of transmitter in SISO Fixed WiMAX type system.

Equalizer[][]Equalizer BSC[][]BSC SBC[][]SBC Alamouti[][]Alamouti Decision[][]Decision
Conv. MAP[][]Log-MAP Decoder[][]Decoder Channel[][]Channel Estimation[][]Estimator
p(uk2;o)[][]$\lambda(v_n; O)$ p(uk;i)[][]$\lambda(u_n; I)$ SN0,SN1[][]y_n/r'_n p(uk;o)[][]$\lambda(u_n; O)$
p(bk,i)[][]$\lambda(b_n; I)$ p(bk;o)[][]$\lambda(b_n; O)$ r[][]r_n L(xn)+jL(yn)[][] $\lambda(s_n)$ h[][]$\hat{h}$ bk[][]$\hat{b}_n$
sgn[][]sgn$(\cdot)$ uk2[][]$\lambda(v_n; I)$ Pi[][]Π Pi1[][]Π^{-1}

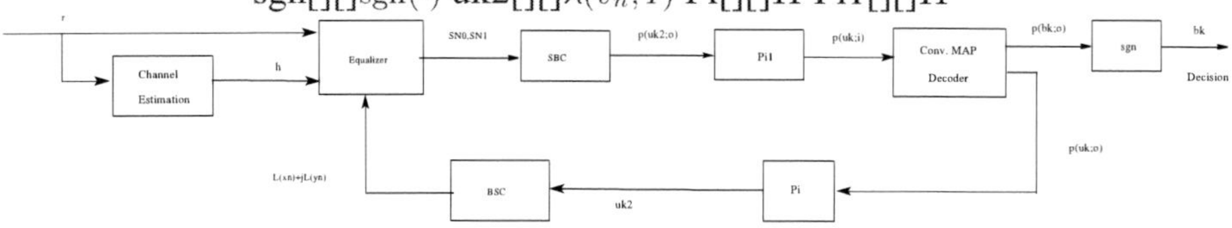

Figure 17. Block diagram for the time-domain turbo equalization.

are subsequently interleaved and each block of two coded and interleaved bits $\{v_n^0, v_n^1\}$ are mapped into one of the four QPSK symbols denoted as s_n. The symbols are transmitted over the SUI-3 channel, which has a tap spacing of 0.5μs, and maximum tap delay of 1.0μs. In the simplest scenario, when the transmitted data rate is 2×10^6 symbols/sec or 2 Msymbols/sec (the symbol duration is $T_s = 0.5\mu s$), the multipath fading is modeled as a tapped-delay line with adjacent taps equally spaced at the symbol period. The received signal is formed as $r_n = h_0 s_n + h_1 s_{n-1} + h_2 s_{n-2} + w_n$, where w_n is complex additive white Gaussian noise (AWGN) with zero mean and variance N_0, and h_0, h_1, h_2 are channel coefficients which are complex Gaussian distributed and assumed to remain constant during the transmission of one block of data. They, however, vary from block to block. The amplitude of the first tap $|h_0|$ is characterized by a Ricean distribution due to the presence of line of sight (LOS) propagation. The amplitudes of the taps $|h_1|, |h_2|$ are Rayleigh distributed. Now, suppose we would like to achieve the data rate of $q \times 2$ Msymbols/sec over the SUI-3 channel, the received signal is then reformed as

$$r_n = h_0 s_n + h_1 s_{n-q} + h_2 s_{n-2q} + w_n. \tag{31}$$

For the symbol of interest s_n, the interfering symbols are s_{n-q} and s_{n-2q}. Therefore, the ISI does not come from its next neighbouring symbols as in the previous case. The ISI span in terms of the number of symbol periods increases from 2 symbols to $2q$ symbols.

The task of the receiver is to detect the transmitted information bits $\{b_n\}$ given the received observation $\{r_n\}$. To this end, we need first to detect the transmitted symbols $\{s_n\}$ which are corrupted with ISI and AWGN noise. A time-domain equalizer can be applied to remove the detrimental effect of ISI. The estimated symbols are then converted to coded bits, which are subsequently deinterleaved and decoded to obtain an estimate of the information sequence. Here, we focus on the turbo equalization algorithm which combines equalizer and channel decoder in an iterative fashion. The existing techniques can be broadly classified into trellis MAP based [5] and filter based [7, 8, 10, 9] approaches. However, they are not designed for high speed wireless links since their complexity grows drastically with data rate as explained in Section 1.. A turbo equalization scheme suitable for reliable transmission in broadband Fixed WiMAX type systems will be introduced next.

The proposed algorithm is illustrated in Fig. 17. First, an estimate of the channel coefficients h_i is obtained using a training sequence. Based on the equalizer output y_n, the symbol-to-bit converter (SBC) computes the LLR values of the coded and interleaved bits v_n, denoted by $\lambda(v_n; O)$, which are deinterleaved to $\lambda(u_n; I)$. Based on the soft input $\lambda(u_n; I)$, a SiSo outer channel decoder computes the LLR of each information bit $\lambda(b_n; O)$ and each coded bit $\lambda(u_n; O)$. The former is used to make a decision on the transmitted information bit at the final iteration, and the latter is interleaved and passed through a bit-to-symbol converter (BSC) to derive a soft symbol estimate $\lambda(s_n)$, which is used for equalization at the next iteration. The performance is improved by repeating this process in an iterative manner. The equalization block will be described in detail next. The focus of this study is the broadband high data rate Fixed WiMAX applications, for which the received signal is expressed by (31).

2.1.1. Initial stage

We obtain an initial estimate of the transmitted symbols at the first equalization stage. To this end, a conventional MMSE based linear equalizer or a decision feedback equalizer (DFE) can be applied. However, as shown in (31), the ISI spans for a larger number of symbol periods when the data rate increases, and the length of equalizer has to be increased accordingly to accommodate the ISI. For the SUI-3 channel with 60 symbols of dispersion, the length of the linear FIR equalizer has to be over a hundred taps in order to achieve satisfactory performance, which is not feasible in practice. A simple solution is to use coherent detection, which has a poor performance for the SUI-3 channel as demonstrated in [26]. Here, we apply a coherent combining approach to tackle this problem. Let us take 3 received samples which are q symbols apart, i.e.,

$$\underbrace{\begin{bmatrix} r_n \\ r_{n+q} \\ r_{n+2q} \end{bmatrix}}_{\mathbf{r}_n} = \underbrace{\begin{bmatrix} s_n & s_{n-q} & s_{n-2q} \\ s_{n+q} & s_n & s_{n-q} \\ s_{n+2q} & s_{n+q} & s_n \end{bmatrix}}_{\mathbf{S}_n} \underbrace{\begin{bmatrix} h_0 \\ h_1 \\ h_2 \end{bmatrix}}_{\mathbf{h}} + \underbrace{\begin{bmatrix} w_n \\ w_{n+q} \\ w_{n+2q} \end{bmatrix}}_{\mathbf{w}}. \tag{32}$$

One can see from these equations that these three received samples all contain the desired symbol s_n, which not only appears in the first-tap term ($h_0 s_n$ in r_n), but also appears in the second-tap term ($h_1 s_n$ in r_{n+q}), as well as the third-tap term ($h_2 s_n$ in r_{n+2q}). When the channel taps are not equally spaced, which is likely to be the case in practice, we need adaptively allocate the received samples so that each of them contains the desired symbol. In order to yield multipath diversity gain, we should combine all of signals expressed by (32) together, i.e.,

$$y_n = \hat{\mathbf{h}}^* \mathbf{r}_n = \hat{h}_0^* r_n + \hat{h}_1^* r_{n+q} + \hat{h}_2^* r_{n+2q} = \gamma s_n + \eta_n, \tag{33}$$

where $\hat{\mathbf{h}} = \begin{bmatrix} \hat{h}_0 & \hat{h}_1 & \hat{h}_2 \end{bmatrix}^\top$ is an estimate of the channel vector $\mathbf{h} = \begin{bmatrix} h_0 & h_1 & h_2 \end{bmatrix}^\top$. The combined noise and ISI term is denoted as $\eta \sim \mathcal{CN}(0, N_\eta)$. Assuming accurate channel estimation, i.e., $\hat{h}_i \approx h_i$, its variance can be computed as $N_\eta = \gamma_0(\gamma_1 + \gamma_2 + N_0) + \gamma_1(\gamma_0 + \gamma_2 + N_0) + \gamma_2(\gamma_0 + \gamma_1 + N_0)$, where $\gamma_i = |h_i|^2$, and $\gamma = \sum_{i=0}^{2} \gamma_i$ is total received power from different paths. The conditional PDF of y_n given s_m can be derived as

$$f(y_n|s_m) = \frac{1}{\pi N_\eta} \exp\left(-\frac{|y_n - \gamma s_m|^2}{N_\eta} \right). \tag{34}$$

Based on (34), the LLR value of v_n^0 can be computed as

$$\lambda(v_n^0; O) = \ln \frac{f(y_n|v_n^0 = +1)}{f(y_n|v_n^0 = -1)} \approx \ln \frac{\exp\left(-|y_n - \gamma s_+|^2/N_\eta\right)}{\exp\left(-|y_n - \gamma s_-|^2/N_\eta\right)}$$
$$= \frac{2\gamma}{N_\eta} \mathrm{Re}\left\{s_+^* y_n - s_-^* y_n\right\}, \tag{35}$$

where s_+ and s_- are defined in the same way as in (17) for a QPSK constellation with Gray-mapping, and $\lambda(v_n^1; O)$ can be derived similarly.

2.1.2. Subsequent Stages

With a data estimate from previous stages, the contribution of ISI can be subtracted from the received samples so that the signal is less contaminated given correct decision feedback. In [7, 8, 9, 10], an MMSE filter is applied to suppress the interference. However, as demonstrated previously, when the data rate increases, the MMSE filter length has to be increased considerably in order to accommodate the ISI, which leads to prohibitive computational complexity. Instead of increasing the equalizer length to capture ISI, we only take 3 received samples which are q symbols apart (they all contain the desired symbol s_n), and cancel the contribution of interfering symbols using decision feedback, i.e.,

$$\mathbf{r}'_n = \mathbf{r}_n - \bar{\mathbf{S}}_n \hat{\mathbf{h}}, \tag{36}$$

where the matrix $\bar{\mathbf{S}}_n$ denotes the soft estimate of $\mathbf{S}_n$. Its elements are computed according to the LLR values as $\bar{s}_n = \tanh[\lambda(v_n^0)/2]/\sqrt{2} + j\tanh[\lambda(v_n^1)/2]/\sqrt{2}$. To simplify the notation, the iteration (stage) index is omitted whenever no ambiguity arises. As with the initial equalization stage expressed by (33), the complexity of the above procedure is only determined by the number of non-zero taps in the channel (which is 3 for the SUI-3 channel). When the data rate increases, the interfering symbols become further apart from the desired symbols. However, canceling the symbols that are far away is no more complex than canceling the neighboring symbols. Consequently, the complexity of the proposed scheme does not increase linearly with data rate. The above formulas can be written in vector form as

$$\mathbf{r}'_n = \begin{bmatrix} r'_n \\ r'_{n+q} \\ r'_{n+2q} \end{bmatrix} = \begin{bmatrix} h_0 \\ h_1 \\ h_2 \end{bmatrix} s_n + \begin{bmatrix} z_n \\ z_{n+q} \\ z_{n+2q} \end{bmatrix} = \mathbf{h}s_n + \mathbf{z}_n,$$

where $\mathbf{z}_n$ stands for the combined noise and interference cancellation residual vector. In order to ease the algorithm derivation, we assume perfect cancellation, $\mathbf{z}_n = \mathbf{w}_n$, and each element of $\mathbf{w}_n$ is a Gaussian random variable, i.e., $w_n \sim \mathcal{N}(0, N_0)$. Due to the assumption of perfect cancellation, this scheme is sub-optimum during the initial stages of turbo equalization, but will approach optimality when the ISI is effectively canceled as the iterative process proceeds. The same assumption has been made, e.g., in [9]. The conditional PDF of $\mathbf{r}'_n$ given s_m is thus derived as

$$f(\mathbf{r}'_n|s_m) = \frac{1}{(\pi N_0)^L} \exp\left(-\frac{\|\mathbf{r}'_n - \mathbf{h}s_m\|^2}{N_0}\right). \tag{37}$$

Based on (37), the LLR value of v_n^0 can be computed as

$$\lambda(v_n^0; O) = \ln \frac{f(\mathbf{r}'_n | v_n^0 = 0)}{f(\mathbf{r}'_n | v_n^0 = 1)} \approx \ln \frac{\exp\left(-\|\mathbf{r}'_n - \mathbf{h}s_+\|^2/N_0\right)}{\exp\left(-\|\mathbf{r}'_n - \mathbf{h}s_-\|^2/N_0\right)}$$
$$= \frac{1}{N_0}\left\{\|\mathbf{r}'_n - \mathbf{h}s_-\|^2 - \|\mathbf{r}'_n - \mathbf{h}s_+\|^2\right\} = \frac{2}{N_0}\mathrm{Re}\left\{(\mathbf{h}s_+)^*\mathbf{r}'_n - (\mathbf{h}s_-)^*\mathbf{r}'_n\right\},$$

where s_+ and s_- are defined in the same way as in (17), and $\lambda(v_n^1; O)$ can be derived similarly.

2.1.3. Complexity Comparison

Here, we compare the complexity of OFDM/SC-FDE to that of the proposed turbo equalizer. OFDM and SC-FDE have essentially the same computational complexity, both require $Nlog_2N + N$ complex multiplications for a FFT block with length N [20, 12]; whereas the turbo equalizer requires LN and $(L^2 - L)N$ complex multiplication operations for the initial stage and the subsequent stages[2], respectively, where L denotes the number of non-zero taps in the channel impulse response. Given $N = 1024, L = 3$, 11264 complex multiplications are required for OFDM/SC-FDE; while the cumulative number of complex multiplications at the first three stages of the turbo equalizer are 3072, 9216, 15360, which are comparable to that of OFDM/SC-FDE (it is shown in Section 2.3. that only three stages are required for the turbo equalizer to outperform OFDM/SC-FDE). In contrast to the traditional time-domain turbo equalization which requires a number of multiplications per symbol that is proportional to the channel delay spread, the complexity of proposed scheme only depends on L. For the WiMAX type systems where the channel only contains a few non-zero discrete taps, the difference in complexity between the proposed equalizer and OFDM/SC-FDE scheme is not significant. On the other hand, the complexity of the OFDM/SC-FDE scheme grows more rapidly with the FFT block length N (for OFDM/SC-FDE the complexity is $O(N \log_2 N)$; for turbo equalizer is $O(N)$), and N should be increased as the data rate goes higher in order to minimize the fraction of overhead due to the insertion of a cyclic prefix. Consequently, the complexity gap between OFDM/SC-FDE and the turbo equalization scheme becomes smaller as data rate increases.

Next, we consider the complexity of the channel decoder. In the field of coding theory, a conventional approach to characterize the computational complexity of a decoding algorithm is to enumerate all executed basic operations [17, 18]. The basic operations performed by the Viterbi, max log-MAP and exact log-MAP decoding algorithms include addition (`ADD`), subtraction (`SUB`), multiplication by ± 1 (`MUL`), division by 2 (`DIV`), maximum -or minimum- of two numbers (`MAX`) and table look-up (`LKUP`).

In order to evaluate the computational complexity of the Viterbi decoding algorithm for a rate-k/n convolutional code with memory order ν (note that memory order = constraint length – 1 for binary codes), we split the algorithm into three different phases [4]. First, the branch metrics of the trellis diagram are calculated (phase V1) and then the path metrics are updated (phase V2). The algorithm is completed (phase V3) when hard decisions are made and estimates of the source information bits are generated. Calculation of a branch metric

[2]These figures do not include the complexity of soft demapping which is the same for the turbo equalizer and OFDM/SC-FDE.

Table 1. Computational complexity of Viterbi, max log-MAP and exact log-MAP decoding algorithms.

Viterbi decoding algorithm						
phase	add	sub	mul	div	max	lkup
V1	$(n-1)\cdot 2^{k+\nu}$	-	$n\cdot 2^{k+\nu}$	-	-	-
V2	$2^{k+\nu}$	-	-	-	$(2^k-1)\cdot 2^\nu$	-
V3	-	-	-	-	-	k
Max log-MAP decoding algorithm						
phase	add	sub	mul	div	max	lkup
L1a	$(n+k-1)\cdot 2^{k+\nu}$	-	$(n+k)\cdot 2^{k+\nu}$	$2^{k+\nu}$	-	-
L1b	$(n-1)\cdot 2^{k+\nu}$	-	$n\cdot 2^{k+\nu}$	$2^{k+\nu}$	-	-
L2	$2^{k+\nu}$	-	-	-	$(2^k-1)\cdot 2^\nu$	-
L3	$2^{k+\nu}$	-	-	-	$(2^k-1)\cdot 2^\nu$	-
L4a	$2\cdot 2^{k+\nu}$	n	-	-	$n\cdot(2^{k+\nu}-2)$	-
L4b	$2\cdot 2^{k+\nu}$	k	-	-	$k\cdot(2^{k+\nu}-2)$	-
Exact log-MAP decoding algorithm						
phase	add	sub	mul	div	max	lkup
L1a	$(n+k-1)\cdot 2^{k+\nu}$	-	$(n+k)\cdot 2^{k+\nu}$	$2^{k+\nu}$	-	-
L1b	$(n-1)\cdot 2^{k+\nu}$	-	$n\cdot 2^{k+\nu}$	$2^{k+\nu}$	-	-
L2	$(2^{k+1}-1)\cdot 2^\nu$	$(2^k-1)\cdot 2^\nu$	-	-	$(2^k-1)\cdot 2^\nu$	$(2^k-1)\cdot 2^\nu$
L3	$(2^{k+1}-1)\cdot 2^\nu$	$(2^k-1)\cdot 2^\nu$	-	-	$(2^k-1)\cdot 2^\nu$	$(2^k-1)\cdot 2^\nu$
L4a	$(n+2)\cdot 2^{k+\nu}-2$	$n\cdot(2^{k+\nu}-1)$	-	-	$n\cdot(2^{k+\nu}-2)$	$n\cdot(2^{k+\nu}-2)$
L4b	$(k+2)\cdot 2^{k+\nu}-2$	$k\cdot(2^{k+\nu}-1)$	-	-	$k\cdot(2^{k+\nu}-2)$	$k\cdot(2^{k+\nu}-2)$

during phase V1 requires n MUL operations for the computation of the n inner products between the codeword bits associated with the branch and the received soft bits, and $(n-1)$ ADD operations for the summation of the n products. Hence, phase V1 requires $(n-1)\cdot 2^{k+\nu}$ MUL and $n\cdot 2^{k+\nu}$ ADD operations, since 2^k branches emerge from each of the 2^ν states per trellis step. Note that no a-priori information is exploited during phase V1. During the next phase, the best paths among all competing paths merging in each state are determined. Computation of the metric of a path ending in a state requires 2 ADD operations (i.e., previous path metric plus branch metric), thus computation of the metrics of all 2^k paths merging in that state requires 2^k ADD operations. Furthermore, selection of the surviving path among all competing paths in that state involves (2^k-1) MAX operations. Taking into account that the same operations are repeated for each of the 2^ν states of the trellis, we conclude that phase V2 requires a total of $2^{k+\nu}$ ADD and $(2^k-1)\cdot 2^\nu$ MAX operations per trellis step. Finally, k hard bits are obtained after k LKUP operations per trellis step [27], at the end of phase V3. The computational requirements of the Viterbi algorithm are summarized in Table 1.

The exact log-MAP and max-log-MAP algorithms can also be split into the following phases [17, 28]: branch metrics calculation with or without a-priori information for the source bits (phase L1a or L1b, respectively), forward metrics calculation (phase L2), backward metrics calculation (phase L3) and soft decision for either the codeword bits (phase L4a) or the source information bits (phase L4b). The number of basic operations for each phase are depicted in Table 1. Note that the max log-MAP algorithm uses the MAX operation to approximate the forward and backward metrics as well as the soft output values in phases L2, L3 and L4, respectively, whilst the exact log-MAP corrects this approximation using additional operations.

A fair comparison between two decoding algorithms can only be made if all operations are broken down into unit operations, namely *equivalent additions* [27]. This approach delivers results with wider applicability, since the complexity measure is not tied to specific hardware implementations. Let us introduce the notation $\mathcal{C}(\cdot)$ to represent the computational complexity of a scheme, algorithm, phase of an algorithm or operation, expressed in equivalent additions. As explained in [27, 29], the complexity of the basic operations is $\mathcal{C}(\texttt{ADD}) = \mathcal{C}(\texttt{SUB}) = \mathcal{C}(\texttt{MUL}) = \mathcal{C}(\texttt{DIV}) = 1$, whilst $\mathcal{C}(\texttt{MAX}) = 2$ and $\mathcal{C}(\texttt{LKUP}) = 3$. The aggregate computational complexity of an algorithm can be evaluated by summing up the complexities of the constituent phases. For example, based upon the data in Table 1, the computational complexity of the Viterbi algorithm can be obtained as follows

$$\begin{aligned}\mathcal{C}(\text{Viterbi}) &= \mathcal{C}(\text{V1}) + \mathcal{C}(\text{V2}) + \mathcal{C}(\text{V3}) \\ &= \big((n+1)\cdot 2^k - 1\big)\cdot 2^{\nu+1} + 3k\end{aligned} \tag{38}$$

since

$$\begin{aligned}\mathcal{C}(\text{V1}) &= (n-1)\cdot 2^{k+\nu}\cdot\mathcal{C}(\texttt{ADD}) + n\cdot 2^{k+\nu}\cdot\mathcal{C}(\texttt{MUL}) &&= (2n-1)\cdot 2^{k+\nu}, \\ \mathcal{C}(\text{V2}) &= 2^{k+\nu}\cdot\mathcal{C}(\texttt{ADD}) + (2^k-1)\cdot 2^{\nu}\cdot\mathcal{C}(\texttt{MAX}) &&= 3\cdot 2^{k+\nu} - 2^{\nu+1}, \\ \mathcal{C}(\text{V3}) &= k\cdot\mathcal{C}(\texttt{LKUP}) &&= 3k.\end{aligned}$$

The computational complexity of the max log-MAP and exact log-MAP algorithms, which both produce soft outputs, can be evaluated in a similar fashion. However, the number of iterations between the components that exchange soft information at the receiver also needs to be taken into account. For example, let us consider a scheme that employs turbo equalization (TE), according to which the equalizer and a soft-output convolutional decoder exchange LLR values in an iterative manner. Note that only the equalizer exploits a-priori knowledge of the codeword bits. If I_o denotes the number of iterations between the two components, the decoder will produce soft values for the codeword bits at the end of each of the $(I_o - 1)$ first iterations, to assist the equalizer in generating better estimates of the received symbols. However, at the end of the last iteration the decoder outputs soft values for the information bits before the decoding algorithm terminates. Consequently, the overall decoding computational complexity of a turbo equalization scheme employing a soft-output algorithm, either exact log-MAP or max log-MAP, can be expressed as follows

$$\begin{aligned}\mathcal{C}(\text{TE}) = (I_o - 1)&\cdot\big(\mathcal{C}(\text{L1b}) + \mathcal{C}(\text{L2}) + \mathcal{C}(\text{L3}) + \mathcal{C}(\text{L4a})\big) \\ &+ \mathcal{C}(\text{L1b}) + \mathcal{C}(\text{L2}) + \mathcal{C}(\text{L3}) + \mathcal{C}(\text{L4b}) \\ = I_o&\cdot\big(\mathcal{C}(\text{L1b}) + \mathcal{C}(\text{L2}) + \mathcal{C}(\text{L3})\big) + (I_o - 1)\cdot\mathcal{C}(\text{L4a}) + \mathcal{C}(\text{L4b}).\end{aligned} \tag{39}$$

Fig. 18 compares the computational complexity of a system using non-iterative equalization and Viterbi decoding for a rate-1/3 convolutional code to that of a system employing turbo equalization. In both cases, it has been assumed that the complexity of the equalizer is not significant, hence the complexity of the overall system is mainly determined by the complexity of the decoding algorithm. As an example, we observe in Fig. 18 that a scheme

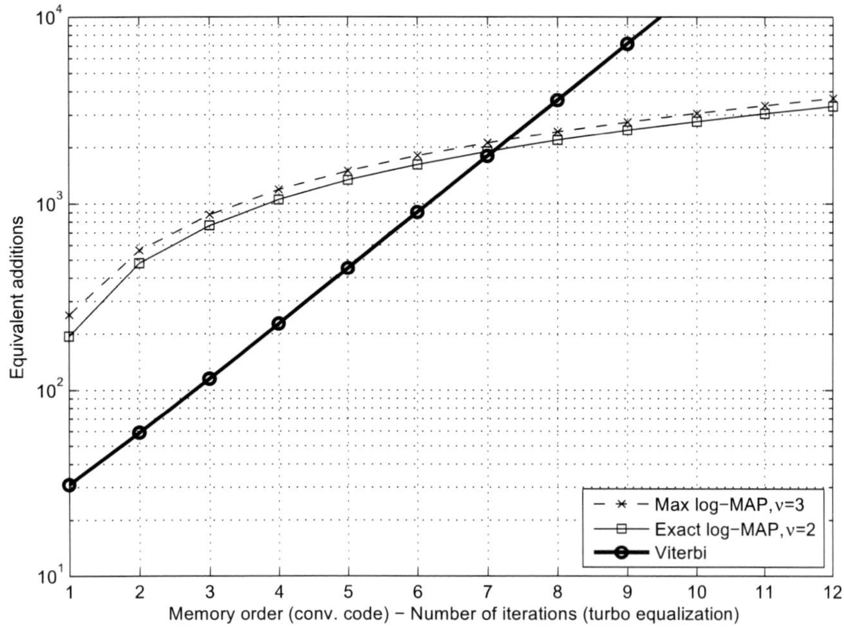

Figure 18. Computational complexity of systems employing rate-1/3 convolutional coding combined with iterative or non-iterative equalization. In the case of iterative equalization, the channel decoder uses either the exact log-MAP or the max log-MAP decoding algorithm. The Viterbi algorithm is used when there is no information exchange between the decoder and the equalizer. The horizontal axis represents the memory order of the convolutional code in a non-iterative scheme or the number of iterations in a scheme employing turbo equalization.

employing the exact log-MAP algorithm for a code of memory order $\nu = 2$ yields similar computational complexity after 7 iterations to that of a non-iterative scheme using the conventional Viterbi algorithm for a code of memory order $\nu = 7$. Furthermore, it is obvious to see that increasing the number of iterations makes the decoding process more computational expensive, while latency builds up; consequently, it is not worthwhile to increase the number of iterations above a value beyond which the performance improvement is insignificant.

A similar complexity analysis for schemes using turbo decoding was carried out in [29]. The authors considered various MIMO configurations using OFDM/QPSK over the SUI-3 channel and presented a performance comparison between turbo-coded and convolutional-coded Fixed WiMAX systems, under the condition of identical computational complexity. The results are reproduced in Fig. 19. It was demonstrated that, in contrast to the AWGN case, turbo coding does not offer a significant performance advantage over convolutional coding in Fixed WiMAX systems without antenna diversity or with limited antenna diversity. As the number of antennas is increased, turbo codes eventually achieve a substantial performance gain over convolutional codes [29].

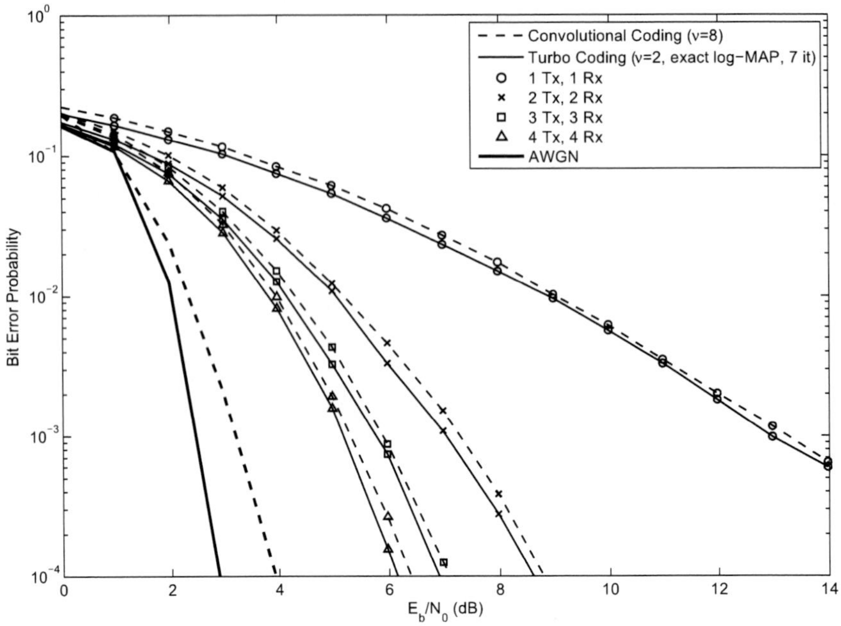

Figure 19. Error rates for turbo-coded and convolutional-coded OFDM systems on an SUI-3 channel for both SISO and MIMO antenna configurations. The input frame size is 1021 bits [29].

2.2. Time-Domain Turbo Equalization for MIMO Systems

Successful WiMAX systems should offer high bit rates and high spectrum efficiency. MIMO technology offers significant leverages to enable such features. Use of multiple antennas along with efficient coding and equalization techniques will greatly increase power and spectrum efficiencies. The challenge is to develop and deliver a well designed WiMAX system that captures the capabilities of MIMO technology without sacrificing robustness, simplicity and cost.

In this section, we extend the aforementioned turbo equalization algorithm to MIMO systems and derive a space-time turbo equalization algorithm which is well-suited to broadband Fixed WiMAX applications. The transmission system under study is shown in Fig. 20. The information sequence $\{b_n\}$ is encoded into coded bits $\{u_n\}$, which are subsequently interleaved and each block of two coded and interleaved bits $\{v_n^0, v_n^1\}$ is mapped into one of the four QPSK symbols. We use the space-time coding scheme proposed by Alamouti [14]. The transmitted symbols are space-time encoded according to the generator matrix $\mathbf{G} = \begin{bmatrix} s_n^0 & s_n^1 \\ -s_n^{1*} & s_n^{0*} \end{bmatrix}$, where s_n^0, s_n^1 denote modulation symbols. The encoding and transmission mechanism is shown in Table 2. The transmitted symbols are grouped into blocks of 2 symbols at each antenna. At a given time, two symbols are simultaneously transmitted from the two antennas. At time instance t, the symbol transmitted from antenna zero is denoted as s_n^0, and the symbol transmitted from antenna one is denoted as s_n^1. During the next symbol period $t+T$, symbol $-s_n^{1*}$ is transmitted from antenna zero, and s_n^{0*} is transmitted from antenna one.

bn[][]b_n rn[][]r_n vn[][]v_n wn[][]w_n yn[][]y_n un[][]u_n un$_p$[][]v$_n$ Int[][]$\prod$ ISI[][]SUI-3 Channel[][]Channel QPSK[][]QPSK Modu.[][]Modu. Encoder[][]Encoder Conv.[][] Channel sn0 -sn1*[][]$s_n^0 \quad - s_n^{1*}$ sn1 sn0*[][]$s_n^1 \quad s_n^{0*}$

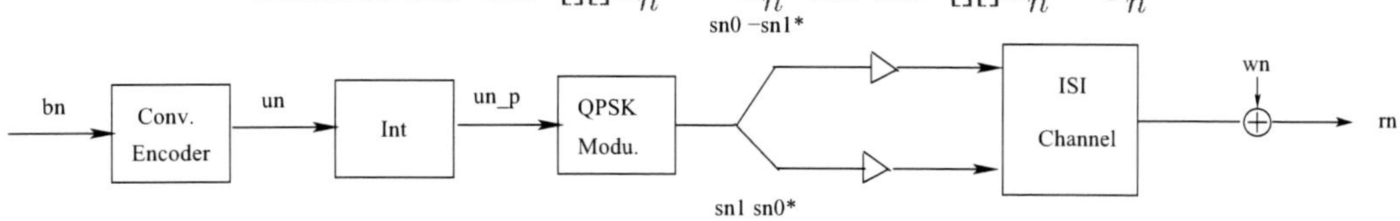

Figure 20. Block diagram of transmitter for single-carrier STBC coded Fixed WiMAX type system.

Table 2. The encoding and transmission sequence for the 2 Tx antenna system.

	$t-2T$	$t-T$	t	$t+T$	$t+2T$	$t+3T$
antenna 0	s_{n-1}^0	$-s_{n-1}^{1*}$	s_n^0	$-s_n^{1*}$	s_{n+1}^0	$-s_{n+1}^{1*}$
antenna 1	s_{n-1}^1	s_{n-1}^{0*}	s_n^1	s_n^{0*}	s_{n+1}^1	s_{n+1}^{0*}

For simplicity, we assume two transmit antennas and one receive antenna in the derivation of the proposed algorithm (the full rate STBC with complex constellation do not exist for more than two transmit antennas). However, its extension to a $2 \times N_R$ system with multiple receive antennas is straightforward. Each complex channel coefficient is denoted as h_{ij}^l where the first (second) subscript $i(j)$ is the index of the transmit (receive) antenna, the superscript l refers to the number of the channel tap.

In the case where the transmitted data rate is 2 Msymbols/s, the multipath fading is modeled as a tapped-delay line with adjacent taps spaced equally at the symbol duration. The signals at receive antenna during the two symbol periods can be formed according to Table 2 as $r_n^0 = h_{00}^2 s_{n-1}^0 - h_{00}^1 s_{n-1}^{1*} + h_{00}^0 s_n^0 + h_{10}^2 s_{n-1}^1 + h_{10}^1 s_{n-1}^{0*} + h_{10}^0 s_n^1 + w_n^0$;and $r_n^1 = -h_{00}^2 s_{n-1}^{1*} + h_{00}^1 s_n^0 - h_{00}^0 s_n^{1*} + h_{10}^2 s_{n-1}^{0*} + h_{10}^1 s_n^1 + h_{10}^0 s_n^{0*} + w_n^1$, where w_n^0, w_n^1 are the complex additive white Gaussian noise with zero mean and variance N_0. When the data rate is increased to $q \times 2$ Msymbols/s, the received signals are reformed as

$$r_n^0 = \begin{cases} h_{00}^2 s_{n-q}^0 - h_{00}^1 s_{n-\frac{q+1}{2}}^{1*} + \underline{h_{00}^0 s_n^0} + h_{10}^2 s_{n-q}^1 + h_{10}^1 s_{n-\frac{q+1}{2}}^{0*} + \underline{h_{10}^0 s_n^1} + w_n^0, & \text{q is odd;} \\ h_{00}^2 s_{n-q}^0 + h_{00}^1 s_{n-\frac{q}{2}}^0 + \underline{h_{00}^0 s_n^0} + h_{10}^2 s_{n-q}^1 + h_{10}^1 s_{n-\frac{q}{2}}^1 + \underline{h_{10}^0 s_n^1} + w_n^0, & \text{q is even;} \end{cases}$$

$$r_n^1 = \begin{cases} -h_{00}^2 s_{n-q}^{1*} + h_{00}^1 s_{n-\frac{q-1}{2}}^0 - \underline{h_{00}^0 s_n^{1*}} + h_{10}^2 s_{n-q}^{0*} + h_{10}^1 s_{n-\frac{q-1}{2}}^1 + \underline{h_{10}^0 s_n^{0*}} + w_n^1, & \text{q is odd;} \\ -h_{00}^2 s_{n-q}^{1*} - h_{00}^1 s_{n-\frac{q}{2}}^{1*} - \underline{h_{00}^0 s_n^{1*}} + h_{10}^2 s_{n-q}^{0*} + h_{10}^1 s_{n-\frac{q}{2}}^{0*} + \underline{h_{10}^0 s_n^{0*}} + w_n^1, & \text{q is even;} \end{cases} \tag{40}$$

For the symbols of interest s_n^0 and s_n^1 (which are underlined in the above equations so that they can be distinguished from the interference and the noise), the interfering symbols are s_{n-q}^0, $s_{n-\frac{q}{2}}^0$, s_{n-q}^1, etc. The higher the data rate is, the larger is the number of symbol intervals spanned by the ISI, i.e., the value of q becomes greater.

The space-time turbo equalization procedure is basically the same as in the SISO case illustrated in Fig. 17 except the mechanism inside the equalizer block is somewhat different. The inputs to the equalizer are channel estimate $\hat{h}^l_{ij}$ and symbol estimate in the form of LLR $\lambda(s_n)$ where s_n denotes either s^0_n or s^1_n. The output of equalizer is denoted as $\widetilde{s_n}$, which is the soft decision of s_n. The soft estimate of the symbol is then mapped to the LLR values of coded bits $\lambda(v_n; O)$ by the SBC, and deinterleaved to yield $\lambda(u_n; I)$. The decoding process is the same as in the SISO case. The proposed space-time turbo equalization algorithm will be described in detail next. The focus of this study is the broadband Fixed WiMAX applications, for which the received signal is expressed by (40).

2.2.1. Initial Stage

The Alamouti algorithm was originally developed for flat fading channels and so does not take into consideration the ISI introduced by frequency-selective fading channels. Some modifications have to be made in order to combat ISI and obtain multipath diversity gain. Let us take pairs of received samples that are $q/2$-symbol interval apart (assume q is even)

$$\begin{bmatrix} r^0_n \\ r^1_n \\ r^0_{n+q/2} \\ r^1_{n+q/2} \\ r^0_{n+q} \\ r^1_{n+q} \end{bmatrix} = \begin{bmatrix} s^0_{n-q} & s^0_{n-\frac{q}{2}} & \underline{s^0_n} & s^1_{n-q} & s^1_{n-\frac{q}{2}} & \underline{s^1_n} \\ -s^{1*}_{n-q} & -s^{1*}_{n-\frac{q}{2}} & \underline{-s^{1*}_n} & s^{0*}_{n-q} & s^{0*}_{n-\frac{q}{2}} & \underline{s^{0*}_n} \\ s^0_{n-q/2} & \underline{\underline{s^0_n}} & s^0_{n+q/2} & s^1_{n-q/2} & \underline{\underline{s^1_n}} & s^1_{n+q/2} \\ -s^{1*}_{n-q/2} & \underline{\underline{-s^{1*}_n}} & -s^{1*}_{n+q/2} & s^{0*}_{n-q/2} & \underline{\underline{s^{0*}_n}} & s^{0*}_{n+q/2} \\ \underbrace{s^0_n} & s^0_{n+\frac{q}{2}} & s^0_{n+q} & \underbrace{s^1_n} & s^1_{n+\frac{q}{2}} & s^1_{n+q} \\ \underbrace{-s^{1*}_n} & -s^{1*}_{n+\frac{q}{2}} & -s^{1*}_{n+q} & \underbrace{s^{0*}_n} & s^{0*}_{n+\frac{q}{2}} & s^{0*}_{n+q} \end{bmatrix} \begin{bmatrix} h^2_{00} \\ h^1_{00} \\ h^0_{00} \\ h^2_{10} \\ h^1_{10} \\ h^0_{10} \end{bmatrix} + \begin{bmatrix} w^0_n \\ w^1_n \\ w^0_{n+q/2} \\ w^1_{n+q/2} \\ w^0_{n+q} \\ w^1_{n+q} \end{bmatrix}. \tag{41}$$

From (40) and (41), one can see that the desired symbols s^0_n, s^1_n not only appear in the first-tap terms (with one line underneath), but also appear in the second-tap terms (with two lines underneath), as well as in the third-tap terms (with underbrace). In order to take advantage of multipath propagation and obtain diversity gain, we should apply the Alamouti detection scheme on all the three taps and combine the desired signals from different taps. The three-tap combining scheme is expressed as

$$\begin{aligned} \widetilde{s^0_n} &= \hat{h}^{0*}_{00} r^0_n + \hat{h}^0_{10} r^{1*}_n + \hat{h}^{1*}_{00} r^0_{n+q/2} + \hat{h}^1_{10} r^{1*}_{n+q/2} + \hat{h}^{2*}_{00} r^0_{n+q} + \hat{h}^2_{10} r^{1*}_{n+q} \\ &= \sum_{i,l} \hat{h}^{l*}_{i0} h^l_{i0} s^0_n + \xi^0_n = \gamma s^0_n + \xi^0_n; \\ \widetilde{s^1_n} &= \ldots = \sum_{i,l} \hat{h}^{l*}_{i0} h^l_{i0} s^1_n + \xi^1_n, \end{aligned} \tag{42}$$

where $\gamma = \sum_{i,l} \hat{h}^{l*}_{i0} h^l_{i0}$ is the total received power from different paths, and $\xi^0_n, \xi^1_n \sim \mathcal{CN}(0, N_\xi)$. The variance N_ξ is computed with noise variance and channel coefficients. The conditional PDFs of $\widetilde{s^0_n}$ and $\widetilde{s^1_n}$ given s_m are thus derived as

$$f(\widetilde{s^0_n}|s_m) = \frac{1}{\pi N_\xi} \exp\left(-\frac{|\widetilde{s^0_n} - \gamma s_m|^2}{N_\xi}\right); \quad f(\widetilde{s^1_n}|s_m) = \frac{1}{\pi N_\xi} \exp\left(-\frac{|\widetilde{s^1_n} - \gamma s_m|^2}{N_\xi}\right).$$

The LLR values can be derived based on the PDFs as shown in (35).

2.2.2. Subsequent Stages

The summation in (42) is carried out over all possible values of $i \in \{0,1\}$, and $l \in \{0,1,2\}$. Therefore, this scheme also leads to temporal diversity gain in addition to the spatial diversity gain obtained by the original Alamouti scheme. On the other hand, however, ξ_n^0, ξ_n^1 in (42) also contain a lot of ISI terms, which in turn will have a detrimental effect on the overall system performance. In order to tackle this problem, we employ the multistage interference cancellation technique to cancel the contribution of the ISI terms. Let us denote $\bar{s}_{n-i}^0$, $\bar{s}_{n-i}^1$ as a soft estimate of s_{n-i}^0, s_{n-i}^1 from previous stage. Given a channel estimate $\hat{h}_{ij}^l$ and symbol estimates $\{\bar{s}_{n-i}^0, \bar{s}_{n-i}^1\}$, the ISI canceled version of the received signal r_n^0, denoted as $\bar{r}_n^0$ can be written according to (40) as

$$\bar{r}_n^0 = (h_{00}^2 s_{n-q}^0 - \hat{h}_{00}^2 \bar{s}_{n-q}^0) - (h_{00}^1 s_{n-\frac{q}{2}}^0 - \hat{h}_{00}^1 \bar{s}_{n-\frac{q}{2}}^0) + \underline{h_{00}^0 s_n^0}$$
$$+ (h_{10}^2 s_{n-q}^1 - \hat{h}_{10}^2 \bar{s}_{n-q}^1) + (h_{10}^1 s_{n-\frac{q}{2}}^1 - \hat{h}_{10}^1 \bar{s}_{n-\frac{q}{2}}^1) + \underline{h_{10}^0 s_n^1} + w_n^0. \tag{43}$$

At the beginning of the iterative process, the symbol estimates $\{\bar{s}_{n-i}^0, \bar{s}_{n-i}^1\}$ needed for interference cancellation are derived by the three-tap combining algorithm expressed by (42). In the following stages, they are obtained from the output of the Log-MAP decoder. Other ISI canceled versions of the received signals, e.g., $\bar{r}_n^1, \bar{r}_{n+q/2}^0, \bar{r}_{n+q/2}^1, \bar{r}_{n+q}^0, \bar{r}_{n+q}^1$ can be formed similarly, i.e., by canceling the contribution from the symbols other than s_n^0, s_n^1. Using the aforementioned combining technique, the soft decisions of s_n^0, s_n^1 can now be formed based upon the ISI canceled signals as

$$\widetilde{s_n^0} = \hat{h}_{00}^{0*}\bar{r}_n^0 + \hat{h}_{10}^0\bar{r}_n^{1*} + \hat{h}_{00}^{1*}\bar{r}_{n+q/2}^0 + \hat{h}_{10}^1\bar{r}_{n+q/2}^{1*} + \hat{h}_{00}^{2*}\bar{r}_{n+q}^0 + \hat{h}_{10}^2\bar{r}_{n+q}^{1*} \tag{44}$$
$$\epsilon_n^0 = \gamma s_n^0 + \epsilon_n^0;$$
$$\widetilde{s_n^1} = \ldots = \sum_{i,l} \hat{h}_{i0}^{l*} h_{i0}^l s_n^1 + \epsilon_n^1 = \gamma s_n^1 + \epsilon_n^1,$$

where ϵ_n^0, ϵ_n^1 denote the noise plus cancellation residual. Given correct decision feedback, all the ISI terms will be eliminated. The variance of ϵ_n^0, ϵ_n^1 will be much smaller than that of ξ_n^0 ξ_n^1 in (42), consequently, the BER performance will be greatly improved.

As indicated by (43) and (**??**), the proposed scheme only requires linear processing at the receiver (most linear MMSE based equalizers would require matrix inversion), and the complexity of the above procedure is not affected by the value of q. As analyzed previously, the complexity of the equalizer is comparable to that of OFDM/SC-FDE for channels having a small number of non-zero taps, and the increase in complexity is mainly due to the Log-MAP decoding.

Assuming perfect cancellation, ϵ_n^0, ϵ_n^1 only contain the noise component, and can be denoted as $\epsilon_n^0, \epsilon_n^1 \sim \mathcal{CN}(0, \gamma N_0)$. The conditional PDF is thus derived as

$$f(\widetilde{s_n^0}|s_m) = \frac{1}{\pi\gamma N_0}\exp\left(-\frac{|\widetilde{s_n^0} - \gamma s_m|^2}{\gamma N_0}\right); \quad f(\widetilde{s_n^1}|s_m) = \frac{1}{\pi\gamma N_0}\exp\left(-\frac{|\widetilde{s_n^1} - \gamma s_m|^2}{\gamma N_0}\right). \tag{45}$$

The LLR value $\lambda(v_n^0; O), \lambda(v_n^1; O)$ can be derived from $f(\widetilde{s_n^0}|s_m)$; the LLRs $\lambda(v_{n+1}^0; O), \lambda(v_{n+1}^1; O)$ can be computed based on $f(\widetilde{s_n^1}|s_m)$ as shown in (17).

2.3. Performance Comparison

Performance comparison between frequency and time-domain solutions is provided in this section. We employ a rate $1/3$ convolutional code with constraint length 5 and generator polynomials $(25, 33, 37)_8$. During each Monte-Carlo run, the block size is set to 1360 information bits followed by 4 tails bits to terminate the trellis. Four zeros are appended at the end of the bit sequence to make the total number of transmitted bits equal to 2^{12}. For simplicity, we assume perfect channel estimation. It was shown in [30] that the channel estimation error can be made arbitrarily small provided that the training sequence is sufficiently long. The coded bits are interleaved by a random interleaver and mapped into QPSK symbols, which are transmitted over the SUI-3 channel. The simulation curves are obtained by averaging the simulation results over a minimum of 1000 blocks of transmitted data and after at least 100 errors have occurred. SC-FDE is implemented using the MMSE criterion. The parameters for the OFDM/SC-FDE systems are: the data block duration $T = 12.8\mu s$, cyclic prefix duration $T_{\text{CP}} = 1.6\mu s$, and $N = 1024$ sub-carriers[3]. Therefore, each FFT/IFFT sample duration is $T_s = \frac{T}{N} = 0.0125\mu s$. For the studied system with rate $1/3$ convolutional code and QPSK modulation, the corresponding information sequence data rate is approximately 54 Mbps, and the rate is kept the same for the system with OFDM/SC-DFE and the one with time-domain turbo equalization (TD-TE).

The performance comparison between OFDM, SC-FDE and TD-TE with QPSK Gray mapping is given in Fig. 21 and Fig. 22 for the 1Tx-1Rx and 2Tx-2Rx systems, respectively. The Log-MAP algorithm is used for channel decoding. It is observed that the turbo equalization algorithm converges after 3 or 4 iterations, beyond which the performance improvement is negligible. This indicates that the latency introduced by the iterative process is moderate, since the most significant gains are obtained within 3 iterations. One can see from both figures that the TD-TE outperforms the non-iterative OFDM/SC-FDE schemes at the 3rd iteration, and the performance gain is more significant in the 2Tx-2Rx (MIMO) system than in the 1Tx-1Rx (SISO) system. Comparing to the SISO system, the performance gain by adding one receive antenna and one transmit antenna is $8 \sim 9$ dB. Here, E_b refers to the transmitted bit energy (power loss due to cyclic prefix is taken into account), and is not affected by the number of receive antennas. The gain would be $5 \sim 6$ dB if we define E_b as the received bit energy. As indicated by Fig. 21 and Fig. 22, OFDM performs better than SC-FDE by a very small margin in both SISO and MIMO systems, which concurs with the results given in Section 1.4..

In Fig. 22, we also compare the performance of STBC/QPSK (using either TD-TE, OFDM or SC-FDE) to that of spatial multiplexing (SM) [31], also referred to as V-BLAST [32], which is a technique aiming at increasing the bit rates in wireless radio links. The latter is BPSK modulated in order to keep the same throughput as the STBC coded systems. Unfortunately, the V-BLAST system cannot operate at such low SNR values as shown by the topmost curve. Although V-BLAST offers high data transmission rates, STBC can

[3]256-FFT is specified in IEEE 802.16 standard. The value of N is chosen to be much larger here in order to reduce the SNR loss incurred by the cyclic prefix.

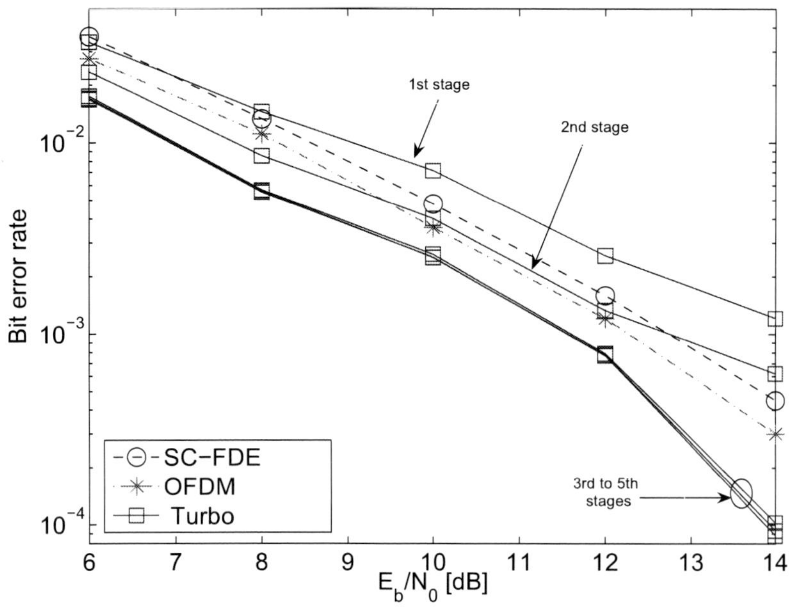

Figure 21. Performance comparison of different schemes for 1Tx-1Rx (SISO) system.

achieve the same data rates by implementing higher signal constellation, and at the same time, offering considerably better reliability.

With the parameters setting in our simulations, the channel delay spread spans over 80 symbols for a single-carrier system. This would mean a transversal filter with over one hundred taps, and at least several hundreds multiplication operations per data symbol if conventional time-domain equalization schemes are to be applied. The complexity of proposed turbo scheme only depends on the number of non-zero components in the channel, it is particularly suited to the Fixed WiMAX applications since the channel conditions are relatively static in nature and a 3-tap model can adequately describe the channel [24, 33]. Complications would arise when extending it to more dynamic systems, such as the Mobile WiMAX or a cellular network in which the channels including the number of channel taps are time-varying. The presented turbo scheme would still be applicable if a fixed number of non-zero taps in the equalizer can be allocated adaptively to capture the ISI of the channel, i.e., an equalizer is adaptively assigned with a limited number of taps in order to yield a reduction in complexity in comparison with a conventional sample spaced equalizer. However, it might be more advantageous to apply OFDM/SC-FDE under such circumstances since the frequency-domain solutions would generally work in spite of the variations in the channel conditions.

Fig. 23 compares the performance of the time-domain turbo equalization for 2Tx-2Rx systems, one with QPSK anti-Gray mapping to one with Gray-mapping. The TD-TE with anti-Gray mapping is designed similarly as shown in Section 1.2.2. for the iterative OFDM system. Here, we use a convolutional code with code rate $R_c = 1/2$, constraint length

OFDM[][]OFDM VBLAST[][]VBLAST SC-FDE[][]SC-FDE Turbo[][]TD-TE

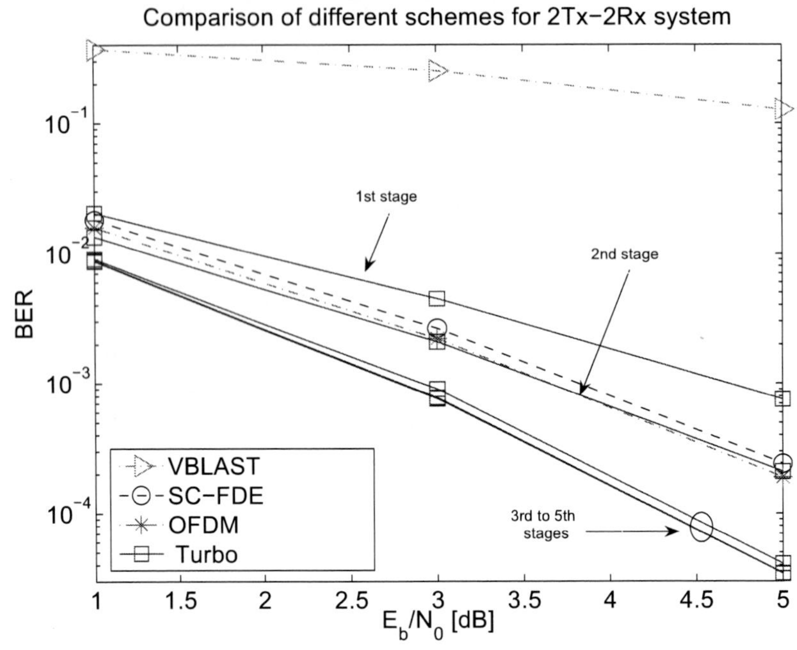

Figure 22. Performance comparison of different schemes for 2Tx-2Rx (MIMO) system.

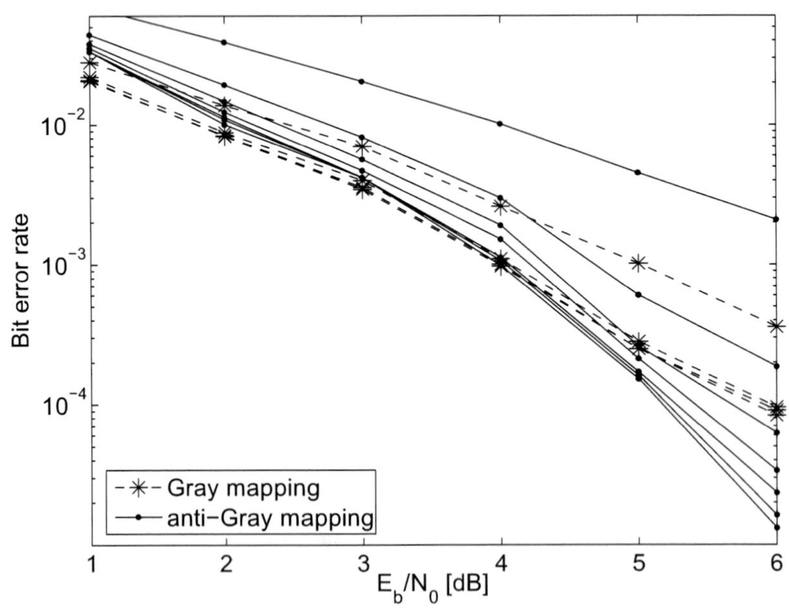

Figure 23. Performance of the time-domain turbo equalization for 2Tx-2Rx (MIMO) system. The employed convolutional code has rate $1/2$ and memory 2. The curves with the same marker represent the performance of the same system at different iterations.

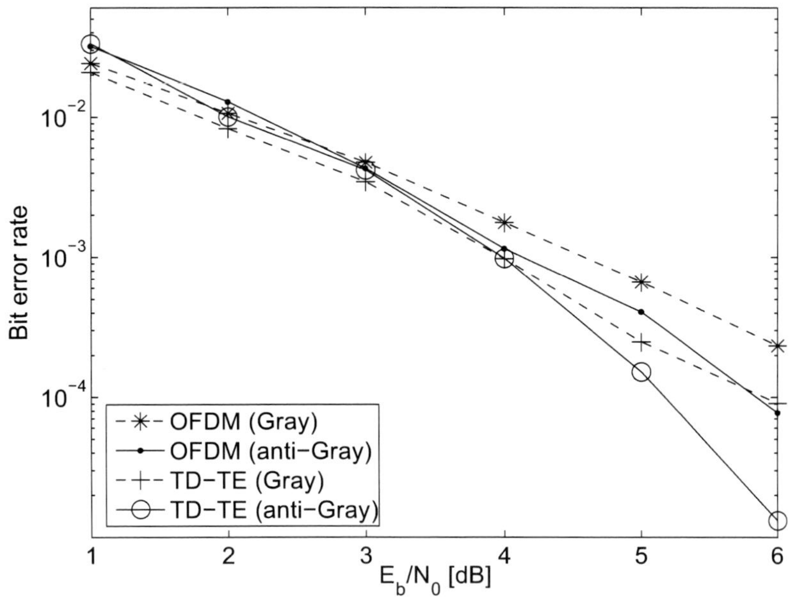

Figure 24. TD-TE vs. OFDM; Gray mapping vs. anti-Gray mapping for 2Tx-2Rx (MIMO) system. The employed convolutional code has rate $1/2$ and memory 2.

$L_c = 3$, generator ploynomial $(5, 7)$ in octal form, and the Max-Log-MAP algorithm for channel decoding. As can be seen from the figure, in the high SNR region ($E_b/N_0 > 4$ dB), the performance of the proposed turbo equalizer can be further improved by applying anti-Gray mapping. A performance gain over Gray mapping of up to 0.8 dB is observed at a target BER$= 10^{-4}$. However, the system with anti-Gray mapping has to iterate 5-7 times to reach convergence depending upon the SNR level, comparing to 2-3 times for the system with Gray-mapping. The performance improvement comes at the expense of higher computational complexity and longer detection latency caused by the slower convergence of the system with anti-Gray mapping.

Fig. 24 shows the performance comparison between TD-TE and OFDM with both Gray and anti-Gray mappings. The curves are plotted after the systems have reached convergence. The parameter settings are the same as in Fig. 23 and Fig. 15 in Section 1.4.. All the systems have comparable performance in the low SNR region (when $E_b/N_0 < 4$ dB); wheareas noticeable performance gains can be obtained in the high SNR region (when $E_b/N_0 > 4$ dB) by applying anti-Gray mapping to both TD-TE and OFDM. At a target BER$= 10^{-4}$, TD-TE with anti-Gray mapping outperforms the other systems by at least 1 dB.

3. Conclusions

Different alternatives for achieving high data rate transmission in Fixed WiMAX type systems have been studied and compared in this book chapter. Numerical results show that, with the deployment of MIMO technique and properly designed receiver algorithms, both frequency and time-domain based WiMAX systems ensure reliable transmission at a high data rate. Compared with OFDM/SC-FDE, the proposed time-domain turbo equalizers achieve better performance at the cost of higher computational complexity incurred by the iterative process. We also show that anti-Gray bit-to-symbol mapping can be employed in both approaches to improve the system performance. In contrast to the existing time-domain equalization algorithms, the complexity of which increases drastically with data rate, the proposed time-domain equalization schemes are particularly suited for broadband wireless applications. By comparison, OFDM/SC-FDE represents a more conservative solution, that is it is not necessarily optimum, but guarantees operation in most environments; whereas TD-TE maximizes performance by exploiting both the spatial diversity offered by multiple antenna systems and temporal diversity obtained from multipath propagation. Evolution of WiMAX systems will continue for many years to come and we believe that the results from this work could provide a valuable source of information for future versions of the IEEE 802.16 standard.

References

[1] H. Bolcskei, A. Paulraj, K. Hari, R. Nabar, W. Lu. "Fixed broadband wireless access: state of the art, challenges, and future directions". *IEEE Communication Magazine*, pp. 100–108, Jan. 2001.

[2] IEEE 802.16 Working Group on Broadband Wireless Access Standards. Available at *http://grouper.ieee.org/groups/802/16/*.

[3] C. Eklund. "IEEE Standard 802.16: A technical overview of the wirelessMAN air interface for broadband wireless access". *IEEE Communications Magazine*, pp. 98–107, June, 2002.

[4] J. Proakis. *Digital Communications*, 4th edition, McGraw-Hill, 2000.

[5] C. Douillard. "Iterative correction of intersymbol interference: turbo-equalization". *European Transactions on Telecommunications*, pp. 507-511, Sept. 1995.

[6] C. Berrou, A. Glavieux, P. Thitimajshima. "Near Shannon limit error-correcting coding and decoding". *Proc. IEEE International Conference on Communications*, pp. 1064–1070, June 1993.

[7] K. Narayanan. " Turbo equalization". In *Wiley Encyclopedia of Telecommunications*, vol. 5, pp. 2716–2727, 2002.

[8] M. Tuchler, R. Koetter, A. Singer. "Turbo equalization: principles and new results", *IEEE Transactions on Communications*, vol. 50, pp. 754–767, May 2002.

[9] C. Laot, A. Glavieux, Joel Labat. "Turbo equalization: adaptive equalization and channel decoding jointly optimized". *IEEE Journal on Selected Areas in Communications*, vol. 19, no. 9, Sept. 2001.

[10] M. Tuchler, A. Singer, R. Koetter. "Minumum mean square error equalization using a priori information", *IEEE Transactions on Signal Processing*, vol. 50, no. 3, March 2002.

[11] I. Koffman. "Broadband fixed wireless access solutions based on OFDM access in IEEE802.16". *IEEE Communication Magazine*, vol. 40, pp. 96–103, April, 2002.

[12] D. Falconer, S. Ariavisitakul, A. Benyamin-Seeyar, B. Edison. "Frequency domain equalization for single-carrier broadband wireless systems". *IEEE Communications Magazine*, pp. 58–66, April 2002.

[13] H. Sari, G. Karam, I. Jeanclaude. "Transmission techniques for digital terrestrial TV broadcasting". *IEEE Commun. Mag.*, pp. 100–109, Feb. 1995.

[14] A. Alamouti. "A simple transmit diversity technqiue for wireless communications". *IEEE Journal on Selected Areas in Communications*, vol. 16, no. 8, pp. 1451–1458, Oct. 1998.

[15] V. Tarokh, H. Jafarkhani and A. R. Calderbank. "Space-time block codes from orthogonal designs". *IEEE Transactions on Information Theory*, vol. 45, no. 5, pp. 1456–1467, July 1999.

[16] D. Agrawal, V. Tarokh, A. Naguib and N. Seshadri. "Space-time coded OFDM for high data-rate wireless communication over wideband channels". Proc. *IEEE Vehicular Technology Conference*, vol. 3, pp. 2232-2236, May 1998.

[17] P. Robertson, E. Villebrun, P. Hoeher. "A comparison of optimal and sub-optimal MAP decoding algorithms operating in the log domain". *Proc. IEEE International Conference on Communications*, pp. 1009-1013, 1995.

[18] M. P. C. Fossorier. "Iterative Reliability-Based Decoding for Low Density Parity Check Codes". *IEEE Journal on Selected Areas in Communications*, vol. 19, no. 5, pp. 908–917, May 2001.

[19] S. Brink, J. Speidel, R. Yan. "Iterative demapping for QPSK modulation". *IEE Electronic Letters*, vol. 34, no. 15, pp. 1459–1460, July 1998.

[20] D. Falconer. "Frequency domain equalization for 2-11 GHz broadband wireless systems". Available at *http://www.ieee802.org/16/tutorial/*80216*t* – 01_01.*pdf*

[21] N. AlDhahir. "Single carrier frequency domain equalization for space-time block coded transmissions over frequency selective fading channels". *IEEE Comm. Letters*, vol. 5, pp. 304–306, July 2001.

[22] L. Dong, Y. Zhao. "Frequency-domain turbo equalization for single carrier mobile broadband systems". *Proc. Milcom'06*, pp. 1–7, Oct. 2006.

[23] Part 16: Air interface for fixed broadband wireless access systems, amendment 2: medium access control modifications and additional physical layer specifications for 211GHz, 802.16a IEEE standard for local and metropolitan area networks, April 2003.

[24] V. Erceg et al. "Channel models for fixed wireless applications". IEEE 802.16a cont. IEEE 802.16.3c-01/29r4, June 2003.

[25] J. Hagenauer, E. Offer, and L. Papke. "Iterative decoding of binary block and convolutional codes". *IEEE Transactions on Information Theory*, vol. 42, no. 2, pp. 429–445, Mar. 1996.

[26] P. Xiao, R. Carrasco, I. Wassell. "Performance Analysis of Conventional Detection in BFWA Systems". In *Proc. Second IFIP International Conference on Wireless and Optical Communications Networks, WOCN'2005*, pp. 447-452, March 2005.

[27] P. Wu. "On the complexity of turbo decoding algorithms". *Proc. IEEE Vehicular Technology Conference*, vol. 2, pp. 1439–1443, May 2001.

[28] L. Bahl, J. Cocke, F. Jelinek, J. Raviv. "Optimal decoding of linear codes for minimizing symbol error rate". *IEEE Transactions on Information Theory*, vol. 20, pp. 284–287, March 1974.

[29] I. Chatzigeorgiou, M. Rodrigues, I. Wassell, R. Carrasco. "Comparison of convolutional and turbo coding for broadband FWA systems". *IEEE Transaction on Broadcasting*, vol. 53, no. 2, pp. 494–503, June 2007.

[30] P. Xiao, R. Carrasco, I. Wassell. "Estimation of FWA MIMO channels". In proc. *IEEE Information Theory Workshop, ITW'2006*, pp. 641–645, Oct. 2006.

[31] A. Paulraj, et al. "Increased capacity in wireless broadcast system using distributed transmission/directional reception". *US Patent*, no. 5,345,599, 1994.

[32] G. Foschini, M. Gans. "On limits of wireless communications in a fading environment when using multiple antennas". *Wireless Personal Communications*, vol. 6, no. 3, pp. 311–335 March 1998.

[33] C. Hong, I. Wassell, G. Athanasiadou, S. Greaves, M. Sellars. "Wideband tapped delay line channel model at 3.5GHz for broadband fixed wireless access system as a function of subscriber antenna height in suburban environment". *The Fourth International Conference on Information, Communications & Signal Processing (ICICS-PCM 2003)*, Dec. 2003.

In: VLSI and Computer Architecture
Editor: Kenzo Watanable

ISBN: 978-1-60692-075-6

Chapter 2

VLSI INTERCONNECTS AND THEIR DELAY PERFORMANCE

Brajesh Kumar Kaushik
Department of Electronics and Electrical Engineering,
G.B.Pant Engineering College, Pauri Garhwal-246001, INDIA
R.P. Agarwal, R.C. Joshi
Department of Electronics and Computer Engineering,
Indian Institute of Technology-Roorkee, Roorkee-247667, INDIA
Sankar Sarkar
Department of Electronics and Communication Engineering
Mody Institute of Technology and Science, Sikar, INDIA

Abstract

The feature size of integrated circuits has been aggressively reduced in the pursuit of improved speed, power, silicon area and cost characteristics. Semiconductor technologies with feature sizes of several tens of nanometers are currently in development. As per, International Technology Roadmap for Semiconductors (ITRS), the future nanometer scale circuits will contain more than a billion transistors and operate at clock speeds well over 10GHz. Distributing robust and reliable power and ground lines; clock; data and address; and other control signals through interconnects in such a high-speed, high-complexity environment, is a challenging task.

The performance of a high-speed chip is highly dependent on the interconnects, which connect different macro cells within a VLSI/ULSI chip. With ever-growing length of interconnects and clock frequency on a chip, the effects of interconnects cannot be restricted to *RC* models. The importance of on-chip inductance is continuously increasing with faster on-chip rise times, wider wires, and the introduction of new materials for low resistance interconnects. It has become well accepted that interconnect delay dominates gate delay in current deep sub micrometer VLSI circuits. With the continuous scaling of technology and increased die area, this behavior is expected to continue.

To avoid prohibitively large latencies, designers scale down global wire dimensions more slowly than the transistors dimensions and this causes a rapid growth in gap between transistors and interconnects densities on a chip. Thereby, as technology advances to Giga-

Scale Integration (GSI), global interconnect resource becomes more and more valuable and it is essential to use global interconnects optimally.

Propagation delay in global interconnects have become a core research problem. Therefore, a lot of work is being carried out to address these problems. Various models have been suggested in literature to analyze interconnects. The present chapter reviews in detail the works carried out in various research aspects and associated problems of VLSI interconnects. A case study is undertaken to understand the methodology to analytically calculate the output waveform and propagation delay. For this case study the effect of short circuit current on the propagation delay is also considered. Furthermore, this chapter also discusses various state-of-the-art delay minimization techniques.

1. Introduction

The semiconductor industry has been fueled by enhancements in integrated circuit (IC) density and performance, resulting in information revolution for over four decades and is expected to continue in future. The periodic improvement in density (as per Moore's Law) and performance has, been mainly achieved through aggressive device scaling and/or increase in chip size. Aggressive scaling of semiconductor process technology over the last several decades has resulted in creation of many new products, such as computers, camera, cell phones and information appliances. The trend is expected to continue for the coming years and create countless opportunities and challenges. Recent developments in semiconductor industry show a rapid increase in chip frequency and design complexity. Introduction of newer technologies is now moving towards a two year cycle as compared to traditional three year cycle. Though technology scaling helps in addressing design complexity and performance trends, it opens up a whole new spectrum of design validation challenges. As far as MOS transistor scaling is concerned, device performance improves as gate length, gate dielectric thickness, and junction depth are scaled. In sharp contrast, scaled chip wiring (interconnect) suffers from increased resistance due to decrease in conductor cross-sectional area and may also suffer from increased capacitance if metal height is not reduced with conductor spacing. As operating frequencies continue to spiral upward, parasitic inductive effects must also be considered [46, 57, 76, 109, 110, 127, 130, 147, 165]. Thus, interconnect parasitics play an increasing role in overall chip performance as feature size scales [17, 18, 58, 107, 108, 122]. Every system implemented either through ASIC design [131, 132, 133, 142, 145, 146, 151] or on FPGA [5, 158] are prone to the effects of the parasitic components of interconnect impedance.

The function of interconnects or wiring systems is to distribute clock and other signals and to provide power/ground to and among the various circuits/systems functions on the chip. The performance *s.a.* time delay and power dissipation of a high-speed chip is highly dependent on the interconnects, which connect different macro cells within a VLSI chip [29, 90]. To escape prohibitively large delays, designers scale down global wire dimensions more sluggishly than the transistor dimensions [137, 153]. As technology advances, interconnects have turned out to be more and more important than the transistor resource, and it is essential to use global interconnects optimally. For high-density high-speed submicron-geometry chips, it is mostly the interconnection rather than the device performance that determines the chip performance [138, 139].

Distribution of the clock and signal functions is accomplished on three types of wiring (local, intermediate, and global) as shown in Figure 1. An interconnect depending on its length, can be classified as local, semi-global and global [130]. Local wiring, consisting of very thin lines, connects gates and transistors within an execution unit or a functional block (such as embedded logic, cache memory, or address adder) on the chip. Local wires usually span a few gates and occupy first and sometimes second metal layers in a multi-level system. The length of a local interconnect wire approximately scales with scaling of technology, as the increased packing density of the devices make it possible to similarly reduce the wire lengths. Intermediate wiring provides clock and signal distribution within a functional block with typical lengths up to 3–4 mm. Intermediate wires are wider and taller than local wires to provide lower resistance signal/clock paths. Global wiring provides clock and signal distribution between the functional blocks, and it delivers power/ground to all functions on a chip. Global wires, which occupy the top one or two layers, are longer than 4mm and can be as long as half of the chip perimeter. The length of global interconnect wires grow proportionally to the die size. The length of semi-global interconnect behaves intermediately. The global interconnects are much wider than local and semi-global interconnects. Thus resistance of global interconnects is small and therefore their behavior resembles that of lossless transmission lines.

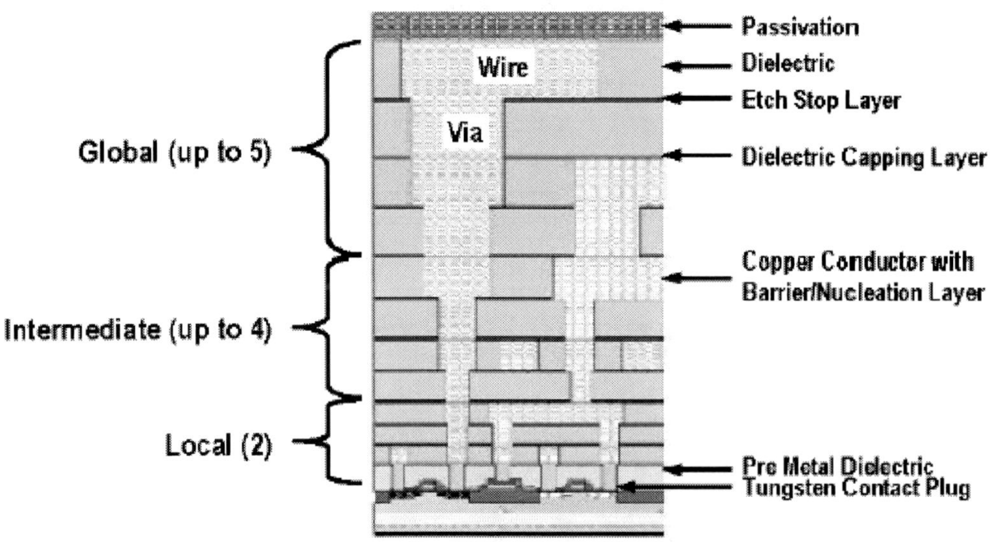

Source: ITRS [69].

Figure 1. Placement of various interconnects (Local, Intermediate and Global).

Wide wires are frequently encountered in clock distribution networks, power and ground lines, and other global interconnects such as data bus and control lines in upper metal layers (Figure 1). These wires are low resistive lines that can exhibit significant inductive effects. Due to presence of these inductive effects, the new generation VLSI designers have been forced to model the interconnects as distributed *RLC* transmission lines, rather than simple *RC*–ones. Modeling interconnects as distributed *RLC* transmission line, has posed many challenges in terms of accurately determining the signal propagation delay; power dissipation through an interconnect; crosstalk between co-planar interconnects and interconnects on different planes due to capacitive and inductive coupling; and optimal repeater insertion.

At lower frequencies, interconnects cause no difficulty for signal propagation. However, at higher frequencies, they cause severe signal degradation such as signal delay, crosstalk, ringing, reflections and distortions which must be addressed by VLSI designers. It has been predicted since long time that interconnect wiring delays rather than transistor logic delays would be the major contributors to the overall global path delays for ICs fabricated by the deep submicron CMOS processes [11, 12]. The increasing dominance of the interconnect coupled with the aggressive scaling in operating frequencies to increase chip performance has fundamentally changed the nature of IC design. The impact of the interconnect, therefore, needs to be considered during all stages of design and at all levels of design hierarchy. Even during process development, interconnect performance is an important consideration in the design of the on-chip metal system. With the upgrade of technology from micron to nanometer regime, technological, device and interconnect challenges are closely examined by different researchers.

On-chip global interconnects are among the top challenges in CMOS technology scaling due to rapidly increasing operating frequencies and growing chip size[69]. The clock signal has already been brought into the multi-gigahertz range [94, 163, 166] where inductance and other transmission line effects of on-chip long lines become important-- [46, 155]. For higher operating frequencies, dispersion [155] and skin effects [46] are among the new concerns. The use of reverse scaling methodology [102] will decrease the line resistance, but the line inductance effects will become more prominent. The global clock network, which was already power hungry, is likely to consume more power and hence become even more difficult to design [25]. Particularly, the delay induced by word lines, bit lines, clock lines, and bus lines in memory [88,123] or logic VLSI will remain the key concerns while designing the interconnects.

2. Modeling Interconnect as *RC* & *RLC* Circuits

During, early phase of VLSI design, the gate parasitic impedances had been much larger than the interconnect parasitic impedances, since size of gate (width and length) were quite large. For example, 5μm was a typical minimum feature size in 1980. Therefore, interconnect parasitic impedances were modeled as short circuit. With the conventional technology that used a feature size of 1μm or above, interconnects resistance was negligible compared to the driver resistance. Thus, the interconnect and loading gates were modeled as a lumped loading capacitance. The interconnect delay was determined by the driver resistance times the total loading capacitance. However, with the scaling of technology and increased chip sizes (submicron VLSI technology), the cross-sectional area of interconnects had been scaled down while their lengths increased. Therefore, the interconnect resistance was comparable to the driver resistance and the interconnect capacitances became comparable to the gate parasitic capacitances, which forced the designers to model interconnects as *RC* line. With the introduction of *RC* models interconnection delay, power consumption and repeater insertions became important in realizing high performance VLSI's. Almost every aspect of the design and analysis was affected by new interconnect model.

With ever-growing length of interconnects and clock frequency on a chip, the effects of interconnects cannot be restricted to *RC* models. The importance of on-chip inductance is continuously increasing with faster on-chip rise times, wider wires, and the introduction of new materials for low resistance interconnects. The usage of higher operating frequencies increases the value of $|j\omega L|$, which plays a role in interconnect delay calculation and its design construction [63].

Wide wires are frequently encountered in clock distribution networks, power and ground lines, and other global interconnects such as data bus and control lines in upper metal layers. These wires are low resistive lines that can exhibit significant inductive effects. Due to presence of these inductive effects, the new generation VLSI designers have been forced to model the interconnects as *RLC* lines.

2.1. Lumped and Distributed Models

Depending on the operating frequency, signal rise times, and nature of its structure, the analytical models can be broadly categorized as lumped and distributed. At lower frequencies, the interconnect can be modeled as lumped *RC* or *RLC* circuits. *RC* circuit responses are monotonic in nature which fails to account for ringing in signal waveforms. In order to account for ringing in signal waveforms, *RLC* circuit models are required [1].

At high signal-speeds, electrical length of interconnects becomes a significant fraction of the operating wavelength, giving rise to signal distortion that do not exist at lower frequencies. The conventional lumped impedance interconnect models are inadequate in such a situation as they do not adequately exhibit the distortions. The transmission line models based on quasi-transverse electromagnetic mode (TEM) assumptions yield better results. The TEM approximation represents the ideal case, where both electric (E) and magnetic (H) fields are perpendicular to the direction of propagation and it is valid under the condition that the line cross section is much smaller than the wavelength. However, the in-homogeneities in practical wiring configurations give rise to E or H fields in the direction of propagation. If the line cross section or the extent of these non-uniformities are a small fraction of the wavelength in the frequency range of interest, the solution of Maxwell's equations are given by the so-called quasi-TEM modes and are characterized by distributed resistance, inductance, capacitance and conductance per unit length parameters [124]. In realistic situations, due to complex interconnect geometries and varying cross-sectional areas, the interconnects may need to be modeled as non-uniform lines. In this case, the per unit length impedance parameters are functions of the distance, along the length of transmission line [21, 98, 121].

In deep submicron technology, lumped models are no longer capable of satisfying the accuracy requirements and cannot exactly locate the coupling elements in the equivalent circuit of the coupled line. It is well accepted that simulations of a distributed *RC* model of an interconnect matches more accurately the actual behavior in comparison to lumped *RC* model [130]. In similar fashion, a distributed *RLC* model outperforms the lumped *RLC* model in terms of modeling accurately the behavior of a line. A distributed *RLC* model of an interconnect, known as the transmission line model, becomes the most accurate approximation of the actual behavior [130]. The transmission line analogy for an interconnect considers the signal propagation to be a wave propagation over the interconnect medium. This is in contrast to the distributed *RC* model, where the signal diffuses from source to the destination governed by the diffusion equation. In the wave mode, a signal propagates by alternatively transferring energy from the electric to magnetic fields, or equivalently from capacitive to the inductive nodes. Interconnect models must incorporate distributed self and mutual inductance to accurately estimate interconnect time delay, power dissipation, crosstalk and other parameters of significance.

The evolution of various models with time is shown in Figure 2. It is assumed that leakage conductance 'g' equals 0, which is true for most insulating materials such as SiO_2, sapphire etc. Dealing with inductance requires efficient extraction methods. Presence of inductance also increases the processing time of the computer-aided design tools. Usually the interconnect circuits extracted from layouts contain a large number of nodes that make the simulation highly CPU intensive.

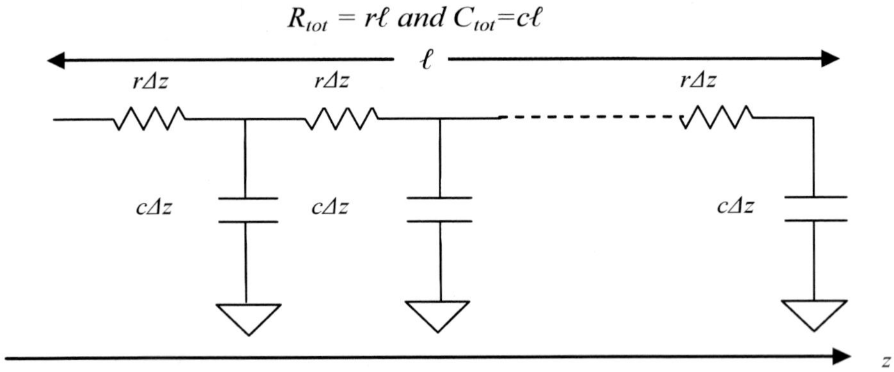

(a)

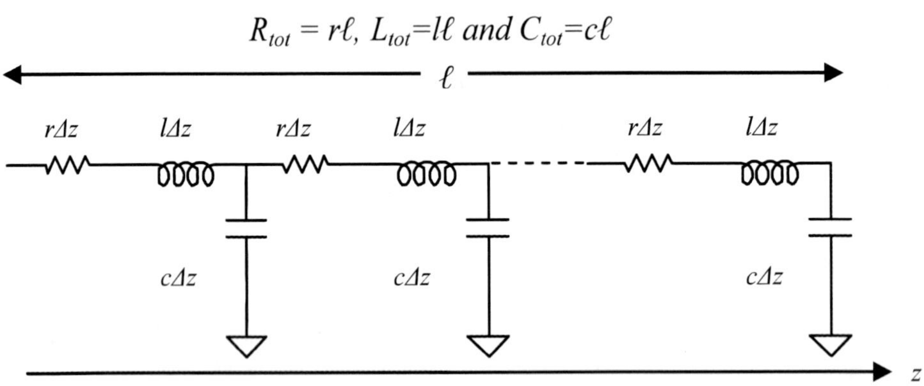

(b)

Figure 2. Development of interconnect models (a) *RC* model; (b) *RLC* model.

Distributed coupled *RLC* models become necessary even for the early design stages, and is suggested in many papers [22, 39, 50]. The distributed π, 3π, and 4π models are used for representing distributed *RLC* interconnects.

Modeling interconnects as distributed *RLC* transmission line, has posed many challenges in terms of accurately determining the signal propagation delay; power dissipation through an interconnect; crosstalk between co-planar interconnects and interconnects on different planes due to capacitive and inductive coupling; and optimal repeater insertion.

3. Extraction of Interconnect Parasitics

For a meticulous interconnect design, it is essential to extract correct values of interconnect parasitic impedances i.e. resistance, capacitance and inductance. In fact, extraction of exact values of capacitance and inductance of an interconnect is a challenging task. For an on-chip interconnect modeling, the peculiar behavior of line capacitance and inductance with neighboring lines has to be taken into consideration. Electrostatic interaction between wires is very short range, and therefore consideration of only nearest neighbors provides sufficient accuracy for capacitance extraction. However, capacitance is a sensitive function of geometry, making any closed-form modeling for general three-dimensional (3-D) wires a complex task. Recently, Quasi 3-D (Q-3-D or 2.5D) modeling has evolved as a reasonably better approach [8, 34, 77] for estimation of interconnect capacitance. The major challenge in Q-3-D model is decomposition of 3-D structure into two-dimensional (2-D) segments, and inclusion of the fringing electric field effects between adjacent 2-D segments. Unlike an electric field, a magnetic field has a long-range interaction. Therefore in inductance extraction, not only the nearest neighbors but also many distant interconnects have to be considered [15]. As a consequence, defining current loops or finding return paths becomes a major challenge in inductance modeling. Extensive research has been conducted in this area, and various techniques have been proposed to extract the values of capacitance and inductance. Traditional approaches for inductance analysis are based on simple loop inductance models [91, 99, 150]. The loop inductance and resistance are extracted by defining ports at the driving gate, and then solving the current distribution for an *RL* model of the circuit using tools such as Fast Henry [81]. The extracted inductance and resistance are then combined with lumped capacitance to construct a net list. While extracting the inductance, current distribution is determined solely by the resistance and inductance of the conductors. This leads to significant inaccuracies, since the interconnect and device decoupling capacitances strongly affect current return paths. Also, defining a port at the driving gate ignores other current paths, such as the short-circuit gate current and the power grid current generated by the switching of other gates in the vicinity of the signal net. However, the simplicity of the loop inductance model reduces simulation time, and can be used as a pre-layout estimation methodology.

An alternative method such as the Partial Equivalent Elements Circuit (PEEC) [135, 136] based on partial inductances for wire segments can be used for extraction. The PEEC method is used to construct a circuit model that does not require the predetermination of current loops. PEEC models have been used to obtain current distribution through signal net and the return currents [64]. The application of this technique is limited to highly simplified structures like coplanar waveguides. Furthermore, it ignores important components that determine current paths, and hence lack accurate estimation capability.

Sim et al. [149] observed quantitatively the effect of random signal lines on the on-chip inductance. An empirical model for high-frequency inductance is developed using an S-parameter-based methodology and a full wave solver. In particular, quasi-TEM wave like propagation mode is observed above 10GHz, revealing a unique relationship between capacitance and inductance of the signal line. The random capacitive coupling effect is incorporated and the frequency-dependent *RLC* model is confirmed to be valid up to 100GHz. Sim et al. [148] further extended their work and proposed a unified model for resistance, inductance and capacitance extraction. The model consisted of two components, a quasi-three-dimensional (3D) capacitance model and an effective loop inductance model. In the capacitance model, a concept of effective width for a 3D wire has been used. This is derived from the combination of an analytical two-dimensional (2D) model and an analytical "wall-to-wall" model. The effective width provides a physics-based approach to decompose any 3-D structure in to a series of 2-D segments, resulting in efficient and accurate capacitance extraction. A highly efficient loop-based interconnect modeling methodology for high frequency clock network design is formulated in [68]. Closed-form loop resistance and inductance models are proposed for fully shielded global clock interconnect structures, which capture high-frequency effects including inductance and proximity effects. The models are validated through comparisons with electromagnetic simulations and measured data taken from a Power 4 chip.

Mezhiba and Friedman [103] analyzed the inductive characteristics of several types of gridded power distribution networks. The inductance extraction program Fast-Henry is used to evaluate the inductive properties of grid structured interconnect. In power distribution grids with alternating power and ground lines, the inductance is shown to vary linearly with grid length and inversely linearly with the number of lines in the grid. The inductance is also relatively constant with frequency in these grid structures. These properties permit the efficient estimation of the inductive characteristics of power distribution grids. To optimize the process of allocating on-chip metal resources, inductance/area/resistance tradeoffs in high speed performance distribution grids are explored. Two tradeoff scenarios in power grids with alternating power and ground lines are considered. In the first scenario, the total area occupied by the grid lines is maintained constant and the grid inductance versus grid resistance tradeoff is evaluated as the width of the grid lines varies. In the second scenario, the metal area of the grid is maintained constant and the grid inductance versus grid area tradeoff is investigated. In both cases, the grid inductance increases virtually linearly with line width, rising more then eightfold for a tenfold increase in line width. The grid resistance and grid area, however, decrease relatively slowly with line width. This decrease in grid resistance and area is limited to a factor of two for particular interconnect characteristics.

Mezhiba and Friedman [104] extended their previous work [103] and analyzed the impedance characteristics of multilayer power distribution grids spanning many layers of interconnect with disparate electrical properties. They show that the electrical characteristics of multilayer grids vary significantly with frequency. As the frequency increases, a large share of the current flow is transferred from the low-resistance upper layers to the low-inductance lower layers. The inductance of a multilayer grid therefore decreases with frequency, while the resistance increases with frequency. The lower layers of multilayer power grids provide a low-inductance current path, significantly reducing the grid impedance at high frequencies. An analytical model is also presented to determine the impedance characteristics of a multilayer grid from the inductive and resistive properties of the comprising individual grid layers.

Eo *et al.* [57] developed a model for high speed and high density VLSI circuit and found that interconnect circuit parameters vary with frequency. The model considers the silicon substrate properties, pad parasitic, fringing effects and frequency variant properties of the circuit parameters. The model parameters are compared to scattering parameter measurements and PISCES-II simulations. A reasonable agreement is obtained with s-parameter measurements. Kopcsay *et al.* [89] proposed a comprehensive 2D inductance modeling approach for VLSI interconnect. Kleveland *et al.*, Sim *et al.*, Ymeri *et al.*, Zagge *et al.* and Zheng *et al.* in references [86, 148, 168, 169, 170] showed that, the equivalent resistance, capacitance and inductance are frequency dependent parameters. With increase in frequency the line resistance per unit length increases, while line inductance per unit length first decreases and finally levels off at very high frequencies. Although inductance decreases with frequency, its reactance increases and makes inductive effect on delay and power to be more significant. In general the interconnect capacitance decreases with frequency, but the decrease depends on the substrate resistivity [169]. For very low substrate resistivity the change in capacitance with frequency may be insignificant [168, 169, 170]. The capacitive reactance also plays an effective role at higher frequencies. A number of researchers have worked to establish the importance of transmission line effects. Models have also been given to extract interconnect parasitic impedance parameters in references [45,106 162]. Huang *et al.* [68] carried out interconnect modeling for multi-gigahertz clock. Sylvester *et al.* [153] considered the characterization of on-chip interconnect with particular attention to ultra-small capacitance measurement and in-situ noise evaluation techniques An approach called the charge-based capacitance measurement technique, to measure femto-farad level wiring capacitances, has the advantage of being compact, having high-resolution and being very simple.

In spite of recent developments, accurate estimation and modeling of inductance in VLSI interconnects still remains a challenging task. Since magnetic fields have a much longer spatial range compared to that of electric fields, in practical high-performance ICs containing several layers of densely packed interconnects, the wire inductance are sensitive to even distant variations in the interconnect topology [130]. Secondly, uncertainties in the termination of the neighboring wires can significantly affect the signal return path and return current distributions and therefore the effective inductance. Many techniques have been employed to extract the capacitance and inductance values, the results are still best approximate for real high-performance VLSI circuits due to the uncertainties in providing valid models of the local physical and electromagnetic environment formed by the orthogonal and parallel interconnects [15]. Also, accurate estimation of the effective inductance values requires details of the 3-D interconnect geometry and layout, technology etc., and the current distributions and switching activities of the wires, which are difficult to predict. Moreover, at high frequencies the line inductance is also dependent on the frequency of operation. These are the added complexities for the designers involved in analyzing the behavior of the interconnects..

4. Propagation Delay through Driver Interconnect Load Model

Existing gate-level static timing analyzers break down the path delay into gate delay and interconnect delay. The two components of the path delay have been studied exhaustively by several researchers. These works are discussed in the subsequent subsections.

4.1. Driver Delay Models

In this chapter gates, buffers, or transistors are collectively referred as drivers. Some commonly used approaches to model the drivers for delay computation with interconnects are presented. An interconnect, in addition to introducing a significant wiring delay in a typical CMOS circuit path, also affects the delay of a CMOS driver. For previous technologies, since the driver resistance dominated the interconnect resistance, the capacitance of the interconnect and the input gate capacitances of the fan-out gates determined the delay of a driver. Essentially, because of the two-pole behavior of the *RC* load the output waveform of long interconnects is not accurately approximated by the typical ramp waveform generally found in digital circuits. As the driver output waveform exhibits a long "tail" portion, approximating it by a ramp leads to large inaccuracy in estimating the interconnect delay. With scaling, the resistance of the *RC* interconnects effectively shields some of the load capacitance, thereby affecting the CMOS driver delays.

Wyatt [164] and Brocco *et al.* [23] developed a switch-resistor model which comprised of an effective linear resistor driven by a voltage source (usually assuming a step input or ramped input). The effective resistance of a driver usually depends on the transition time of the input signal, the loading capacitance, and the size of the driver. For example, one can use a resistor of fixed value R_{eff} to model a driver by selecting an appropriate capacitance load C and matching the 50% delay of the driver driving the load with that of the equivalent *RC* circuit ($0.7R_{eff}C$) under the step-input. The switch-resistor model has the advantage that the coupling with the interconnect can be easily modeled by including the effective driver resistance in the interconnect *RC* tree for delay and/or waveform computation. But it may be difficult to model the non-linear behavior of the driver and determining the required intrinsic delay for step function input.

Ousterhout [120] presented a more accurate model, called the *slope model*, which uses a one-dimensional table to compute the effective driver resistance based on the concept of rise-time ratio. It first uses the output load and transistor size to compute the *intrinsic rise-time* of the driver, which is the rise-time at the output under the step input. The rise-time of the driver input is then divided by the intrinsic rise-time of the driver to produce the *rise-time ratio* of the driver. The effective resistance is represented as a piece-wise linear function of the rise-time ratio and stored in a one dimensional table. Given a driver, one first computes its rise-time ratio and then calculates its effective resistance R_{eff} by interpolation according to its rise-time ratio from the one-dimensional table. Multi-dimensional tables can also be used for computing and storing the effective driver resistance as a function of the input slope, output load, etc.

Another approach for driver modeling characterizes the behavior of a driver (such as the driver delay and the output transition time) using all relevant parameters of the input signal(s) and the output load. This allows for very accurate modeling, but the gate delay and the interconnect delay must be computed separately. For example, one can pre-characterize, the delay (t_d) and output transition times (t_f and t_r) of a driver in terms of the input transition time t_t and the total load capacitance C_L using accurate circuit simulation such as SPICE. The characterized results can then be stored in a *look-up table* where each entry is in the form: $\{t_t, C_L, (t_d, t_f, t_r)\}$. Such a model can be very accurate if one can afford the time and space to generate a detailed multi-dimensional table for each gate. Alternatively, one can store the characterization data much more compactly in the form of *k-factor equations* [161], where $k-$factors are determined based on linear regression or least square fits on the characterization data. The k-factor equations are more accurate in general, but are inherently incompatible with *RC* loads.

Qian *et al.* [129] introduced the concept of 'effective capacitance' to overcome above problems. As a result the driver does not see the total capacitance of the load- it only sees an "effective capacitance". They developed an analytical expression for the 'effective load capacitance" of *RC* interconnect. Using this, they showed that, when there is significant shielding, the response waveforms at the gate output might have a large exponential tail. This in turn can strongly influence the delay of the *RC* interconnect. The concept of effective capacitance is extended to develop an equation on the basis of a two-piece gate output approximation. The equation is solved to obtain response waveform.

Dartu *et al.* [40, 41, 42] employed an empirical fitting to approximate the equivalent resistance value for modeling the CMOS gate. They used a *CRC* π-model of an *RC* interconnect to analyze waveform and delays. The empirical characterization of the CMOS gate required numerous SPICE runs for different input transition times and different load capacitances. The interconnect *CRC* π-circuit was mapped to an 'effective capacitance' that would draw the same average current during the transition period of signal. The resulting non-linear equation was solved for effective capacitance (C_{eff}) using a damped Newton-Raphson approach. After an initial guess for C_{eff}, number of iterations are required for converging to an appropriate C_{eff}. The model worked in terms of pre-characterizing the parameters of a time varying Thevenin voltage source model (in series with a fixed resistor) over a range of effective capacitance load values. The shape of the Thevenin voltage waveshape was inaccurately modeled with a single ramp. In current DSM technology where supply voltage is highly scaled, substantial error is incurred by modeling Thevenin voltage waveform shape as linear. As supply voltage scales down, the shape of Thevenin voltage source becomes more and more non-linear and the single saturated ramp used to model this voltage source produces erroneous results. The work of Dartu et al. fits more in the area of library characterization for micron/sub-micron level technology, where supply voltage is not highly scaled. Arunachalam et al. [9] extended Dartu's work further to include *RLC* loading instead of *RC* loading but still used same approach of inaccurately pre-characterizing gate to a linear resistance and mapping the interconnect to an effective capacitance.

Shichman *et al.* [143] and Shockley [144] developed square law models for MOSFETs in which drain current varies as a square of the effective gate voltage i.e. $(V_{gs} - V_t)$ where V_{gs} is gate to source voltage and V_t is the threshold voltage of a MOSFET. These models have been extensively used in computer-aided analyses of CMOS switching circuits. However, these models lose accuracy as channel length is reduced. Sakurai and Newton [140] developed Alpha power MOS model that defines current voltage characteristics for short channel MOSFETs. In this model, the effects of input waveform slope and parasitic drain/source resistance are included. Sakurai and Newton used the Alpha power law model, to analyze the working of CMOS inverter. They observed that neglecting PMOS is not valid when the rising input ramp is very slow compared to the output waveform. However, the approximation is valid if the input slope exceeds one-third of the out-put slope, which is usually true in VLSI. As velocity saturation effect intensifies, the logic threshold voltage becomes more sensitive to the gate width ratio of PMOS and NMOS. The short-circuit power dissipation increases as the carrier velocity saturation effects get severer in short-channel MOSFETs. However, for relatively slow inputs, Sakurai's model ceases to be valid, thereby leading to inaccuracies.

Hirata *et al.* [66] derived a closed-form formula of propagation delay for static CMOS logic gates considering short-circuit current and current flowing through gate capacitance. They used the nth-power law MOSFET model [141] and represented the short circuit current by a piecewise-linear function for carrying out detailed analysis of the transient behavior of a CMOS inverter. They reported an error of less than 8% from circuit simulation. They numerically demonstrated that the influence of short-circuit power on delay increases with slower input transition time and smaller output load capacitance.

Agarwal et al. [4] presented a library-compatible approach to gate-level timing characterization in the presence of resistive/inductive/capacitive (*RLC*) interconnect loads. They showed that when transmission line effects are significant in *RLC* interconnects, driver output waveforms are non smooth and exhibit inflection points during transition. They proposed a two-ramp model based on transmission-line theory that predicts both the 50% delay and waveform shape (slew rate) at the driver output reasonably well when inductive effects are significant. The approach does not rely on piecewise linear Thevenin voltage sources and is compatible with library characterization methods. They showed that a one-ramp assumption may be sufficient for *RC* and weakly inductive lines, but becomes highly inaccurate for inductively dominated interconnects. The results when compared with SPICE demonstrate errors below 10% for both delay and slew rate. They also proposed a new criterion for evaluating the importance of on-chip inductance by comparing rise time at the driver output with the time of flight.

4.2. Interconnect Delay Models

As VLSI technology reaches deep submicron feature size, the model used to estimate interconnect delay changed from the simple capacitive model for micron technologies to the sophisticated high-order moment matching delay model [32].

Whenever inductance is considered to be negligible, the *RC* model can be viewed as a limiting case of the *RLC* transmission line model. The other limiting case is an inductance-capacitance (*LC*) transmission line where the resistance is negligible. These cases have been thoroughly examined in [72, 130]. Although it is highly improbable that the resistance of on-chip interconnects will become negligible in reality, the *LC* analysis provides an upper limit for analyzing inductive effects in VLSI circuits. The behavior of an *RLC* transmission line can therefore be bounded by analyzing the behavior of the *RC* and *LC* cases [72]. Various techniques have been proposed for the delay analysis of interconnects. Simulation techniques using SPICE tool provides the most accurate insight into arbitrary interconnect structures but are computationally expensive. For reducing simulation time, Lin *et al.* [96] presented transient simulation methods of lossy interconnect based on convolution techniques. Closed-form analytical formulas are mostly preferred over SPICE simulations for faster calculations and gaining overall insight into the factors affecting the results. Zhou *et al.* [171] proposed faster techniques based on moment computations. These methods are too expensive to be used during iterative layout optimization. Therefore, designers began to use the Elmore delay model [52] in the performance-driven design of clock distribution and Steiner global routing topologies. Following paragraphs presents some of the delay models for *RC* & *RLC* interconnects.

The Elmore delay model [52] was applied during early eighties to analyze the delay of transistor-level circuits, which were modeled as *RC* trees [134]. As an extension, the delay model was also used to analyze the wiring delays. Elmore delay approximation represents the first moment of the transfer function. One of the appealing properties of the model as applied to *RC* trees is that it can be easily written by inspection for any node of the tree. For designers, this property enables quick back-of-the envelop delay estimates for driver-net-receiver combinations. In addition, the Elmore delay is calculated in linear time of *RC* trees-only two traversals of the *RC* tree are required in order to calculate the delay for every node of the *RC* tree. This allows for software programs that can complete the delay analysis (rollups) for tens of thousands of nets within minutes. Another important property which can be easily derived from closed-form formula for delay model is its hierarchical nature. A partial net can be compactly represented by its total capacitance to represent its loading effect on the main net. This compact hierarchical representation is especially useful in designs where parts of the same interconnect may belong to different blocks and are generated by different designers or automated routing programs. The hierarchical and linear complexity properties of the Elmore delay model are exploited in several physical design and circuit optimization programs [32]. The model is used to quickly estimate the relative delays of different paths in the circuit, permitting more exhaustive simulations to be performed for only the critical paths. Also, it is widely used as a delay model for the synthesis of VLSI circuits such as repeater insertion in *RC* trees and wire sizing. Elmore delay has high degree of fidelity. An optimal or near optimal solution achieved by a design methodology based on the Elmore delay is also near-optimal based on a more accurate (e.g., SPICE-computed [113]) delay for routing constructions [20] and wire sizing optimization [31]. Simulations [35] have shown that the clock skew derived under the Elmore delay model has a high correlation with SPICE-derived skew data [73].

In spite of all these advantages, one of the main drawbacks of the Elmore delay model is that it can significantly overestimate the actual delay. Gupta *et al.* [61, 62] showed that the Elmore delay is a theoretical upper bound for the 50% delay for a family of input waveform. This is especially true for nodes that are closer to the driver which are driven by sharper signal waveshapes.

Elmore delay has been successfully applied to several aspects of high-performance chip design. Its applicability to the timing verification of high performance circuit designs has reduced accuracy, but efficient timing model for transistor stacks can be effectively used in conjunction with circuit simulation. This was demonstrated in the design of state of the art 600MHz commercial microprocessor [115]. Unfortunately, Elmore model cannot accurately estimate the delays for *RLC* interconnect lines and trees [79]. This is primarily due to the fact that it does not cover non-monotonic responses [52] which can occur in *RLC* circuits [73]. This inaccuracy of Elmore delay is harmful to current performance-driven routing methods, which try to optimize interconnect segment lengths and widths as well as driver and repeater sizes. Previous moment-based approaches (e.g., [96]) compute the delay estimates only from a simulated response but not from an analytical formula as in [79].

Sakurai [138] suggested a means to calculate the response and delay of a distributed *RC* line. Time-domain response is calculated from the transfer function using the Heaviside expression over poles of the transfer function. Furthermore, Sakurai approximates the response using a single pole and observes the variation of delay with respect to source and load parameters. The heuristic delay formula is almost identical to the Elmore delay equation. Hence, it suffers of the same constraints related to Elmore delay model.

With high-speed signals used in modern VLSI circuits, the electrical length of the interconnects becomes a significant fraction of the wavelength. This makes the distributed transmission line model a necessity to be used instead of the conventional lumped-impedance interconnect model. Analytical models of networks which contain coupled lossy transmission lines have been presented by Djordjevic *et al.* [48], where equations describing the transmission line system, and the terminal and interconnecting networks and terminations, are combined in frequency domain. The time-domain response is obtained by applying the inverse fast Fourier transform (FFT). A major difficulty caused by this approach is when the analysis has to span a time interval of several line transient times. Griffith and Nakhla [60] proposed an alternative method for the analysis of lossy coupled transmission lines with arbitrary linear terminal and interconnecting networks. They used the numerical inversion of Laplace transform to obtain the time-domain response. This method has been claimed to be more reliable than the FFT-based methods since it does not suffer from the usual aliasing problems. However, this method is expensive in terms of CPU requirements. Thus the need for a reasonable compromise between accuracy and CPU requirements arose, which made Rubenstein *et al.* [134] to develop delay estimate methods which rely upon an *RC*-tree interconnect model. However, this model is not adequate for high frequency MOS integrated circuits and cannot be used for high speed VLSI design. It means, developing general *RLC* lumped and distributed network models are of importance.

The aggressive scaling of operating frequencies in interconnect-dominated submicron process has necessitated the use of highly efficient interconnect analysis techniques with accuracy comparable to that of a circuit simulator. The conservativeness of first order models is a high price to pay in a design environment where picoseconds matter. Beginning with the application of an Asymptotic Waveform Evaluation (AWE) [126] to the design of the clock tree [49], model order reduction techniques (also referred to as moment-matching techniques) found increasing usage in a variety of interconnects analysis applications. Model order reduction offers excellent accuracy while providing reasonable run times as compared to Elmore-based methods. AWE was developed as an extension to the work of Penfield and Rubinstein [125, 134] which used the Elmore delay model for delay and timing prediction of *RC* trees. AWE is a technique for approximating the dominant poles (time constants) of linear *RLC* circuits. Using the dominant poles one can generate approximate time and/or frequency domain responses. AWE dramatically improved dominant pole/zero analysis for analog circuit design optimization, but it had an even greater impact on time domain analyses of *RLC* interconnect. The approximate dominant poles are generated in AWE by moment matching. Moments, as defined in classical mechanics or probability theory, are related to frequency domain coefficients. The essence of AWE is that these coefficients, or moments, can be calculated easily and efficiently for linear circuits in general, and then mapped to a set of dominant poles and zeros. In general, AWE can be directly applied to any circuit that can be linearized including circuits encountered during noise analysis. Another advantage of model order reduction is that it provides an extremely compact characterization of the interconnect. The large *RC* netlists created by most extraction tools can be compacted significantly. Furthermore, accurate analyses can be carried out later for different input waveforms efficiently avoiding the need for detailed circuit simulation. In fact, for some common input waveshape assumptions like the saturated ramp, the output waveform is available in a closed form manner. This enables solving for the time points of interest (50%, 90% time delays) required for calculating delay and slope, respectively, directly without having to generate intermediate points. Moment matching technique, which is recognized as a Padé-type approximation [10], must be applied with care due to its inherent potential for instability. That is, given the set of moments for a stable *RLC* circuit, generalized moment matching can result in one or more of the approximate dominant poles occurring in the right half plane. AWE suffers due to numerical instability for higher order approximations. This limits the order of the approximations using AWE to less than approximately eight poles (some of them are unstable poles). This limited number of poles is inappropriate for evaluating the transient response of an underdamped *RLC* tree which requires a much greater number of poles to accurately capture the transient response at all the nodes [70].

The success of AWE in dealing with interconnect analysis problems sparked a flurry of research in applying different model order reduction techniques in order to overcome some of the inherent limitations of AWE. Odabasioglu *et al.* [118, 119] described one of the more successful reduced-order modeling techniques called PRIMA (Passive reduced-order interconnect and macromodeling algorithm) in dealing with low-loss transmission-line circuits which present a problem to AWE because of the lack of any clearly dominant poles.

Since the Elmore single-pole delay estimates cannot accurately estimate the delay for *RLC* interconnects, Zhou *et al.* [171] proposed a two-pole approximation for the transfer function to compute the response at the load for *RLC* interconnect trees. However, the response computation does not provide any analytical expression for delay; it is also too time consuming to be used in iterative optimization of layout. In 1995, Krauter *et al.* [92] proposed to improve the Elmore delay model by using higher order moments; this work led to a heuristic net delay model equal to the sum of the first moment and its standard deviation.

Kahng and Muddu [79] showed that Krauter *et al.*'s [92] delay model is not accurate for various source and load parameters. They studied various combinations of first and second order moments, from which they developed a fast delay analytical model of *RLC* interconnects assuming a step input. The solutions developed by Kahng and Muddu [79], composed of three different formulas for the cases of real, complex, and multiple poles. At that time, there were no closed-form solutions for the moments of a tree that could be directly incorporated into the delay model. Kahng and Muddu [79] showed that Elmore delay estimates can be as much as 50% away from the SPICE computed delays, while their proposed analytical delay model estimates are within 15% of the SPICE delays. Also, they extended their delay model to estimate source-sink delays in arbitrary interconnect trees.

Ismail *et al.* [73] introduced a simple tractable delay formula for *RLC* trees. They tried to preserve the useful characteristics of the Elmore delay model while maintaining the same accuracy characteristics. This delay model with the closed-form expressions considers all damping conditions of an *RLC* circuit including the underdamped response, which was not considered by the Elmore delay due to the non-monotonic nature of the response. These solutions are presented for the 50% delay, rise time, overshoots, and settling time of signals in an *RLC* tree. Their generated delay expressions for an *RLC* tree have the same accuracy characteristics as the Elmore approximation for *RC* trees. These expressions consider both monotonic and non-monotonic signal responses. Due to this continuity, this delay model [73] is claimed to be always stable with arbitrary inputs and computationally efficient since the number of multiplication operations required to evaluate the approximation at all of the nodes of an *RLC* tree is linearly proportional to the number of branches in the tree.

Ismail and Friedman [74] derived a closed form solution of the output response of a CMOS inverter driving an open-circuited *RLC* transmission line using Alpha power law model of transistors for DSM technologies. Figures of merit have been developed that determine the relative accuracy of *RC* impedance to model on-chip interconnects. The range of length of an interconnect where transmission-line model becomes necessary is shown to be based on parasitic impedances of the line (*R*, *L* and *C*) and the rise time of the input signal at the gate driving the line. They used a more realistic approach in reference [71] by terminating the *RLC* transmission line by a load capacitance C_L. But replacement of an Alpha power law transistor model for driving gate, by a linear resistor has made analysis impractical. However, curve-fitted closed-form solution for the propagation delay are presented, which shows that neglecting inductance can cause large errors (over 35%) in the propagation delay for current on-chip inductance. It is also shown that the traditional quadratic dependence of the propagation delay on the length of interconnect for *RC* lines, tends to a linear dependence as inductance effects increase. Thereby, for a same length of interconnect, an *RC* model will overestimate the propagation delay as compared to a more realistic *RLC* transmission line model.

El-Moursy and Friedman [54, 55] presented a model of the effective capacitance of an *RLC* load driven by a CMOS inverter. They claimed that interconnect inductance introduces a shielding effect which decreases the effective capacitance seen by the driver of a circuit, reducing the gate delay. The interconnect inductance decreases the gate delay and increases the time required for the signal to propagate across an interconnect, thus reducing the overall delay to drive an *RLC* load. They showed that ignoring line inductance would overestimate the circuit delay, and consequently inefficiently oversize the circuit driver. It is claimed that considering line inductance in the design process reduces gate area and dynamic power dissipation.

Davis et al. [43] significantly extended Sakurai's work [139] by including the self and mutual inductance in models of high-speed Giga Scale Integration (GSI) interconnects. The distributed *RLC* model is driven by a gate represented by a linear resistor and terminated by an open-circuit at the load end. The transient response of a single semi-infinite distributed *RLC* interconnect driven by a step input voltage is obtained. Furthermore considering the reflections at source and load ends, the transient response of output voltage at the end of an interconnect of finite length is expressed. This expression is compared to HSPICE simulation of an interconnect using 1, 10, 50 and 500 lumped *RLC* elements. Simulation results show that as the number of lumped elements is increased the results converge to the distributed *RLC* solutions. The transient response expression of the finite *RLC* interconnect contains summation of later reflection terms even when the source resistance is equal to the lossless characteristic impedance. This differs from traditional lossless transmission theory in which a matched source absorbs all power from the transmission line leaving only the first reflection. It is shown that a distributed *RLC* interconnect, however, prevents this type of perfect matching because the voltage and current ratio are out of phase; and their ratio changes with time. In lossless transmission line theory, the ratio of the voltage and the current is always a constant equal to the lossless characteristic impedance, which allows perfect impedance matching.

Cao et al. [24] developed two-pole and second-order waveform models, while Banerjee and Mehrotra [13, 14, 15] used second- and fourth-order Padè approximations for the transfer function, the solutions of which require numerical iterations to calculate line characteristics such as time delay and crosstalk. Coulibaly and Kadim [38] presented an analytical method for estimation of delay through distributed *RLC* line. This method is based on a fourth order transfer function when the input is a ramp with finite rise time.

Venkatesan et al. [156] improved the model presented by Davis et al. [43] by including a capacitive load termination to a distributed *RLC* line, which accurately models on-chip and off-chip high-speed global wires that drive large capacitive loads. This work differs from the previous work [43, 44] in the manner that it is focused on the state-of-the-art global interconnect structures such as those described in [7] and [90], where one or two signal interconnects are flanked by shielding power/ground lines and sandwiched between ground planes. The Laplace domain transfer function of the interconnect circuit is rigorously solved using a convergent series of modified Bessel functions to obtain an explicit solution for the transient response to step and ramp inputs. These solutions are used to develop unified models for time delay, crosstalk, and repeater insertion *RC* and *RLC* interconnects [157]. The solutions are verified against HSPICE simulations. Due to mathematical complexity involved in multiple reflections, it has been assumed by Davis et al. [43, 44] and Venkatesan et al. [156, 157] that the first reflection provides significant information about the transient characteristics. This approximated expression of first reflection provides the detailed transient response of an *RLC*-modeled interconnect in a range upto three times of flight of the signal. When compared to a distributed *RC* model, the later significantly underestimates the 50% time delay and significantly overestimates the 90% response time of interconnect. Finally, approximated closed form expressions for overshoot and time delay are derived, and compared to the exact (SPICE) results using *RLC* model, for various interconnect lengths and driver impedance. The approximated expressions provide less than 5% error when ratio of gate impedance to characteristic impedance is less than 0.2 or ratio of total line resistance to characteristic impedance is greater than 2.3.

4.3. Composite Driver-Interconnect-Load Model-A Case Study

Predicting accurately the waveform shape and propagation delay in a driver-interconnect load model has been an important design perspective since long time. Previously, Chatzigeorgiou et al. [27], analyzed distributed *RC* interconnect load represented as *CRC* π-model driven by CMOS gate, but inaccurately neglected inductive effects. As discussed earlier, Kahng and Muddu [80] proposed a π-model for distributed *RLC* interconnects to estimate the driving point admittance at the output of a CMOS gate. With the help of this model, a good approximation of interconnect line characteristics can be obtained. The π-*RLC* model has a better accuracy in estimating output waveform and delay calculations than the previously proposed ones. The π-model becomes more accurate as the resistance, capacitance and inductance of the distributed *RLC* line increases. An attempt to model the interconnect line by distributed *RLC* line was made in [9, 13, 14, 15, 43, 44, 71, 156, 157], where the CMOS drivers are replaced by simple resistors. When such an equivalent linear resistor models the nonlinear CMOS transistors, it leads to discrepancy in results. The consequence is an overestimation of the inductance effects [71]. This behavior can be understood by noting that a transistor in a CMOS gate operates partially in the linear region and partially in the saturation region during switching. In the linear region, the transistor can be accurately approximated by a resistor. However, in the saturation region, the transistor is more accurately modeled as a current source with a parallel high resistance. The Thevenin equivalent of this circuit is a voltage source with a high resistance in series. This high resistance in series with an interconnect line overrides the series resistance and inductance of the line. Then a predominantly capacitive (*RC*) behavior is shown by interconnect. Since a transistor operates partially in saturation region and partially in linear region, the metrics presented by previous works are not accurate enough for all operating conditions of the transistors.

The case study undertaken discusses the work proposed in [87]. This work includes a more accurate transistor model *viz.*, Alpha power law [140], for analyzing output-waveform and determining propagation delay in driver-*RLC* interconnect load model in a VLSI chip. Representing transistor and distributed *RLC* interconnect by appropriate models results in higher accuracy [87]. Accurate analytical expressions for the output waveform and the propagation delay is found by solving the corresponding system equations of an inverter driving a π-circuit. The parasitic capacitance due to long interconnect is large because of which the gate-to-drain coupling capacitance has negligible effect on the output. Therefore, the gate-to-drain coupling capacitance is ignored in this analysis [87]. In brief, the equivalent π-model is used to capture with higher accuracy the performance of CMOS gates driving *RLC* interconnect loads. The proposed analysis determines the output waveform evolution accurately while keeping the complexity low. In addition, the output waveforms at both ends of an interconnect line are efficiently approximated by piecewise linear waveforms enabling the calculation of the voltage waveform at each point of the interconnect line. The analysis provides good results for two different cases of input ramp conditions i.e. fast and slow.

The driver-interconnect model consists of a π-model for interconnect and Alpha power law model for the CMOS gate transistor. The distributed *RLC* interconnect line is represented in Figure 3. The total line resistance, capacitance and inductance of the interconnect is R_{tot}, C_{tot} and L_{tot} respectively.

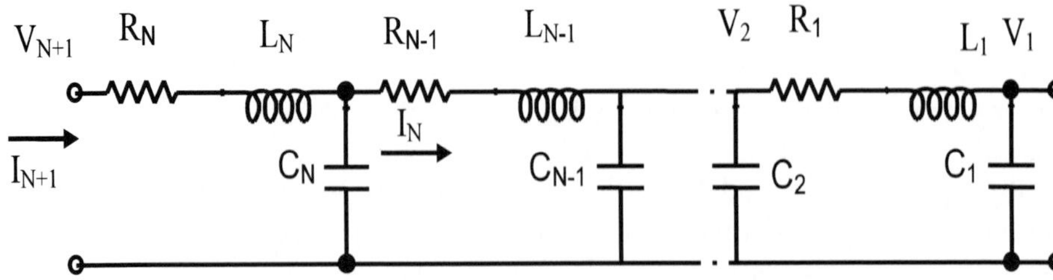

Figure 3. N-Segment *RLC* interconnect.

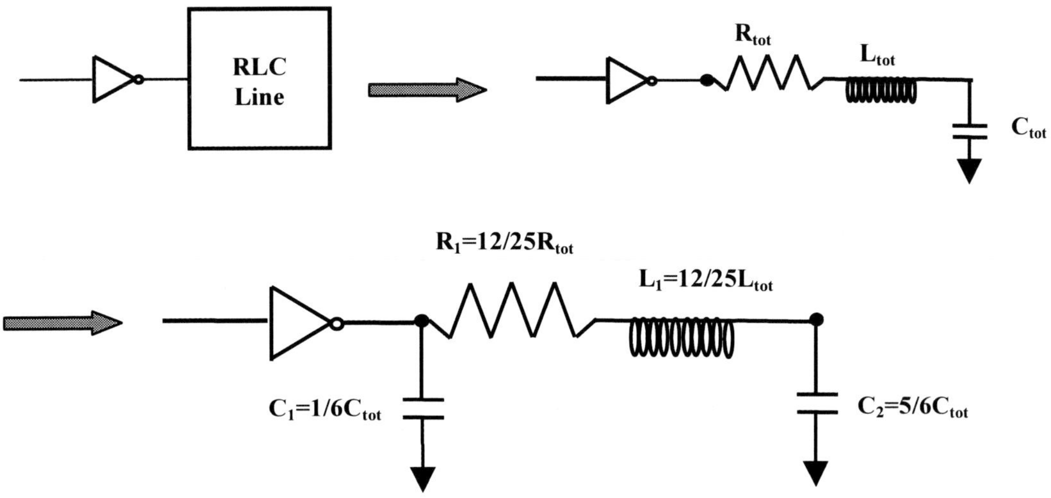

Figure 4. An equivalent π-model of the *RLC* distributed interconnect.

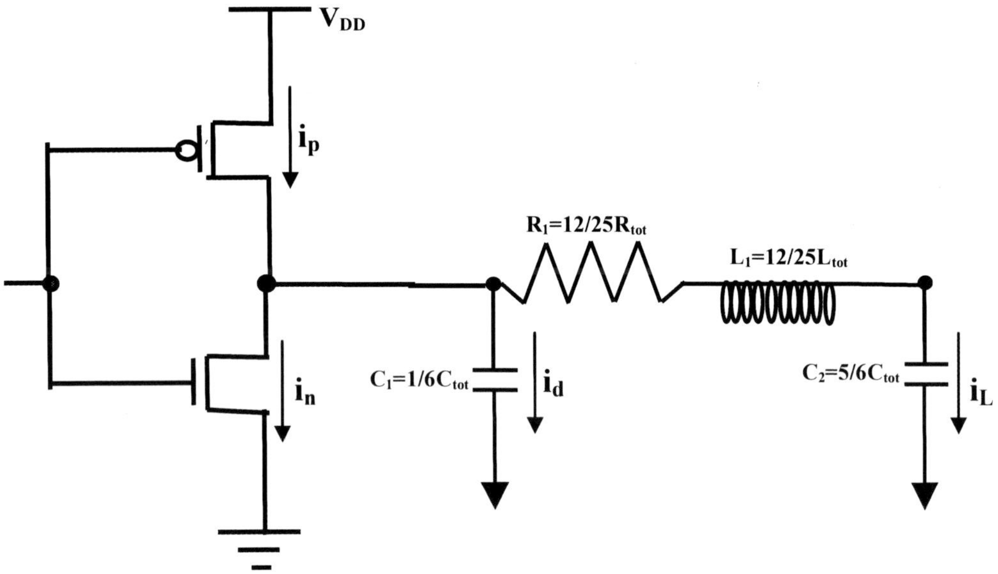

Figure 5. Model of the CMOS gate driving π-model of an *RLC* interconnect line.

The equivalent π-model as proposed by [80] is shown in Figure 4, where R_1, L_1, C_1 and C_2 represents equivalent *RLC* values of the π-model. Figure 5 shows CMOS transistors driving the interconnect load.

In order to solve the differential equation that describes the operation of the circuit in Fig. 5, the parasitic current through the PMOS transistor is considered negligible. This is a reasonable assumption since long interconnects present a high capacitance thus reducing the maximum value of the short-circuit current [140].

Two main cases for input ramps are considered: for fast (slow) inputs, the NMOS device is in saturation (in the linear region) when the input voltage reaches its final value. An input ramp is categorized to fast or slow ramp depending on the state NMOS device attains when the input voltage reaches its final value. If the NMOS continue to operate in saturation region when the input ramp has reached its final value, the ramp is called fast. On the other hand, if the NMOS transistor switches to linear region of operation before the input ramp attains its final value, the input ramp is said to be slow. Thus depending on the status of NMOS device the ramp is called fast or slow. The region of operation for a fixed size NMOS device invariably depends on the load characteristics. Depending on the length of interconnect, a ramp with fixed slew rate can therefore be called either fast or slow. If the interconnect is long, the discharging through the NMOS device will be slow enough to compel the NMOS to continue its operation in saturation region while the input ramp has reached its final value, this input ramp is therefore categorized as fast ramp. Contrary to this condition, keeping all other parameters same, if the length of interconnect is short, the NMOS device discharges the load capacitance quick enough to enter linear region of operation by the time the input ramp reaches its final value. Under this condition the input ramp is treated as slow ramp. For illustration, assume a condition where the input ramp is having a transition time of 50ps, but has a short interconnect (i.e. small parasitic resistance, capacitance and inductance) loaded to the driver transistor. Under this condition the driver transistor will enter linear region of operation by the time the input ramp has reached its final value. Thus this ramp will still be considered slow. Contrary to this, if the input ramp has a slew rate of 500ps but the driver has to drive a long interconnect (i.e. large parasitic resistance, capacitance and inductance), wherein the transistor will remain in saturation region of operation till the time the input ramp has reached its final value. Under this condition the input ramp is called fast ramp.

In order to obtain the output voltage expression analytically, four regions of operation are considered for each of the input ramps condition i.e. fast and slow. The regions of operation of NMOS transistor for a rising input fast and slow ramps are summarized in Table-1.

Table 1. NMOS operation and Input Ramp condition for Fast and Slow Ramps in different regions of operation.

Region	Fast Input Ramp		Slow Input Ramp	
	NMOS operation	Input Ramp Transition	NMOS operation	Input Ramp Transition
I	Cut off	Rising	Cut off	Rising
II	Saturation	Rising	Saturation	Rising. Extends to a time less than rise time of the input signal
III	NMOS is still in saturation.	Input ramp has reached its final value.	Linear	Input is still rising.
IV	Linear	Input signal has constant value.	Linear	Input ramp has reached its final value.

The analytical piecewise output waveform and the SPICE output waveform for fast input ramp are compared in Figure 6. The simulation set-up utilizes TSMC 0.18μm technology transistor with $W_n = 10.08\mu m$. The total interconnect resistance, inductance and capacitance are taken as 833Ω, 37.48nH and 6pF. The input ramp has rise time $\tau = 0.2\text{ns}$. The agreement observed between analytical and SPICE waveforms is good. It is observed that the piecewise analytical results very well replicates the SPICE waveforms.

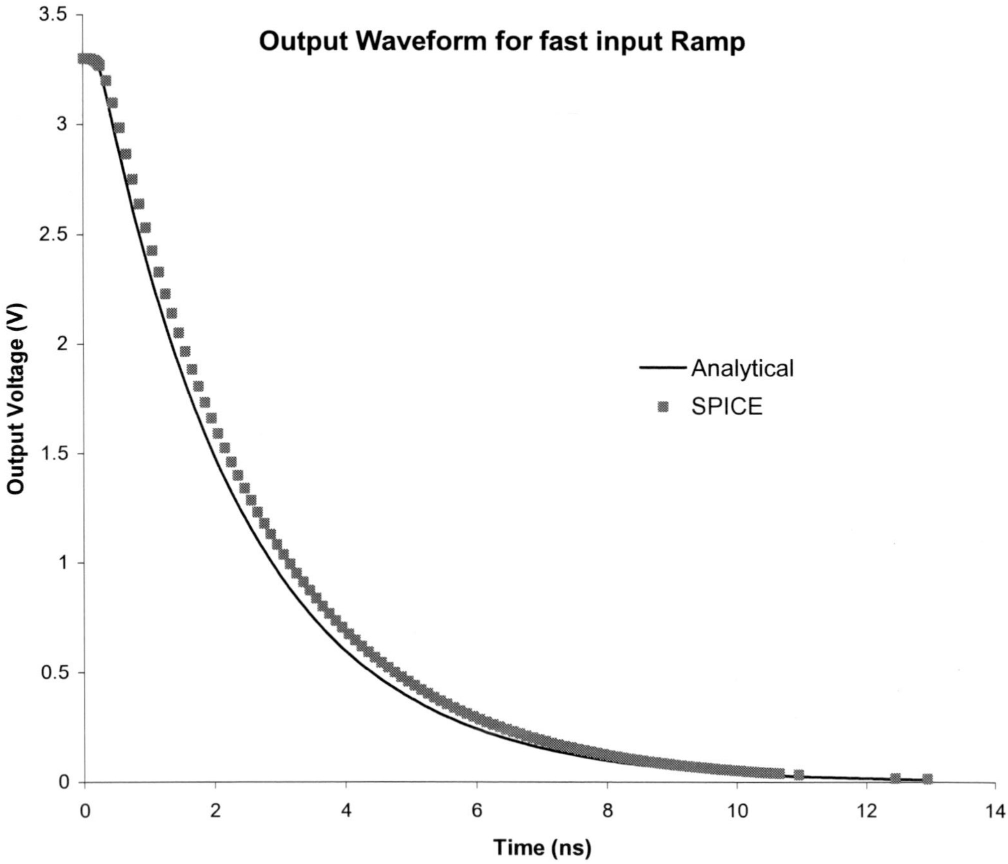

Figure 6. Comparison of output waveforms generated by analytical calculation and SPICE simulation for fast input ramp.

Piecewise analytical and the SPICE output waveforms for slow input ramp are compared in Figure 7. Again the experimental set-up utilizes TSMC 0.18μm technology transistor with $W_n = 10.08\mu\text{m}$. The total interconnect resistance, inductance and capacitance are taken as 833Ω, 37.48nH and 6pF. The input ramp has rise time of $\tau = 0.5\text{ns}$. Some small deviation is seen in the middle of the curve which corresponds to region-III. This is due to two reasons, firstly, because during analytical calculations an assumption is made that the input is an average of initial input voltage and final input voltage V_{DD}; secondly, due to the fact that the α-power law model is not as accurate in the linear region as it is in saturation region [140].

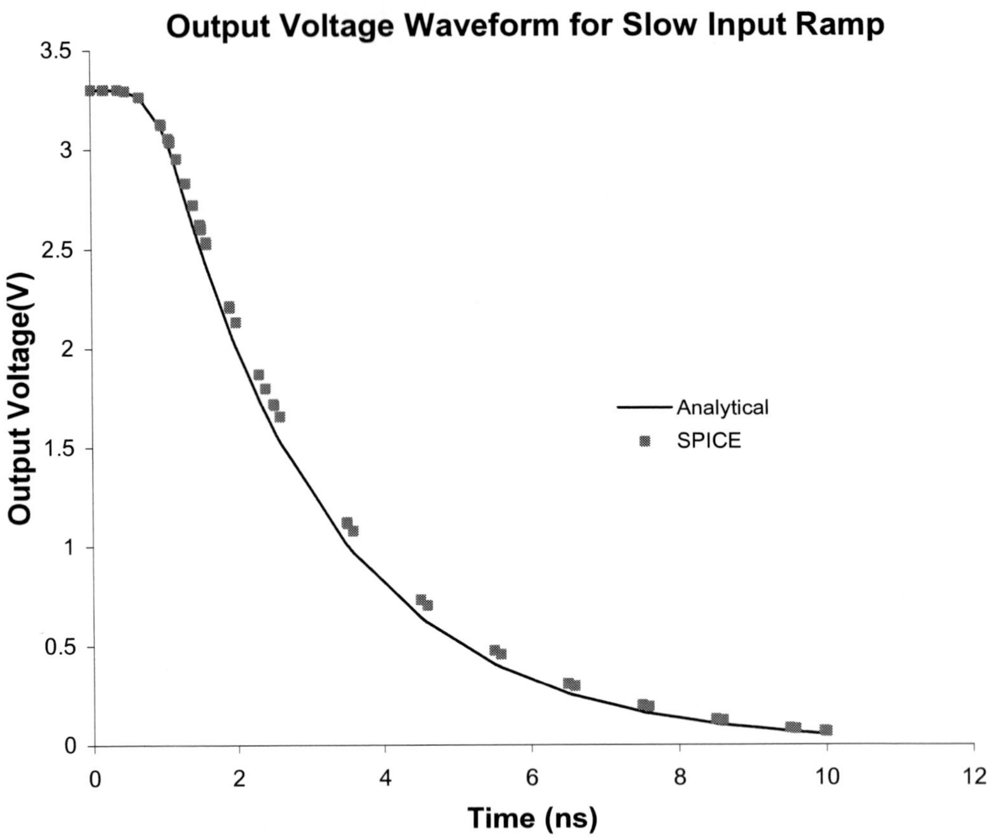

Figure 7. Comparison of output waveforms generated by analytical calculation and SPICE simulation for slow input ramp.

4.3.1. Effect of Short-Circuit Current on Propagation Delay

In the analysis upto this stage, the current through the PMOS transistor was considered negligible. Generally, this is a valid assumption because the capacitive load in long interconnect lines is large enough so that the output voltage does not change significantly until the time the PMOS transistor turns off. This means that the drain-to-source voltage of the PMOS transistor remains small and its current also takes small values. However, the value of the short-circuit current also depends on the width of the driving transistors and the input slope and it may become significant for the case of large drivers and for large input transition times. Therefore, a method given by Chatzigeorgiou et al. [27], for taking into account its influence on the estimation of the propagation delay is used.

The short-circuit current through the PMOS transistor exists in the interval $[0;t_p]$ where t_p is the time when the PMOS transistor turns off (when $V_{in} = V_{DD} - |V_{TP}|$). A simplified representation of the PMOS current during this period is shown in Figure 8. During the output voltage overshoot, which ends at time t_{ov}, the PMOS current is negative which means that the current is flowing toward V_{DD}. The minimum value of the PMOS current occurs at time

t_1 when the NMOS transistor starts conducting. The maximum value occurs when the PMOS transistor enters saturation at time point t_{s-p}. The existence of the PMOS current after time point t_{ov} results in a decrease of the output load discharging current and thus in an increase of the propagation delay. It acts like an amount of charge initially stored at the output node and which has to be removed through the NMOS transistor. On the contrary, the PMOS current before t_{ov} acts as an amount of charge that is being removed from the output load thus it speeds up the output evolution. Consequently, the total equivalent charge Q_e can be calculated by integrating the current of the PMOS device from time 0 to time t_p. If the maximum and minimum value for the PMOS current is calculated and the PMOS current is approximated by linear functions of time [32], then this equivalent charge could be simply obtained as the sum of the area of the two triangles which are set up below and above the time axis.

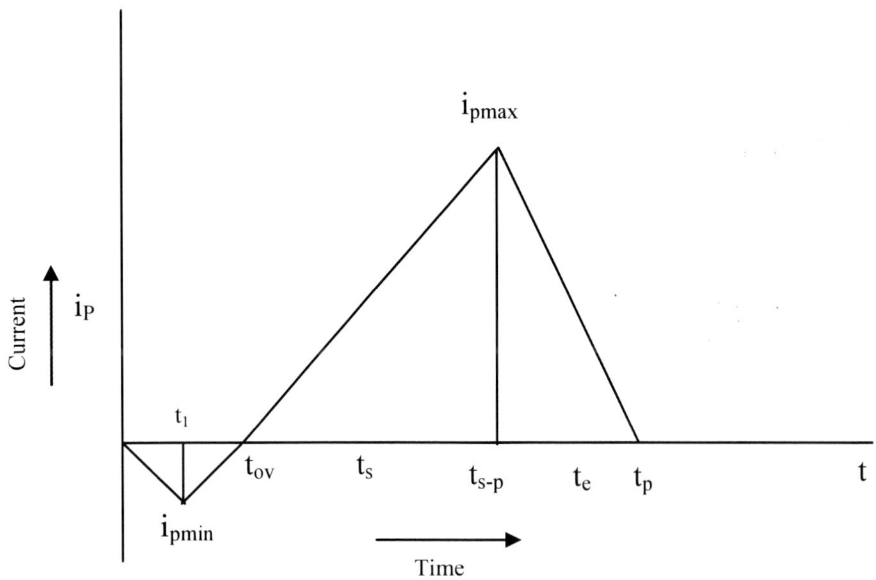

Figure 8. Representation of PMOS short circuit current.

The effect of short circuit current that increases propagation delay by time duration t_{ad} has been taken into account while calculating effective propagation delay presented in Tables 2 to 6. It is observed that the effect of short-circuit current significantly rises, as the transition time of the signal increases. However, for fast rising signals short circuit current have smaller effect on effective propagation delay.

4.3.2. Fifty Percent Propagation Delay Evaluation

The propagation delay for a CMOS gate can be calculated as the time from the half-V_{DD} point of the input to the half-V_{DD} point of the output (V_o). Using this definition, the average error in the calculation of propagation delay for several realistic interconnect load configurations is found to be around 2%.

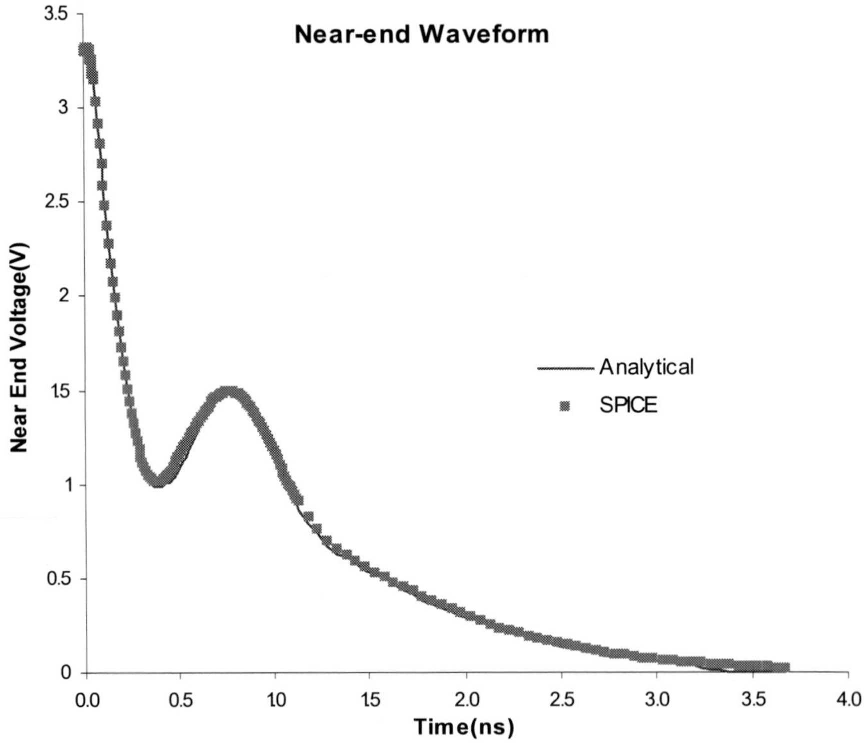

Figure 9. Near-end waveform of an interconnect driven by an input ramp having transition time of 50ps.

Tables 2, 3, 4, 5 and 6 shows fifty percent propagation delay found through SPICE simulation and through analytical calculation while taking into account t_{ad} and without t_{ad} for input slew rates of 50ps, 100ps, 200ps, 500ps and 1500ps. Different values of resistance, capacitance and inductance of the interconnect line are considered. The propagation delay is found analytically initially without taking into account the time factor t_{ad} due to short circuit current through PMOS transistor. Thereafter, the time t_{ad} is calculated for corresponding slew rates and then added to the analytically calculated propagation delay. It is observed that as the value of resistance decreases, the interconnect changes its state from resistive to inductive. An inductive interconnect produces certain undesired effects such as overshoots, undershoot and oscillations. These undesired ringing effects are more prominent in two conditions *viz.* either in case when the input ramp has very small transition time (i.e. high frequency signal) or in case when interconnect line is highly inductive (i.e. ratio of line's resistance to inductance is small enough). Near-end of the interconnect shows ringing more explicitly as compared to the far-end of interconnect. This is because of the reason that the ringing effect dies out as the signal travels to the far-end of the interconnect. One such case is considered where the transition time of input ramp is 50ps and the line parameters are

$R_{tot} = 208.33\Omega$, $L_{tot} = 74.96\text{nH}$ and $C_{tot} = 6\text{pF}$. The analytical piecewise waveform and the SPICE waveform are compared for near end and far end of the interconnect in Figure 9 and 10 respectively.

Due to voltage oscillations at the drain of the NMOS, the transistor switches its state back and forth between saturation and linear region of operation. Thus, accurate determination of propagation delay becomes tedious. Because of this reason, the percent error with respect to SPICE results is higher for extremely inductive interconnect. However, such high ratios of (L_{tot}/R_{tot}) are not found in practically oriented on-chip interconnects. For realistic values of interconnect parasitic resistance, inductance and capacitance the average error observed is nominal.

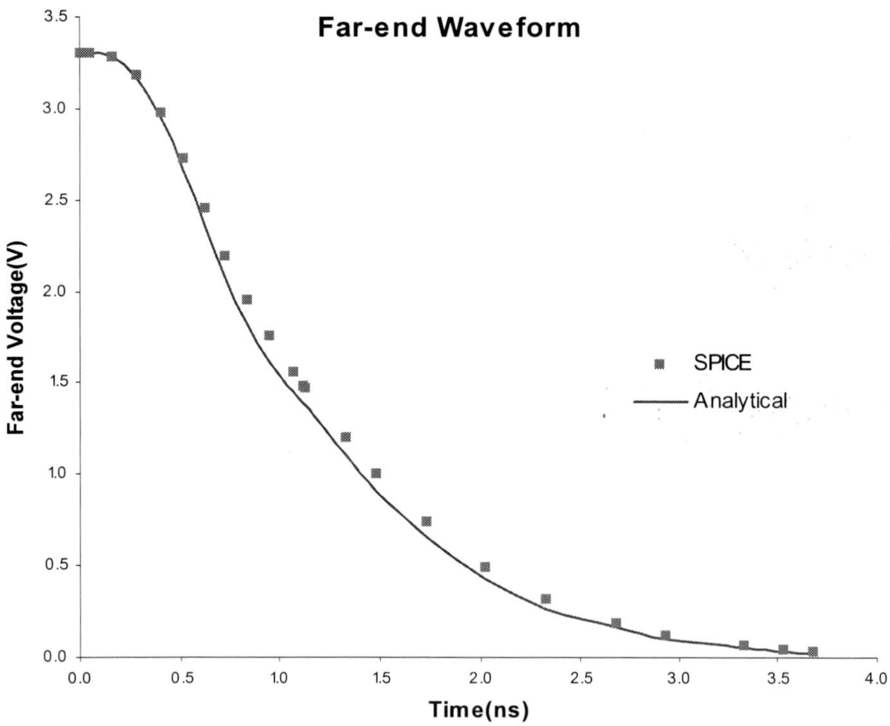

Figure 10. Far-end waveform of an interconnect driven by an input ramp having transition time of 50ps.

The last column of Tables 2 to 6 shows the type of input ramp i.e. fast or slow and in same column the second entry shows the region of operation in which the output reaches 50% of V_{DD}. It is observed that as the line resistance reduces, the output reaches 50% of V_{DD} in region-III (saturation region of NMOS) itself. Thus assuming that the transistor operates in linear region is quite incorrect. Table 5 (500ps) shows that the type of input ramp is considered fast for larger loads and slow for smaller loads. Thus an input ramp with same transition time of 500ps is fast for one set of interconnect load and slow for other set. The percentage error in the table corresponds to the error between analytically calculated propagation delay including time t_{ad} and propagation delay obtained through SPICE simulation.

Table 2. Fifty percent propagation delay evaluation through analytical calculation and SPICE simulation for an input having transition time of 50ps.

R_{tot} (Ω)	L_{tot} (nH)	C_{tot} (pF)	Prop. Delay by SPICE (ns)	Analytically Calculated Prop. Delay without t_{ad} (ns)	Analytically Calculated Prop. Delay with t_{ad} (ns)	Error (%)	Input ramp type, Reaches 50% of V_{DD} in Region
1666.67	37.48	6	3.2134	3.101	3.281	2.06	Fast, Regn.-IV
		12	6.4068	6.1925	6.3425	-1.01	Fast, Regn.-IV
	74.96	6	3.2183	3.105	3.285	2.03	Fast, Regn.-IV
		12	6.4113	6.197	6.347	-1.01	Fast, Regn.-IV
833.33	37.48	6	1.8624	1.729	1.909	2.44	Fast, Regn.-IV
		12	3.7034	3.4485	3.5985	-2.92	Fast, Regn.-IV
	74.96	6	1.881	1.737	1.917	1.88	Fast, Regn.-IV
		12	3.7092	3.452	3.602	-2.98	Fast, Regn.-IV
416.67	37.48	6	1.2338	1.085	1.265	2.47	Fast, Regn.-IV
		12	2.45	2.162	2.312	-5.97	Fast, Regn.-IV
	74.96	6	1.2463	1.0855	1.2655	1.52	Fast, Regn.-IV
		12	2.4517	2.143	2.293	-6.92	Fast, Regn.-IV
208.33	37.48	6	1.0002	0.9279	1.1079	9.72	Fast, Regn.-III
		12	1.9845	1.8497	1.9997	0.76	Fast Regn.-III
	74.96	6	0.98567	0.8777	1.0577	6.81	Fast, Regn.-III
		12	1.9848	1.85	2	0.76	Fast, Regn.-III

Table 3. Fifty percent propagation delay evaluation through analytical calculation and SPICE simulation for an input having transition time of 100ps.

R_{tot} (Ω)	L_{tot} (nH)	C_{tot} (pF)	Prop. Delay by SPICE (ns)	Analytically Calculated Prop. Delay without t_{ad} (ns)	Analytically Calculated Prop. Delay with t_{ad} (ns)	Error (%)	Input ramp type, Reaches 50% of V_{DD} in Region
1666.67	37.48	6	3.2167	3.094	3.294	2.347	Fast, Regn.-IV
		12	6.4101	6.185	6.365	-0.709	Fast, Regn.-IV
	74.96	6	3.2217	3.099	3.299	2.343	Fast, Regn.-IV
		12	6.4145	6.189	6.369	-0.714	Fast, Regn.-IV
833.33	37.48	6	1.866	1.722	1.922	2.914	Fast, Regn.-IV
		12	3.7068	3.4413	3.6213	-2.361	Fast, Regn.-IV
	74.96	6	1.8744	1.731	1.931	2.931	Fast, Regn.-IV
		12	3.7125	3.445	3.625	-2.414	Fast, Regn.-IV

Table 3. Continued

R_{tot} (Ω)	L_{tot} (nH)	C_{tot} (pF)	Prop. Delay by SPICE (ns)	Analytically Calculated Prop. Delay without t_{ad} (ns)	Analytically Calculated Prop. Delay with t_{ad} (ns)	Error (%)	Input ramp type, Reaches 50% of V_{DD} in Region
416.67	37.48	6	1.2375	1.067	1.267	2.328	Fast, Regn.-IV
		12	2.4534	2.154	2.334	-5.116	Fast, Regn.-IV
	74.96	6	1.2499	1.0785	1.2785	2.237	Fast, Regn.-IV
		12	2.4551	2.1345	2.3145	-6.075	Fast, Regn.-IV
208.33	37.48	6	1.0039	0.9208	1.1208	10.430	Fast, Regn.-III
		12	1.988	1.8425	2.0225	1.706	Fast, Regn.-III
	74.96	6	0.98928	0.86925	1.06925	7.479	Fast, Regn.-III
		12	1.9882	1.842	2.022	1.672	Fast, Regn.-III

Table 4. Fifty percent propagation delay evaluation through analytical calculation and SPICE simulation for an input having transition time of 200ps.

R_{tot} (Ω)	L_{tot} (nH)	C_{tot} (pF)	Prop. Delay by SPICE (ns)	Analytically Calculated Prop. Delay without t_{ad} (ns)	Analytically Calculated Prop. Delay with t_{ad} (ns)	Error (%)	Input ramp type, Reaches 50% of V_{DD} in Region
1666.67	37.48	6	3.2273	3.083	3.303	2.37	Fast, Regn.-IV
		12	6.4200	6.1715	6.3715	-0.72	Fast, Regn.-IV
	74.96	6	3.2322	3.088	3.308	2.37	Fast, Regn.-IV
		12	6.4245	6.175	6.375	-0.74	Fast, Regn.-IV
833.33	37.48	6	1.8772	1.7095	1.9295	2.86	Fast, Regn.-IV
		12	3.7172	3.4265	3.6265	-2.42	Fast, Regn.-IV
	74.96	6	1.8854	1.7185	1.9385	2.88	Fast, Regn.-IV
		12	3.7229	3.43	3.63	-2.48	Fast, Regn.-IV
416.67	37.48	6	1.2496	1.051	1.271	1.91	Fast, Regn.-IV
		12	2.4645	2.1385	2.3385	-5.26	Fast, Regn.-IV
	74.96	6	1.2613	1.065	1.285	2.05	Fast, Regn.-IV
		12	2.4660	2.119	2.319	-6.21	Fast, Regn.-IV
208.33	37.48	6	1.0001	0.905	1.125	9.96	Fast, Regn.-III
		12	1.9962	1.826	2.026	1.47	Fast, Regn.-IV
	74.96	6	1.0157	0.851	1.071	6.80	Fast, Regn.-III
		12	1.9962	1.8265	2.0265	1.50	Fast, Regn.-III

Table 5. Fifty percent propagation delay evaluation through analytical calculation and SPICE simulation for an input having transition time of 500ps.

R_{tot} (Ω)	L_{tot} (nH)	C_{tot} (pF)	Prop. Delay by SPICE (ns)	Analytically Calculated Prop. Delay without t_{ad} (ns)	Analytically Calculated Prop. Delay with t_{ad} (ns)	Error (%)	Input ramp type, Reaches 50% of V_{DD} in Region
1666.67	37.48	6	3.2507	3.078	3.378	3.7685	Slow,Regn.-IV
		12	6.4428	6.142	6.392	-0.7947	Fast, Regn.-IV
	74.96	6	3.2554	3.082	3.382	3.7433	Slow,Regn.-IV
		12	6.4472	6.146	6.396	-0.8005	Fast, Regn.-IV
833.33	37.48	6	1.9038	1.7	2	4.8100	Slow,Regn.-IV
		12	3.7418	3.3925	3.6425	-2.7261	Fast, Regn.-IV
	74.96	6	1.9114	1.7115	2.0115	4.9764	Slow,Regn.-IV
		12	3.7473	3.397	3.647	-2.7502	Fast, Regn.-IV
416.67	37.48	6	1.2808	1.023	1.323	3.1897	Slow,Regn.-IV
		12	2.4916	2.093	2.343	-6.3423	Fast, Regn.-IV
	74.96	6	1.2906	1.055	1.355	4.7528	Slow,Regn.-IV
		12	2.4929	2.072	2.322	-7.3600	Fast, Regn.-IV
208.33	37.48	6	1.047	0.8484	1.1484	8.8297	Fast, Regn.-III
		12	2.0271	1.7803	2.0303	0.1576	Fast, Regn.-III
	74.96	6	1.0317	0.791	1.091	5.4354	Fast, Regn.-III
		12	2.0256	1.7804	2.0304	0.2364	Fast, Regn.-III

Table 6. Fifty percent propagation delay evaluation through analytical calculation and SPICE simulation for an input having transition time of 1500ps.

R_{tot} (Ω)	L_{tot} (nH)	C_{tot} (pF)	Prop. Delay by SPICE (ns)	Analytically Calculated Prop. Delay without t_{ad} (ns)	Analytically Calculated Prop. Delay with t_{ad} (ns)	Error between (%)	Input ramp type, Reaches 50% of V_{DD} in Region
1666.67	37.48	6	3.3182	3.0285	3.4285	3.217	Slow, Regn.-IV
		12	6.5276	6.137	6.487	-0.626	Slow, Regn.-IV
	74.96	6	3.3221	3.032	3.432	3.202	Slow, Regn.-IV
		12	6.5318	6.1415	6.4915	-0.621	Slow, Regn.-IV
833.33	37.48	6	1.9941	1.682	2.082	4.222	Slow, Regn.-IV
		12	3.8423	3.3925	3.7425	-2.667	Slow, Regn.-IV
	74.96	6	1.9993	1.685	2.085	4.110	Slow, Regn.-IV
		12	3.8472	3.3995	3.7495	-2.606	Slow, Regn.-IV
416.67	37.48	6	1.4033	1.054	1.454	3.487	Slow, Regn.-IV
		12	2.6124	2.053	2.403	-8.714	Slow, Regn.-IV
	74.96	6	1.4086	1.0665	1.4665	3.948	Slow, Regn.-IV
		12	2.6136	2.0635	2.4135	-8.291	Slow, Regn.-IV

Table 6. Continued

R_{tot} (Ω)	L_{tot} (nH)	C_{tot} (pF)	Prop. Delay by SPICE (ns)	Analytically Calculated Prop. Delay without t_{ad} (ns)	Analytically Calculated Prop. Delay with t_{ad} (ns)	Error between (%)	Input ramp type, Reaches 50% of V_{DD} in Region
208.33	37.48	6	1.1814	0.8125	1.2125	2.565	Slow, Regn.-IV
		12	2.1519	1.6184	1.9684	-9.322	Fast,Regn.-III
	74.96	6	1.1826	0.841	1.241	4.706	Slow, Regn.-IV
		12	2.1473	1.4625	1.8125	-18.472	Slow, Regn.-IV

Thus, this case study introduced an analytical method for the calculation of the output waveform and propagation delay of a CMOS inverter driving a distributed *RLC* interconnect load, modeled as a *CRLC* π-circuit. It is observed that the output waveforms generated by the SPICE and the analytical equations closely match each other. The analytical driver-interconnect load model gives sufficiently good results for different cases of slow and fast input ramps. For each case of stimulations the model gives an insight to four regions of operation of the CMOS gate. The voltage waveform at the end of an interconnect line is obtained for each region of operation. Since the method leads to the output waveforms at both ends of an interconnect line, a further step is introduced in order to calculate the voltage waveform at each point of the line. It can be concluded that analytically calculated 50% propagation delays are in very good agreement with SPICE simulation results.

5. Delay Minimization Techniques

The length and organization of the interconnects, the communication paths, place a lower bound on the area and time delay of the system operations. Mead and Rem [100, 101], showed that an optimum design is possible using the area time product as a cost function. Conditions are also outlined by them; under which propagation delays in VLSI circuits are logarithmic functions of interconnect lengths. These conditions are imposed by area requirements and the velocity of light.

Mohsen et al. [105] examined minimization of delays associated with driving and sensing signals from large capacitance paths by optimizing the fan-out factor of the driver stages, the gain of the input sensing stages and the path voltage swing. It is concluded that minimum delay time is achieved when the delay times of the successive stages of the driver chain, the high capacitance path and the input sensing stage are comparable. Wann *et al.* [159] identified asynchronous and clocked control structures for VLSI interconnects delay minimization. Munch et al. [107] proposed an analytical model for simultaneous placement and binding in a single, unified mixed integer linear programming to optimize the overall interconnect length and delay in a linear bit-slice data path.

Bakoglu et al. [12] reduced interconnection delays by replacing the passive interconnections on a chip by the active interconnections, that is, by inserting inverters or "repeaters" at appropriate spacing depending on the preferred driving mechanism. In *RC*-model interconnect without repeaters, the propagation delay increases proportionally to the square of the interconnection length because both capacitance and resistance increase linearly with length. The use of repeaters, by dividing the interconnect wire into smaller subsections, makes the time delay dependence on length to be linear. The additional delay due to repeaters in a long interconnect is marginal in comparison to large amount of reduction in interconnect delay. Thus, the overall propagation delay through repeater inserted interconnect is reduced. Several methods of repeater insertion for reducing delays which include driving interconnection using minimum size inverters, optimum size inverters, and cascaded (tapered and non-tapered) inverters have been discussed in [2, 6, 11,12, 47, 59]. Bakoglu et al. [11, 12] have shown that for an *RC*-model the most appropriate methodology of minimizing propagation delay through an on-chip interconnect is optimal repeater insertion, in which the repeaters are uniform CMOS buffers, each of same size and placed at equal interval. An optimum number of sections and size of the buffer is derived so that minimum propagation delay is achieved [12]. In current scenario, with growing importance of inductive effects, optimal repeaters are inserted in *RLC*-model as shown in Figure 11.

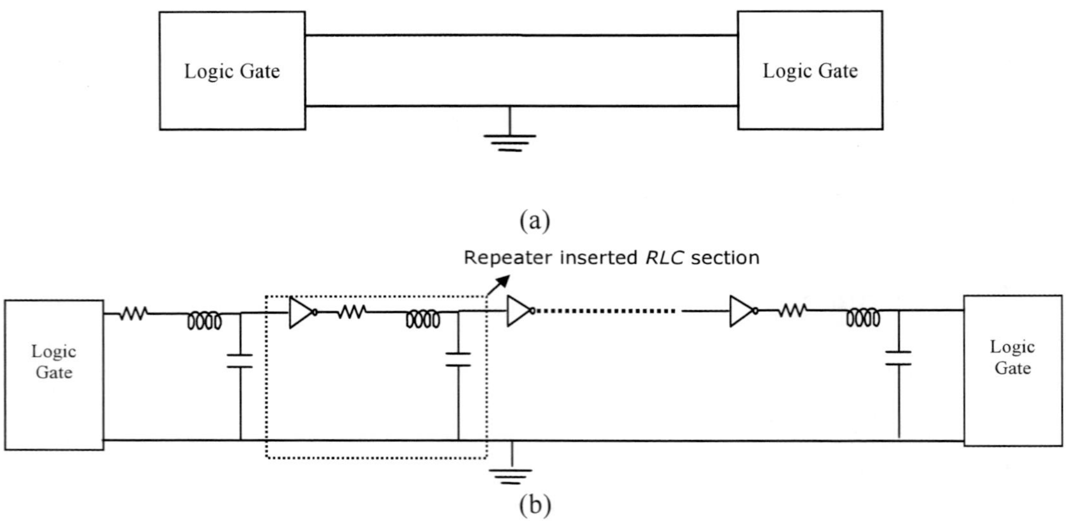

Figure 11. (a) Logic gates connected by an interconnect wire (b) Optimal repeater insertion in *RLC* model of an interconnect.

Tam et al. [154] outlined clock design for the first implementation of the IA-64 microprocessor. They showed that a clock distribution with an active distributed deskewing technique achieves a low skew of 28ps. Their work describes the global, regional and clock distributions.

Nakagome *et al.* [114] discussed low swing signaling, in an effort to reduce the insertion of repeater logic, reduce power dissipation, and potentially drive longer distances. This technique uses two wires to transmit the data differentially across the entire length of the bus (as used in SRAM). A sense amplifier (SA) detects the small voltage differential developed across the distant terminals. This kind of technique is also often used for long buses with multiple drivers. Implementing such a bus with repeaters (or buffers) may require their locations to be fixed at each driver. The locations of repeaters may require modifications to multiplex the data from either the source driver or the rest of the bus. This can significantly increase delay. The differential bus enables bidirectional transmission of the data to multiple receivers. The implementation of this technique comes with significant cost. The differential signaling requires two routing tracks instead of one, the low voltage swing developed at the receiver is highly sensitive to noise injection (even accepting common mode components) and requires extensive design management. The technique enables data transmission in just one phase of the clock, the other being needed for the precharge phase. Although another phase of data transmission could follow, but that would result in the intermediate SA adding delay, logic intrusion, and may still be subject to full clock skew and jitter variations. All these subtract from the time available to develop the differential. Techniques to enable greater routing density through single-ended low swing signaling have been proposed [65,114]. These techniques effectively halve the differential available through dual-rail signaling and are especially error-prone to noise, due to either reduced supply voltages or the removal of common-mode components. Various charge-sharing techniques have also been proposed to effect the low swing signaling and hence reduce power [85, 167]. These techniques are only valid for very short buses for which the resistance of the interconnect is negligible; otherwise, at the high frequencies of VLSI chip the differential developed at the point of charge sharing cannot extend to the distant SA.

Nekili and Savaria [116] proposed active regeneration technique for improving transmission delay in which a number of regenerator stations are dispersed across the length of the bus. These regenerators sense when a voltage transition is occurring and provide an additional current boost to speed up the transition [117]. One advantage of this technique is that the regenerators can be located almost anywhere along the bus and are therefore less constrained by underlying signal routes and processing logic. Furthermore, both bidirectional and multisource behavior of a wire are directly supported by this technique, and differential signaling is not required. The disadvantage of this technique is primarily in its power dissipation, since the regenerators have a relatively slow response time and significant crossover current. The slow response time is a design requirement of the technique to prevent the amplification of unwanted noise injected onto the bus.

Izumikawa *et al.* [75] proposed to sense the direction of current instead of differential voltage on wires for minimizing delay. An advantage of this technique is that very long interconnect lengths could be driven since the *L/R* time constants is now of relevance rather than the *RC* time constant, for which the former is length independent and very small. This technique is therefore primarily limited by the signal's time of flight and device switching speeds. Disadvantages include high power dissipation due to the large potential difference across the wire which is sustained for the entire sensing phase. Another problem is noise immunity concerns akin to those of inductively coupled return currents that extend across many conductors.

Wave Pipelined Multiplex (WPM) technique discussed by Chen et al. [28] essentially consists of multiple data signals in flight along the bus nominally separated in time by one clock cycle. This technique is used for minimizing overall delay and area of interconnects. In WPM technique, all intermediate latches/registers are removed that are present in conventional pipelined interconnect between the source and receiver, resulting in a propagation delay which is larger than a clock cycle. To reduce overall delay and clock loading repeaters can be inserted. For synchronized data, the system with WPM design method will use less chip area than that with traditional pipelining design method. There are some fundamental disadvantages with this technique [26]. First, there is the concern that a subsequent wavefront will, through capacitive and inductive crosstalk, propagate much faster than its predecessor and hence corrupt the earlier transmitted data. Furthermore, at a nominal clock frequency there will be at least a two cycle delay between the generation and reception of data; however, when the clock is slowed down below a given frequency during chip debug, this will than become single cycle delay. Without additional control logic to conditionally pipe the data at the receiver, this phenomenon will result in functional errors.

Joshi and Davis [78] proposed an improved low overhead WPM routing technique. The technique intelligently uses the inherent intraclock period interconnect idleness to implement wire sharing throughout the various hierarchical levels of design. They claimed that the WPM network can be readily incorporated into future gigascale integration (GSI) systems to reduce the number of interconnect routing channels in an attempt to contain escalating manufacturing costs. A system level analysis of WPM routing is carried out and is verified on circuit level. They studied a case of 40 million transistor system that uses WPM network. It has been shown that WPM technique could result in an almost 20% decrease in the number of metal layers for less than 4% increase in dynamic power with no loss of communication throughput performance. The key virtues of WPM routing are its flexibility, robustness, implementation simplicity and its low overhead requirements.

Maheshwari et al. [97] adopted a differential current-sensing technique as an alternative to existing circuit techniques for on chip interconnects. On the downside, current sensing requires a special receiver circuit as compared to the CMOS repeaters. Also, this technique is power inefficient due to the presence of static-power dissipation owing to the low-impedance path to the ground. Current sensing is essentially a low swing signaling technique and hence, it is sensitive to full swing aggressor noise. Huang et al. [67] described new circuits called capacitor coupling trigger (CCT) and capacitor coupling accelerator (CCA). Such circuits have been used to reduce long interconnect (*RC*) delay in a sub-100nm process. The circuits under consideration require capacitors to split the output driving paths to eliminate the short-circuit current and thus improve the signal transition time.

Naeemi et al. [112] reported optimal design of global interconnects for GSI. The interconnect width is optimized to achieve a large data flux density and small latency simultaneously. It is shown that the optimum wire width results in delay that is 33% larger than the time of fight for the signal.

Proper wire sizing can effectively reduce the interconnect delay, especially in deep submicron or nanometer designs where the wire resistance becomes significant. Cong, Leung, and Zhou [36, 37], first used the wire sizing technique for the interconnect delay minimization of general nets based on dynamic programming. They developed an optimal wire-sizing algorithm for a single source *RC* interconnect tree to minimize the sum of weighted delays from the source to timing-critical sinks under the Elmore delay model. An efficient approach to perform global interconnects sizing and spacing (GISS) for multiple nets to minimize interconnect delays has been presented in [33]. This method considers coupling capacitance, in addition to area and fringing capacitances. The work proposes an asymmetric wire sizing scheme, which may be widened or narrowed above and below the centerline of the original wire asymmetrically. The optimal wire sizing and spacing problems for a single net with fixed surrounding wire segments can be solved by adapting the bottom-up dynamic programming (DP)-based buffer insertion and wire-sizing algorithm proposed in [95]. Pullela, Menezes, and Pillage [128] used wire sizing for reducing the clock skew caused by the process variation, while Edahiro [51] for the delay reduction of the clock net. These sizing methods are based on *RC* delay model, and suitable for a tree topology. El-Moursy *et al.* [53] considered wire sizing in conjuction with optimal repeater insertion as an efficient way for reducing Power-Delay-Area-Product (PDAP) in *RLC* modeled interconnects.

El-Moursy et al. [56] considered exponential wire tapering more attractive in *RLC* than for *RC* lines due to the presence of inductive effects. The optimum wire shape to produce the minimum signal propagation delay across an *RLC* line is shown to exhibit a general exponential form. For *RLC* lines, optimum wire tapering achieves a greater reduction in the signal propagation delay as compared to uniform wire sizing. Furthermore, it is shown that exponential tapering outperforms uniform repeater insertion for *RLC* lines. As technology advances, wire tapering becomes more effective than repeater insertion, since a greater reduction in the propagation delay is achieved. It is shown that optimum wire tapering achieves a reduction of 36% in the propagation delay in long *RLC* interconnect as compared to uniform repeater insertion. Wire tapering can reduce both the propagation delay and power dissipation. Wire tapering is also shown to reduce the power dissipation of a circuit by up to 65%. Wire tapering can also improve signal integrity by reducing the inductive noise of the interconnect lines. Wire tapering reduces the effect of impedance mismatch in digital circuits. It is shown that an exponentially tapered interconnect minimizes the time-of-flight of an *LC* line.

More recently for reducing delays, optical interconnects are under consideration as potential alternative to the regime of copper interconnects [3, 83]. The critical length beyond which optical interconnects are preferable than the electrical counterparts is approximately one-tenth of the chip edge length at the 22nm technology node [30]. The critical length depends on technology node. For an optical interconnect, large optical transmitter and receiver are located at the two ends of the waveguide; in contrast to electrical repeaters, which are distributed along the interconnect. Via congestion issues, therefore, are avoided in optical interconnects. CMOS-compatible modulator is one of the most challenging elements in the optical data path. An advantage of optical interconnect is the smaller crosstalk noise as compared with electrical interconnect. The primary limitations for optical interconnect are the yield of the modulators; large footprint and power consumption in the optical components and driving circuits. The generation of sufficient optical power to maintain optical operation; on-chip fabrication of optical interconnect; and silicon optoelectronic circuit packaging are also difficult. The packaging issue involves the integration of all on-chip optical components as well as the coupling to off-chip sources. Thus, both efficient light sources and detectors are crucial for the development of future on-chip optical interconnects. Finally, a set of integrated silicon compatible WDM (wavelength-division multiplexing) components needs to be developed in order to fully exploit the inherent advantages of optical interconnects.

Regime of carbon nanotubes (CNTs) in future VLSI interconnect applications is also considered aggressively [16, 84, 93, 111, 152, 160]. The metallic single-walled CNTs with their good intrinsic properties along with encouraging performance, power and thermal reliability of CNT bundles provide sound foundation for further investments in CNT interconnect research. Several challenges remain to be overcome in the areas of fabrication and process integration. Although these challenges are not expected to cause any fundamental problems, lowering of metal nanotube contact resistance would be vital, especially for local interconnect and via applications. Moreover, rigorous characterization and modeling of electromagnetic interactions in CNT bundles; 3-D (metal) to 1-D (CNT) contact resistance;

impact of defects on electrical and thermal properties; and high-frequency effects are seen as additional challenges. Although optical and carbon nanotubes (CNTs) can be safely predicted as future interconnects, but these technologies are still amateur as compared to well established fabrication technique in copper.

A low dielectric constant interlayer insulator reduces both the gate delay and power by reducing wiring capacitance. Low-resistivity copper and low-permittivity dielectric may provide performance and reliability enhancement in DSM global interconnects. Super conducting materials for interconnects and liquid nitrogen is seen ultimate solution to interconnect speed [63].

6. Conclusion

This chapter presented a detailed literature review in the area of interconnects which leads to the conclusion that in current scenario, it is essential to take an in-depth look at propagation delay in inductive long interconnects through improved and accurate analytical approach. This chapter addresses various aspects of modeling and extraction of the parasitics associated with on-chip interconnects. An analytical method for the calculation of output waveform and propagation delay of a CMOS inverter driving a distributed *RLC* interconnect load, modeled as a *CRLC* π-circuit is undertaken through a case study. The study is done through adequate analytical modeling and compared with supporting computer simulations.

At last various techniques are reviewed for delay minimization through interconnects.

References

[1] Achar R. and Nakhla M., "Simulation of high-speed interconnects," *Proceed. IEEE*, vol. 89, no. 5, pp. 693–728, May 2001.

[2] Adler V. and Friedman E.G., "Repeater design to reduce delay and power in resistive interconnect," *IEEE Trans. on Circuits and Systems-II*, vol. 45, no. 5, pp. 607–616, May 1998.

[3] Agarwal D., "Optical interconnects to silicon chips using short pulses", *Phd. Thesis submitted at Stanford University*, 2002.

[4] Agarwal K., Sylvester D. and Blaauw D., "A library compatible driver output model for on-chip *RLC* transmission lines," *IEEE Trans. on Computer-Aided Design of Integrated Circuits and Systems*, vol. 23, no. 1, pp. 128-136, Jan 2004.

[5] Ahuja S., Kothari L., Vishwakarma D. N., and Balasubramanian S. K., "Field programmable gate array based over current relays," *Electric Power Components and Systems*, vol.32, pp.247-255, 2004.

[6] Alpert C. J., "Wire segmenting for improved buffer insertion," in *Proc. IEEE/ACM Design Automation Conf.*, June 1997, pp. 588–593.

[7] Anderson F. E., Wills J. S. and Berta E. Z., "The core clock system on the next generation Itanium™ microprocessor," in *Proc. ISSCC*, San Francisco, 2002, pp. 146–147.

[8] Arora N. D., Raol K. V., Schumann R. and Richardson L. M., "Modeling and extraction of interconnect capacitances for multilayer VLSI circuits," *IEEE Trans. on*

Computer-Aided Design of Integrated Circuits and Systems, vol. 15, pp. 58–67, Jan. 1996.

[9] Arunachalam R., Dartu F. and Pileggi L.T., "CMOS gate delay models for general *RLC* loading," in: *Proc. IEEE Int. Conf. Comput. Design: VLSI Comput. Process.*, 1997, pp. 224–229.

[10] Baker G.A., Jr., *Essentials of Padé Approximants*, Academic Press, Reading, MA, 1975.

[11] Bakoglu H. B., *Circuits, Interconnections, and Packaging for VLSI*, Reading, MA: Addison-Wesley, 1990.

[12] Bakoglu H. B. and Meindl J. D., "Optimal interconnection circuits for VLSI," *IEEE Trans. on Electron Devices*, vol. ED-32, pp. 903–909, May 1985.

[13] Banerjee K. and Mehrotra A., "Accurate analysis of on-chip inductance effects and implications for optimal repeater insertion and technology scaling," in *Proc. IEEE Symp. VLSI Circuits*, Kyoto, Japan, 2001, pp. 195–198.

[14] Banerjee K. and Mehrotra A., "Accurate analysis of on-chip effects using a novel performance optimization methodology for distributed *RLC* interconnnects," in *Proc. Design Automation Conf.*, Las Vegas, NV, 2001, pp. 798–803.

[15] Banerjee K. and Mehrotra A., "Analysis of on-chip inductance effects for distributed *RLC* interconnects," *IEEE Trans. on Computer-Aided Design of Integrated Circuits and Systems*, vol. 21, pp. 904–915, Aug. 2002.

[16] Banerjee, K. and Srivastava, N., "Are carbon nanotubes the future of VLSI interconnections?," in *Proc. ACM/IEEE Design Automation Conf.,* San Francisco, CA, 2006, pp.809-14.

[17] Benini L., Macii A., Macii E., Poncino M. and Scarsi R., "Synthesis of low-overhead interfaces for power-efficient communication over wide buses," in *Proc. ACM/IEEE Design Automation Conf.*, 1999, pp. 128–133.

[18] Bernstein K. and Rohrer N. J., *SOI Circuit Design Concepts*. Norwell, M A: Kluwer, 1999.

[19] Bhaumik B., Pradhan P., Visweswaran G.S., Varambally R. and Hardi A., "A low power 256 KB SRAM design," in *Proc. 12th Int. Conf. VLSI Design*, Jan. 1999, pp. 67-70.

[20] Boese K. D., Kahng A. B., McCoy B. A. and Robins G., "Fidelity and near-optimality of Elmore-based routing constructions," in *Proc. IEEE Int. Conf. Computer Design*, Oct. 1993, pp. 81-84.

[21] Boulejfen N., Kouki A. and Ghannouchi F., "Frequency and time domain analysis of nonuniform lossy coupled transmission lines with linear and nonlinear terminations," *IEEE Trans. on Microwave Theory and Techniques*, vol. 48, no. 3, pp. 367–379, Mar. 2000.

[22] Branin Jr .F. H., "Transient analysis of lossless transmission lines," *Proc. IEEE*, vol. 55, pp. 2012–2013, 1967.

[23] Brocco L.M., McCormick S.P. and Allen J., "Macromodeling CMOS circuits for timing simulation," *IEEE Trans. on Computer-Aided Design of Integrated Circuits and Systems,* vol. CAD-7, pp. 1237-1249, Dec. 1988.

[24] Cao Y., Huang X., Sylvester D., Chang N. and Hu C., "A new analytical delay and noise model for on-chip *RLC* interconnect," in *Proc. IEDM*, San Francisco, CA, 2000, pp. 823–826.

[25] Chandra G., Kapur P. and Saraswat K. C., "Scaling trends for the on chip power dissipation" in *Proc. IEEE Interconnect Technology Conf.*, 2002, pp. 170–172.

[26] Chandrakasan A., Bowhill W.J. and Fox F., *Design of High Performance Microprocessor Circuits*, IEEE Press, ISBN 0-7803-6001-X, 2001.

[27] Chatzigeorgiou A., Nikolaidis S. and Tsoukalas I., "Modeling CMOS gates driving RC interconnect loads," *IEEE Trans. on Circuits Systems—II: Analog Digital Signal Process*, vol. 48, no.4, pp. 413–418, April 2001.

[28] Chen L. and Tang Z. "Design issues of deep submicron wave pipelining technology" in *Proc. 4th International Conference on ASIC*, Oct. 2001, pp. 62 – 66.

[29] Cheng C.-K., Lillis J., Lin S. and Chang N., *Interconnect Analysis and Synthesis*. New York: Wiley, 1999.

[30] Chen G., Chen H., Haurylau M., Nelson N.A., Albonesi D.H., Fauchet P.M. and Friedman E.G., "Predictions of CMOS compatible on-chip optical interconnect," *Integration, the VLSI Journal*, Elsevier Pub., Netherlands, vol. 40, no.4, pp. 434-446, July 2007.

[31] Cong J. and He L., "Optimal wire sizing for interconnects with multiple sources," *Proc. IEEE/ACM Design Automation Conf.*, Nov. 1995, pp. 568-574.

[32] Cong J., He L., Koh C. and Madden P.H., "Performance optimization of VLSI interconnect layout," *Integration, the VLSI Journal*, Elsevier Pub., Netherlands, vol. 21, no.1-2, pp. 1-94, Nov. 1996.

[33] Cong J., He L., Koh C.-K. and Pan Z., "Global interconnect sizing and spacing with consideration of coupling capacitance," in *Proc. ACM/IEEE Int. Conf. Computer-Aided Design*, Nov. 1997, pp.628-633.

[34] Cong J., Lei Het J., Kahng A.B., Noice D., Shirali N. and Yen S.H.-C., "Analysis and justification of a simple, practical 2 1/2-D capacitance extraction methodology," in *Proc. Design Automation Conf.*, 1997, pp. 627–632.

[35] Cong J., Kahng A.B., Koh C., and Tsao C.-W.A. "Bounded-slew clock and Steiner routing under Elmore delay," *Proc. IEEE Int. Conf. Computer-Aided Design*, Jan. 1995, pp. 66-71.

[36] Cong J. and Leung K.-S., "Optimal wire sizing under the distributed Elmore delay model," in *Dig. Tech. Papers IEEE lnt. Conf. Computer- Aided Design,* 1993, pp. 634-639.

[37] Cong J., Leung K.-S. and Zhou D., "Performance-driven interconnect design based on distributed RC delay model," in *Proc. 30th ACMLEEE Design Automation Conf.* 1993, pp. 606-611.

[38] Coulibaly L.M. and Kadim H.J., "Analytical ramp delay model for distributed on-chip *RLC* interconnects" in *Proc. IEEE Midwest Symposium on Circuits and Systems*, July 2004, vol. 1, pp. 457-460.

[39] Cullum J., Ruehli A. and Zhang T., "A method of reduced-order modeling and simulation of large interconnect circuits and its application to PEEC models with retardation," *IEEE Trans. on Circuits and Systems-II: Analog and Digital Signal Processing*, vol. 47, no.4, pp. 261–273, Apr. 2000.

[40] Dartu F., Menezes N. and Pileggi L.T., "Performance computation for precharacterized CMOS gates with RC loads," *IEEE Trans on Computer-Aided Design of Integrated Circuits and Systems*, vol. 15, no.5, pp. 544–553, 1996.

[41] Dartu F., Menezes N., Qian J. and Pillage L.T., "A gate-delay model for high-speed CMOS circuits", in *Proc. IEEE Design Automation Conference*, 1994, pp. 576–580.

[42] Dartu F. and Pileggi L.T., "Modeling signal wave shapes for empirical CMOS gate delay models," 6th *Intl. Workshop PATMOS '96,* Bologna, Italy, 1996, pp. 57-66.

[43] Davis J. A. and Meindl J. D., "Compact distributed *RLC* interconnect models—Part I: Single line transient, time delay and overshoot expressions," *IEEE Trans. on Electron Devices*, vol. 47, pp. 2068–2077, Nov. 2000.

[44] Davis J. A. and Meindl J. D., "Compact distributed *RLC* interconnect models—Part II: Coupled line transient expressions and peak crosstalk in multilevel interconnect networks," *IEEE Trans. on Electron Devices*, vol. 47, pp. 2078–2087, Nov. 2000.

[45] Delorme N., Belleville M. and Chilo J., "Inductance and capacitance analytic formulas for VLSI interconnects", *Electronics Letters*, vol. 32, no. 11, pp. 996-997, 1996.

[46] Deutsch A., Kopcsay G.V., Restle P.J., Smith H.H., Katopis G., Becker W.D., Coteus P.W., Surovic C.W., Rubin B.J., Dunne R.P., Jr., Gallo T., Jenkins K.A., Terman L.M., Dennard R.H., Sai-Halasz G.A., Krauter B.L. and Knebel D.R., "When are transmission-line effects important for on-chip interconnections?," *IEEE Trans. on Microwave Theory and Techniques*, vol. 45, no. 10, part-2, pp.1836 – 1846, Oct. 1997.

[47] Dhar S. and Franklin M. A., "Optimum buffer circuits for driving long uniform lines," *IEEE Jour. Solid-State Circuits*, vol. 26, pp. 32–40, Jan. 1991.

[48] Djordjevic A. R., Sarkar T. K., and Harrington R. F., "Time domain response of multiconductor transmission lines," *Proceedings IEEE*, vol. 75, pp. 743-764, June 1987.

[49] Dobberpuhl D.W., Witek R.T., Allmon R., Anglin R., Bertucci D., Britton S., Chao L., Conrad R.A., Dever D.E., Gieseke B., Hassoun S.M.N., Hoeppner G.W., Kuchler K., Ladd M., Leary B.M., Madden L., McLellan E.J., Meyer D.R., Montanaro J., Priore D.A., Rajagopalan V., Samudrala S. and Santhanam S., "A 200-MHz 64-b dual-issue CMOS microprocessor", *IEEE Jour. of Solid State Circuits,* vol. 27, no. 11, pp. 1555-1567, Nov. 1992.

[50] Dounavis A., Achar R. and Nakhla M., "Efficient passive circuit models for distributed networks with frequency-dependent parameters," *IEEE Trans. on Advanced Packaging*, vol. 23, pp. 382–392, Aug. 2000.

[51] Edahiro M., "Delay minimization for zero-skew routing," in *Dig. Tech. Papers IEEE Int. Conf. Computer-Aided Design,* 1993, pp. 563-566.

[52] Elmore W. C., "The transient response of damped linear networks with particular regard to wideband amplifiers", *Jour. Appl. Phys.*, vol. 19, pp. 55-63, Jan. 1948.

[53] El-Moursy M.A. and Friedman E.G., "Optimum wire sizing of *RLC* interconnect with repeaters", *Integration, the VLSI Journal* , Elsevier Pub., Netherlands, vol. 38, no. 2, pp. 205 – 225, Dec. 2004.

[54] El-Moursy M.A. and Friedman E.G., "Shielding effect of on-chip interconnect inductance," in *Proc. IEEE Great Lakes Symposium on VLSI,* April 2003, pp. 165–170.

[55] El-Moursy M.A. and Friedman E.G., "Shielding effect of on-chip interconnect inductance," *IEEE Trans. on Very Large Scale Integration (VLSI) Syst.*, vol. 13, no.3, pp. 396–400, March 2005.

[56] El-Moursy M.A. and Friedman E.G., "Wire shaping of *RLC* interconnects" *Integration, the VLSI Journal*, Elsevier Pub., Netherlands, vol. 40, no.4, pp. 461-472, July 2007.

[57] Eo Y. and Eisenstadt W. R., "High-speed VLSI interconnect modeling based on S-parameter measurement" *IEEE Trans. on Comp., Hybrids, Manufact. Technol.*, vol. 16, pp. 555–562, Aug. 1993.

[58] Gasteier M., Munch M. and Glesner M., "Generation of interconnect topologies for communication synthesis," *in Proc.. Design, Automation and Test in Europe*, Feb. 1998, pp. 36 – 42.

[59] Ginneken L. P., "Buffer placement in distributed RC-tree networks for minimal Elmore delay," in *Proc. IEEE Int. Symp. Circuits Syst.*, May 1990, pp. 865–868.

[60] Griffith R. and Nakhla M., "Time-domain analysis of lossy coupled transmission lines," *IEEE Trans. on Microwave Theory Tech.*, vol. 38, pp. 1480-1487, Oct. 1990.

[61] Gupta R., Tutuianu B., Krauter B. and Pileggi L.T., "The Elmore delay as a bound for RC trees with generalized input signals", in *Proc.32nd ACM/IEEE Design Automation Conf.*, June 1995, pp. 364-369

[62] Gupta R., Tutuianu B. and Pileggi L.T., "The Elmore delay as a bound for RC trees with generalized input signals", *IEEE Trans. on Computer-Aided Design of Integrated Circuits and Systems,* vol. 16, no. 1, pp. 95-104, Jan. 1997.

[63] Havemann R.H. and Hutchby J.A., "High performance interconnects: An integration overview", *Proceed. IEEE,* vol. 89, no. 5, pp. 586-601, May 2001.

[64] He L., Chang N., Lin S. and Nakgawa O. S., "An efficient inductance modeling for on-chip interconnects", in *Proc. Custom Integrated Circuits Conference*, May 1999, pp. 457-460.

[65] Hiraki M., Kojima H., Misawa H., Akazawa T. and Hatano Y., "Data-dependent logic swing internal bus architecture for ultra low power LSI's," *IEEE J. of Solid State Circuits,* vol. 30, no.4, pp.397-402, April 1995.

[66] Hirata A., Onodera H. and Tamaru K., "Estimation of propagation delay considering short – circuit current for static CMOS gates", *IEEE Trans. on Circuits and Systems-I: Fundamental Theory and Applications,* vol. 45, no. 11, pp. 1194-1198, Nov. 1998.

[67] Huang H.Y. and Chen S.L., "Interconnect accelerating techniques for sub-100-nm gigascale systems," *IEEE Trans. on Very Large Scale Integration (VLSI) Systems,* vol. 12, no. 11, pp. 1192-1200, Nov. 2004.

[68] Huang X., Restle P., Bucelot T., Cao Y., King T.J. and Hu C., "Loop-based interconnect modeling and optimization approach for multigigahertz clock network design," *IEEE Jour. of Solid State Circuits,* vol. 38, no. 3, pp. 457-463, Mar. 2003.

[69] International Technology Roadmap for Semiconductors. Semiconductor Industry Association. [Online]. Available: http://public.itrs.net

[70] Ismail Y. I. and Friedman E. G., "DTT: Direct truncation of the transfer function- an alternative to moment matching for tree structured interconnect," *IEEE Trans. on Computer-Aided Design of Integrated Circuits and Systems*, vol. 21, No. 2, Feb. 2002.

[71] Ismail Y. I. and Friedman E. G., "Effects of inductance on the propagation delay and repeater insertion in VLSI circuits," *IEEE Trans. on Very Large Scale Integration (VLSI) Systems.*, vol. 8, pp. 195–206, Apr. 2000.

[72] Ismail Y. I., Friedman E. G. and Neves J. L., "Dynamic and short circuit power of CMOS gates driving lossless transmission lines," *IEEE Trans. on Circuits and Systems I: Fundamental Theory and Applications*, vol. CAS-46, no. 8, pp. 950-961, August 1999.

[73] Ismail Y. I., Friedman E. G. and Neves J. L., "Equivalent Elmore delay for *RLC* trees," *IEEE Trans. on Computer-Aided Design of Integrated Circuits and Systems*, vol. 19, no. 1, pp.83-97, Jan. 2000.

[74] Ismail Y. I., Friedman E. G. and Neves J. L., "Figures of merit to characterize the importance of on-chip inductance," *IEEE Trans. on Very Large Scale Integration (VLSI) Systems*, vol.7, pp.442-449, Dec 1999

[75] Izumikawa M. and Yamashina M., "A current direction sense technique for multiport SRAM's", *IEEE Jour. of Solid State Circuits,* vol. 31, no. 4, pp. 546–551, April 1996.

[76] Jarvis D. B., "The effects of interconnections on high-speed logic circuits," *IEEE Trans. on Electron. Computers*, vol. EC-10, pp. 476–487, Oct.1963.

[77] Jin W., Eo Y., Eisenstadt W. R. and Shim J., "Fast and accurate quasi three-dimensional capacitance determination of multilayer VLSI interconnects," *IEEE Trans. on Very Large Scale Integration (VLSI) Systems.*, vol. 9, pp. 450–460, June 2001.

[78] Joshi A.J. and Davis J.A., "Wave-pipelined multiplexed (WPM) routing for gigascale integration (GSI)" *IEEE Trans. on Very Large Scale Integration (VLSI) Systems*, vol. 13, no. 8, pp. 899 - 910, Aug. 2005.

[79] Kahng A. B. and Muddu S., "An analytical delay model for *RLC* interconnects," *IEEE Trans. on Computer-Aided Design of Integrated Circuits and Systems*, vol. 16, no. 12, pp. 1507-1514, Dec. 1997.

[80] Kahng A.B. and Muddu S., "Efficient gate delay modeling for large interconnect loads," IEEE Multi-Chip Module Conf., 1996, pp. 202–207.

[81] Kamon, M., Ttsuk M.J. and White J.K., "FASTHENRY: A multipole-accelerated 3-D inductance extraction program," *IEEE Trans. on Microwave Theory and Techniques*, pp 1750-1758, Sept. 1994.

[82] Kang S.M. and Leblebici Y., *CMOS Digital Integrated Circuits— Analysis and Design*, TMH, New York, 2003.

[83] Kapur P. and Saraswat K.C., "Optical interconnects for future high performance integrated circuits", *Physica E: Low-dimensional Systems and Nanostructures*, vol. 16, no.3/4, pp.620-7, 2002.

[84] Kaushik B. K., Goel S. and Rauthan G., "Future VLSI interconnects: optical fiber or carbon nanotube – a Review" *Microelectronics International*, Emerald Pub. U.K., vol. 24, no. 2, pp.53 – 63, 2007.

[85] Khellah M.M. and Elmasry M.I., "Low power design of high capacitive CMOS circuits using a new charge sharing scheme," in *Proc. IEEE International Solid State Circuits Conf.*, Feb 1999, pp. 286-287.

[86] Kleveland B., Qi Z., Madden L., Furusawa T., Dutton R.W., Horowitz M.A. and Wong S.S., "High-frequency characterization of on-chip digital interconnects", *IEEE Jour. of Solid State Circuits,* vol. 37, no. 6, pp. 716-725, June 2002.

[87] Kaushik B.K., Sarkar S. and Agarwal R.P., "Waveform Analysis and Delay Prediction for a CMOS Gate Driving RLC Interconnect Load", *Integration, the VLSI Journal*, Elsevier Pub., Netherlands, vol. 40, no. 4, pp. 394-405, July 2007.

[88] König A. and Glesner M., "VLSI-implementation of associative memory systems for neural information processing" In *VLSI for Neural Networks and Artificial Intelligence*, Jose G. Delgado-Frias, William R.~Moore (Eds.), Plenum Press New~York and London, 1994.

[89] Kopcsay G.V., Krauter B., Widiger D., Deutsch A. Rubin B.J. and Smith H.H., "A comprehensive 2-D inductance modeling approach for VLSI interconnects – frequency – dependent extraction and compact circuit model synthesis", *IEEE Trans. on Very Large Scale Integration (VLSI) Systems,* vol. 10, no. 6, pp. 695-711, Dec. 2002.

[90] Kowalczyk A., Adler V., Amir C., Chiu F., Choon Chug, De Lange, W., Dubler S., Yuefei Ge, Ghosh S., Tan Hoang, Hu R., Baoqing Huang, Kant S., Kao Y.S., Cong Khieu, Kumar S., Chung Lau, Lan Lee, Liebermensch A., Xin Liu, Malur N., Hiep Ngo, Sung-Hun Oh, Orginos I., Pini D., Shih L., Sur B., Tzeng A., Vo D., Zambare S. and Jin Z., "First generation MAJC dual microprocessor," in *Proc. IEEE International Solid State Circuits Conf.*, San Francisco, CA, 2001, pp. 236–237.

[91] Krauter B. and Mehrotra S. "Layout based frequency dependent inductance and resistance extraction for on-chip interconnect timing analysis," in *Proc. IEEE Design Automation Conf.*, June 1998, pp. 303-308.

[92] Krauter B., Gupta R., Willis J. and Pileggi L.T., "Transmission line synthesis," in *Proc. IEEE Design Automation Conf.*, June 1995, pp. 358-363.

[93] Kreupl F., Graham A.P., Duesberg G.S., Steinhögl W., Liebau M., Unger E. and Hönlein W., "Carbon nanotubes in interconnect applications", *Microelectronic Engineering,* Elsevier Pub., Netherlands, vol. 64, no.1-4, pp. 399-408, Oct. 2002.

[94] Kurd N., Barkarullah J. S., Dizon R. O., Fletcher T. D., and Madland P. D.,"A multigigahertz clocking scheme for the pentium® 4 microprocessor," *IEEE Jour. Solid-State Circuits*, vol. 36, pp. 1647–1653, Nov. 2001.

[95] Lillis J., Cheng C.K. and Lin T.T.Y., "Simultaneous routing and buffer insertion for high performance interconnect," in *Proc. 6th Great Lakes Symp. VLSI,* 1996, pp. 7-12.

[96] Lin S. and Kuh. E.S. "Transient simulation of lossy interconnect," in *Proc. 29th ACM/IEEE Design Automation Conf.*, June 1990, pp. 81-86.

[97] Maheshwari A. and Burleson W., "Differential current-sensing for on-chip interconnects," *IEEE Trans. on Very Large Scale Integration (VLSI) Systems,* vol. 12, no.12, pp. 1321-1329, Dec. 2004.

[98] Manney S., Nakhla M. S. and Zhang Q. J., "Time domain analysis of non uniform frequency dependent high-speed interconnects," in *Proc. IEEE Int. Conf. Computer-Aided Design*, 1992, pp. 449–453.

[99] Massoud Y., Majors S., Bustami T. and White J., "Layout techniques for minimizing on-chip interconnect self-inductance," in *Proc. ACM/IEEE Design Automation Conf.*, June 1998, pp 566-571.

[100] Mead C.A. and Rem M., "Cost and performance of VLSI computing structures," *IEEE Jour. of Solid State Circuits,* vol. SC-14, no.2, pp. 455-462, Apr. 1979.

[101] Mead C.A. and Rem M., "Minimum propagation delays in VLSI," *IEEE Jour. of Solid State Circuits,* vol. SC-17, no.4, pp. 773-775, Aug. 1982.

[102] Meindl J. D., Venkatesan R., Davis J. A., Joyner J., Naeemi A., Zarkesh-Ha P., Bakir M., Mule T., Kohl P. A. and Martin K. P., "Interconnecting device opportunities for gigascale integration (GSI)," in *Proc. IEEE Int. Electron Devices Meeting (IEDM) Tech. Dig.*, 2001, pp. 525–528.

[103] Mezhiba A. V. and Friedman E. G., "Inductive properties of high-performance power distribution grids," in *IEEE Trans. on Very Large Scale Integration (VLSI) Systems*, vol. 10, no.6, pp. 762–776, Dec. 2002.

[104] Mezhiba A. V. and Friedman E. G. "Impedance characteristics of power distribution grids in nanoscale integrated circuits" in *IEEE Trans. on Very Large Scale Integration (VLSI) Systems*, vol. 11, Nov. 2004, 1148-1155.

[105] Mohsen A.M., and Mead C.A., "Delay-time optimization for driving and sensing of signals on high-capacitance paths of VLSI systems," *IEEE Jour. of Solid State Circuits,* vol. 14, no.2, pp. 462-470, Apr. 1979.

[106] Moll F., Roca M. and Rubio A., "Inductance in VLSI interconnection modeling," in *IEEE Proc. Circuits Devices Syst.,* vol. 145, no.3, pp. 175-179, June 1998.

[107] Munch M., Wehn N. and Glesner M., "Optimum simultaneous placement and binding for bit-slice architectures" in *Proc. of the ASP-DAC Asian and South Pacific Design Automation Conference, 1995. IFIP International Conference on Hardware Description Languages '95/CHDL '95/ IFIP International Conference on Very Large Scale Integration VLSI '95*, 29 Aug.-1 Sept. 1995, pp. 735 - 740

[108] Murgan T. and Glesner M., "Limits for switching power consumption in encoded interconnects," in Proc. *IEEE International Symposium on Signals, Circuits and Systems*, 14-15 July 2005, vol. 1, pp.27 – 30.

[109] Murgan T., Ortiz A. Garcia, Schlachta C., Zimmer H., Petrov M., and Glesner M., "On timing and power consumption in inductively coupled on-chip interconnects" in *Proc. Lecture Notes in Comp Science (Springer), Intl. Workshop on Power and Timing Modeling, Optimization and Simulation,* Sep. 2004, pp. 819-828

[110] Murugan T., Schlachta C., Petrov M., Indrusiak L., Garcia A., Glesner M. and Reis R., "Accurate capture of timing parameters in inductively-coupled on-chip interconnects" in *Proc. IEEE Symposium on Integrated Circuits and Systems Design,* Sept. 2004, pp.117 – 122.

[111] Naeemi A., Sarvari R. and Meindl J.D., "Performance comparison between carbon nanotube and copper interconnects for gigascale integration (GSI)", *IEEE Electron Device Letters*, vol. 26, no.2, pp. 84-86, Feb. 2005.

[112] Naeemi A., Venkatesan R. and Meindl J.D., "Optimal global interconnects for GSI," in *IEEE Trans. on Electron Devices,* vol. 50, no.4, pp. 980-987, Apr. 2003.

[113] Nagel L. W. "SPICE2: A computer program to simulate semiconductor circuits". Univ. California, Berkeley, CA, Tech. Report ERL-M520, May 1975.

[114] Nakagome Y., Itoh K., Isoda M., Takeuchi K. and Aoki M., "Sub-IV swing internal bus architecture for future low-power ULSI's," *IEEE Jour. of Solid State Circuits,* vol. 28, no.4, pp.414-419, April 1993.

[115] Nassif N., Desai M.P. and Hall D.H., "Robust Elmore delay models suitable for full chip timing verification of a 600MHz CMOS microprocessor", in *Proc. 35th ACM/IEEE Design Automation Conf.*, June 1998, pp 230-235.

[116] Nekili M. and Savaria Y., "Optimal methods of driving interconnections in VLSI circuits", in *Proc. IEEE International Symposium on Circuits and Systems (ISCAS)*, May 1992, pp. 21-23.

[117] Nekili M. and Savaria Y., "Parallel regeneration of interconnection in VLSI & ULSI circuits", in *Proc. IEEE International Symposium on Circuits and Systems*, Chicago, May 1993, vol. 3, pp.2023-2026.

[118] Odabasioglu A., Celik M. and Pileggi L.T., "PRIMA: Passive reduced order interconnect macromodeling algorithm," in *Proc. IEEE International Conference on Computer-Aided Design*, Nov. 1997, pp. 58–65.

[119] Odabasioglu A., Celik M. and Pileggi L.T., "PRIMA: Passive reduced order interconnect macromodeling algorithm," *IEEE Trans. on Computer-Aided Design of Integrated Circuits and Systems*, vol. 17,no.8, pp. 645–654, Aug. 1998.

[120] Ousterhout J. K., "Switch-level delay models for digital MOS VLSI," in *Proc. Design Automation Conf.* , 1984, pp. 542–548.

[121] Palusinski O. and Lee A., "Analysis of transients in nonuniform and uniform multiconductor transmission lines," *IEEE Trans. on Microwave Theory and Technique*, pp. 127–138, Jan. 1989.

[122] Pandey S., Glesner M. and Mühlhäuser M., "Performance aware on-chip communication synthesis and optimization for shared multi-bus based architecture" in *Proc. of* ACM *18th Annual Symposium on Integrated Circuits and System Design SBCCI*, 2005, pp. 230 – 235.

[123] Pandey S., Zimmer H., Glesner M. and Mühlhäuser M., "High level hardware/software communication estimation in shared memory architecture" in *Proc. IEEE International Symposium on Circuits and Systems*, May 2005, vol. 1, pp. 37 – 40.

[124] Paul C., *Analysis of Multiconductor Transmission Lines*. New York: Wiley, 1994.

[125] Penfield P. and Rubinstein J., "Signal delay in RC tree networks", in *Proc of IEEE Design Automation Conf.*, June 1981, pp. 613-617.

[126] Pillage L. T., and Rohrer R. A., "Asymptotic waveform evaluation for timing analysis", in *IEEE Trans. on Computer-Aided Design of Integrated Circuits and Systems*, vol. 9, pp. 352-366, Apr. 1990.

[127] Priore D. A., "Inductance on silicon for sub-micron CMOS VLSI," in *Proc. IEEE Symp. VLSI Circuits*, May 1993, pp. 17–18.

[128] Pullela S., Menezes N. and Pillage L. T., "Reliable nonzero skew clock trees using wire width optimization," in *Proc. 30th ACM/IEEE Design Automation Conf*, 1993, pp. 165-170.

[129] Qian J., Pullela S. and Pillage L.T., "Modeling the effective capacitance for the RC interconnect of CMOS gates," *IEEE Trans. on Computer-Aided Design of Integrated Circuits and Systems,* vol. CAD-13, pp. 1526-1535, December 1994.

[130] Rabaey J. M., *Digital Integrated Circuits, A Design Perspective*. Englewood Cliffs, NJ: Prentice-Hall, 1996.

[131] Rae K.V.S.H., Singh R. and Shekhar C., "Using dedicated functional simulator for ASIC design in Indian context" *in Proc.. 5th Int. Conf. on VLSI Design*, Jan. 1992, pp.334 – 335.

[132] Rajput S.S. and Jamuar S.S., "High current, low voltage current mirrors and applications," in *VLSI: Systems on a Chip* (Book). Edited by L.M. Silveria, S. Devadas and R.Reis, Kluwer Academic Publishers, Netherlands, 2000.

[133] Rao K., Kansal A., Shekhar C., Srinivas M. and Rao V.N.S.N., "An ASIC for high performance stepper motor control" *in Proc.. 5th Int. Conf. on VLSI Design*, Jan. 1992, pp. 328 – 329.

[134] Rubinstein J., Penfield P. and Horowitz M.A., "Signal delay in RC tree networks," *IEEE Trans. on Computer-Aided Design of Integrated Circuits and Systems,* vol. 2, no.3, pp. 202-211, July 1983.

[135] Ruehli A.E., "Inductance calculations in a complex integrated circuit environment," *IBM Journal of Research and Development*, pp 470-481, Sept. 1972.

[136] Ruehli A. E., "Equivalent circuit models for three dimensional multiconductor systems," *IEEE Trans. on Microwave Theory and Techniques,* vol. 22, no.3, pp. 216-221, March 1974.

[137] Sai-Halasz G.A., "Performance trends in high-end processors", *Proceed. of IEEE,* vol.84, pp. 20-36, Jan.1995.

[138] Sakurai T., "Approximation of wiring delay in MOSFET LSI," *IEEE Jour. Solid-State Circuits*, vol. SC-18, pp. 418–426, Aug. 1983.

[139] Sakurai T., "Closed form expressions for interconnection delay, coupling and crosstalk in VLSI's," *IEEE Trans. on Electron Devices*, vol. 40, pp. 118–124, Jan. 1993.

[140] Sakurai T. and Newton A. R., "Alpha-power law MOSFET model and its applications to CMOS inverter delay and other formulas," *IEEE Jour. Solid-State Circuits,* vol. 25, pp. 584–594, Apr. 1990.

[141] Sakurai T. and Newton A.R., "A simple MOSFET model for circuit analysis," *IEEE Tans. on Electron Devices,* vol. ED-38, no.4, pp. 887-894, Apr. 1991.

[142] Shekhar C. and Mandal A.S., "Design of an efficient ASIC for the control of microwave ovens" *in Proc. 5th Int. Conf. on VLSI Design*, Jan. 1992, pp.293 – 296.

[143] Shichman H. and Hodges D.A., "Modeling and simulation of insulated gate field effect transistors switching circuits," *IEEE Jour. of Solid State Circuits,* vol. SC-3, pp. 285-289, Sep. 1968.

[144] Shockley W., "A unipolar field effect transistor," *Proc. Institute of Radio Engineers,* vol. 40, pp. 1365-1376, Nov. 1952.

[145] Shojaei-Baghini M., Lal R.K. and Sharma D.K., "A low-power and compact analog CMOS processing chip for portable ECG recorders" in *Proc. Asian Solid-State Circuits Conference*, Nov. 2005, pp.473 – 476.

[146] Shojaei-Baghini M., Lal R.K. and Sharma D.K., "An ultra low-power CMOS instrumentation amplifier for biomedical applications, "in *Proc. IEEE International Workshop on Biomedical Circuits and Systems*, Dec. 2004, pp.S1/1 - S1-4

[147] Shoji M., *High-Speed Digital Circuits*. Reading, MA: Addison- Wesley, 1996.

[148] Sim S.P., Krishnan S., Petranovic D.M., Arora N.D., Lee K. and Yang C.Y., "A unified *RLC* model for high speed on-chip interconnects," *IEEE Trans. on Electron Devices,* vol. 50, no. 6, pp. 1501-1510, 2003.

[149] Sim S.P., Lee K., Cary Y. and Yang C.Y., "High-frequency on chip inductance model," *IEEE Electron Device Letters,* vol. 23, no. 12, pp. 740-742, Dec. 2002.

[150] Sinha A. and Chowdhury S., "Mesh-structured on-chip power/ground: design for minimum inductance and characterization for fast R, L extraction," *Proc. of IEEE Custom Integrated Circuits*, May 1999, pp 461-464.

[151] Skribanowitz J., Knobloch T., Schreiter J. and König A., "VLSI implementation of an application-specific vision chip for overtake monitoring, real time eye tracking, and visual inspection" in *Proc.* of the *7th Int. Conf. on Microelectronics for Neural, Fuzzy, and Bio-Inspired Systems MicroNeuro'99*, Spain, April 7-9, 1999, pp. 45-52.

[152] Srivastava N. and Banerjee K., "Performance analysis of carbon nanotube interconnects for VLSI applications", in *Proc. IEEE International Conference on Computer-Aided Design*, Nov. 2005, pp.383 – 390.

[153] Sylvester D. and Wu C., "Analytical modeling and characterization of deep-submicrometer interconnect," *Proceed. of IEEE*, vol.89, no.5, pp. 634-664, May 2001.

[154] Tam S., Rusu S., Desai U.N., Kim R., Zhang J. and Young I., "Clock generation and distribution for the first IA-64 microprocessor," *IEEE Jour. Solid State Circuits,* vol. 35, no.11, pp. 1545-1552, Nov. 2000.

[155] Veghte R. L. and Balanis C. A., "Dispersion of transient signals in microstrip transmission lines," *IEEE Trans. on Microwave Theory and Technique*, vol. MTT-34, pp. 1427–1436, Dec. 1986.

[156] Venkatesan R., Davis J. A. and Meindl J. D., "Compact distributed *RLC* interconnect models—Part III: Transients in single and coupled lines with capacitive load termination," *IEEE Trans. on Electron Devices*, vol. 50, pp. 1081–1093, Apr. 2003.

[157] Venkatesan R., Davis J. A. and Meindl J. D., "Compact distributed *RLC* interconnect models—Part IV: Unified models for time delay, crosstalk, and repeater insertion", *IEEE Trans. on Electron Devices*, vol. 50, pp. 1094–1102, Apr. 2003.

[158] Vishwakarma D. N., Balasubramanian S. K., Kothari L. and Ahuja S., "FPGA implementation of directional relays," in *Proc. 3rd International Conference on Power System Protection and Automation, Nov. 17-18*, New Delhi, India.

[159] Wann D.F. and Franklin M.A., "Asynchronous and clocked control structures for VLSI based interconnection networks," *IEEE Trans. on Computers*, vol. C-32, no.3, pp. 284-293, March 1983.

[160] Wei B.Q., Vajtai R. and Ajayan P.M., "Reliability and current carrying capacity of carbon nanotubes", *Appl. Phys. Lett.* vol. 79, no. 8, pp. 1172-1174, 2001.

[161] Weste N.H.E. and Eshraghian K., *Principles of CMOS VLSI Design,* 2nd edition, Addision-Wesley Publishing Company, "Empirical Delay Models", pp. 213, 1992.

[162] Wong S.C., Lee T.G.Y., Ma D.J., and Chao C.J., "An empirical three-dimensional crossover capacitance model for multilevel interconnect VLSI circuits," *IEEE Trans. on Semiconductor Manufacturing,* vol. 13, No.2, pp. 219-227, May 2000.

[163] Wood J., Edwards T. C. and Lipa S., "Rotary traveling-wave oscillator arrays: A new clock technology," *IEEE Jour. of Solid-State Circuits*, vol. 36, pp. 1654–1665, Nov. 2001.

[164] Wyatt Jr. J.L., "Signal delay in RC mesh networks," *IEEE Trans.* on *Circuits and Systems,* vol. CAS-32, no. 5, pp. 507-510, May 1985.

[165] Yacoub G.Y., Pham H. and Friedman E.G., "A system for critical path analysis based on back annotation and distributed interconnect impedance models," *Microelectronic Journal*, Elsevier Pub., Netherlands, vol. 18, no. 3, pp. 21–30, June 1988.

[166] Yamaguchi K., Fukaishi M., Sakamoto T., Akiyama N. and Nakamura K., "A 2.5-GHz four-phase clock generator with scalable no-feedback-loop architecture," *IEEE Jour. Solid-State Circuits*, vol. 36, pp. 1666–1672, Nov. 2001.

[167] Yamauchi H. and Matsuzawa A., "A signal swing suppressing strategy for power and layout area savings using time multiplexed differential data transfer scheme", *IEEE Jour. Solid-State Circuits*, vol..31, no.9, pp.1285-1294, Sep. 1996.

[168] Ymeri H., Nauwelaers B. and Maex K., "On the frequency-dependent line capacitance and conductance of on-chip interconnects on lossy silicon substrate," *Microelectronics International,* Emerald Pub. U.K., vol. 19, no. 1, pp. 11-18, 2002.

[169] Zaage S. and Groteluschen E., "Characterization of the broadband transmission behavior of interconnections on silicon substrates," *IEEE Trans. on Components, Hybrids, and Manufacturing Technology,* vol. 16, no. 7, pp. 686-691, Nov.1993.

[170] Zheng J., Hahm Yeon-Chang, Tripathi V.K. and Weisshaar A., "CAD-oriented equivalent-circuit modeling of on-chip interconnects on lossy silicon substrate" *IEEE Trans. on Microwave Theory and Techniques*, vol. 48, no. 9, pp.1443 –1451, Sept. 2000.

[171] Zhou D., Su S., Tsui F., Gao D.S. and Cong J.S., "A simplified synthesis of transmission lines with a tree structure" *Analog Integrated Circuits and Signal Processing, Kluwer Academic Publishers,* vol. 5, no.1, pp. 19-30, Jan. 1994.

In: VLSI and Computer Architecture
Editor: Kenzo Watanable
ISBN: 978-1-60692-075-6

Chapter 3

DEVELOPMENT, VALIDATION AND EVALUATION OF A SPACE QUALIFIED LONG-LIFE FLIGHT COMPUTER SERVER

***Esaú Vicente-Vivas*[1,*] *and Francisco J. Mendieta Jiménez*[2,†]**
[1]Instituto de Ingeniería, Universidad Nacional Autónoma de México, Cd. Universitaria, Coyoacan, 04510, México
[2]Centro de Investigación Científica y de Educación Superior de Ensenada, Km. 107 Carretera Tijuana-Ensenada, CP. 22860, AP. 2732, Ensenada, B.C., México

Abstract

Several research institutions have worked together in the development of a 55 Kg low Earth orbit (LEO) microsatellite, aiming the development of university technology toward the space field. As it is known, LEO operation implies a dephased orbital dynamics among satellites and our planet, achieving time limited communications either for downloading high bandwidth scientific telemetry or for uploading command and/or missions for space vehicles. By this reason, among others, long-life space computer architectures constitute an important research field oriented to preserve satellite operations, satellite autonomy and communications among space vehicles and their control Earth stations. Under this scenario, few years of research efforts were dedicated towards the development of a reconfigurable space qualified long-life computer server (SQLLCS) that integrates cold spare redundancies in single points of failure from the architecture to improve hardware reliability. The computer architecture aims for the extension of satellite life even in the presence of important SQLLCS failures. This chapter describes the SQLLCS hardware architecture and underlines the features integrated in the hardware that allow the computer to withstand the harsh space environment. Moreover, it explains some software operations for both the space and ground segments. In addition, the chapter outlines the hardware and software tools specially created for SQLLCS validation purposes, as well as results of a reliability study to predict the server behaviour based on the exponential failure law, the military standard MIL-HDBK217f notice 2 and MatLab software.

---

* Tel.: (5255)5623-3600, ext. 8815, evv@servidor.unam.mx
† mendieta@enst.fr

1. Introduction

Electronic equipment developed for space applications has to accomplish a set of fundamental requirements in order to withstand vibration during the launching phase, as well as the harsh space environment in terms of radiation, extreme temperatures and vacuum during its operative life. For this purpose, space projects have employed approaches both for failure avoidance as well as for fault tolerance [Cardarilli, 2003; Wimmer, 1997; Johnson, 1988]. The former includes the selection of qualified components, enforcing design rules and the periodical review of designs. The last handle hardware failures and software errors when they occur, through the help of redundant hardware and fault diagnosis, as well as fault detection and reconfiguration techniques [Vicente, 2004].

Electronic equipment for space applications, especially those projected for small space vehicles, has strong limitations in terms of weight, volume, and electrical power. Furthermore, the small satellite field is also characterized by the adoption of "faster, cheaper and probably better" approaches to easily access space. In this sense, the use of commercial off-the-shelf (COTS) components is extremely attractive [Caldwell, 2000] and is becoming a common practice [Cibola, 2007; Renaudie, 2007; Guldager, 2005; Elfving, 2003], which in turn has enabled the launching of state-of-the-art electronics into LEO orbits. The big picture is represented by the successful missions developed by the University of Surrey, UK, a worldwide recognized institution both in the small satellite field as well as in the use of COTS components in space platforms [Underwood, 2003; Sweeting, 2001].

In addition, some important space institutions such as the Jet Propulsion Laboratory of NASA have made research efforts towards reusable avionics computer architectures to be used in multiple missions, aiming to reduce the development and production costs of flight projects [Chau, 2001].

On the other hand, the Satex project (figure 1) aimed for the integration of microsatellite generic subsystems (figure 2) with capabilities to adapt them to progressive missions. In this sense the project demanded the development of a redundant and reusable flight computer with capabilities to apply automatic maintenance after the detection of failures. It is important to highlight that around the world very few small satellite missions make use of redundant flight computers [SSTL MicroSat, 2008; SSTL MiniSat, 2008; Cibola, 2007; Bretschneider, 2005; Sperber, 1996]. In other words, most small satellite missions employ centralized computer architectures to automate operations and payloads.

The Satex mission projected five payloads, few of them with dedicated control requirements. This goal demanded the development of a satellite-distributed computing system for automation purposes where satellite operations are governed by the flight computer and where some payloads contain a microcomputer for local control purposes. Furthermore, communications inside the vehicle are accomplished by a fault tolerant redundant local area network (FTRLAN) that achieves safe communications through real time accomplishment of fault diagnosis, fault detection and fault reconfiguration processes [Vicente, 2004].

Figure 1. Satex Microsatellite project..

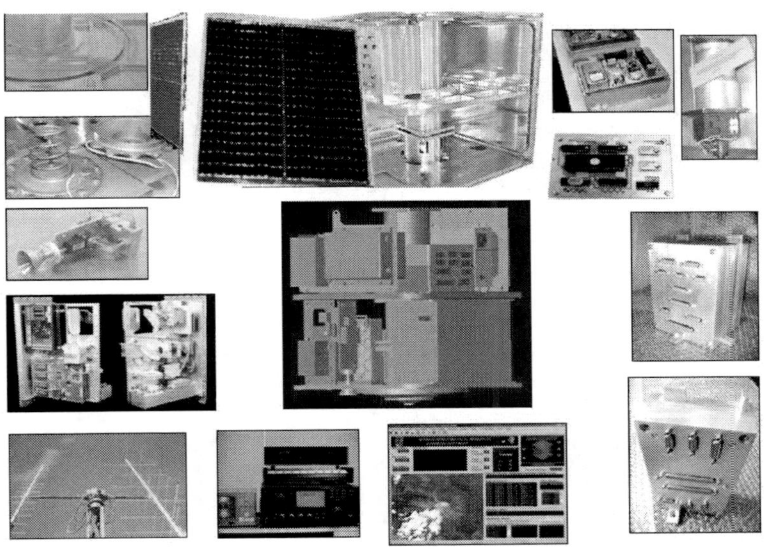

Figure 2. Subsystems developed for Satex Microsatellite.

The operations delegated to the SQLLCS were the following:

- Starting operations for the satellite after deployment into space
- Telemetry acquisition and telemetry packaging
- Communications and protocol handling with the Earth Station (ES)
- Computer server functions among satellite and ES
- Communications with payload microcomputers
- First-stage vehicle stabilization to allow gravity gradient deployment
- Second-stage stabilization process to allow payload pointing to Earth

As realized from the previous list of functions, the SQLLCS represents a single point of failure whose malfunction leads to a failure of the whole satellite system. To overcome this risky situation, a reconfigurable SQLLCS architecture was developed, validated and evaluated. The design integrates up to three single board microcomputers (SBM), each one with enough hardware capabilities to fulfill the requirements of satellite instrumentation. SBM full characteristics are given in a later paragraph.

In order to generate a cost-effective computer, the architecture employs three SBMs with identical printed circuit boards (PCB). This led to the design of a single PCB that employs jumpers to program the SBM identity as well as to provide separate energization paths for three different processor configurations: main, first backup and second backup (see figure 3).

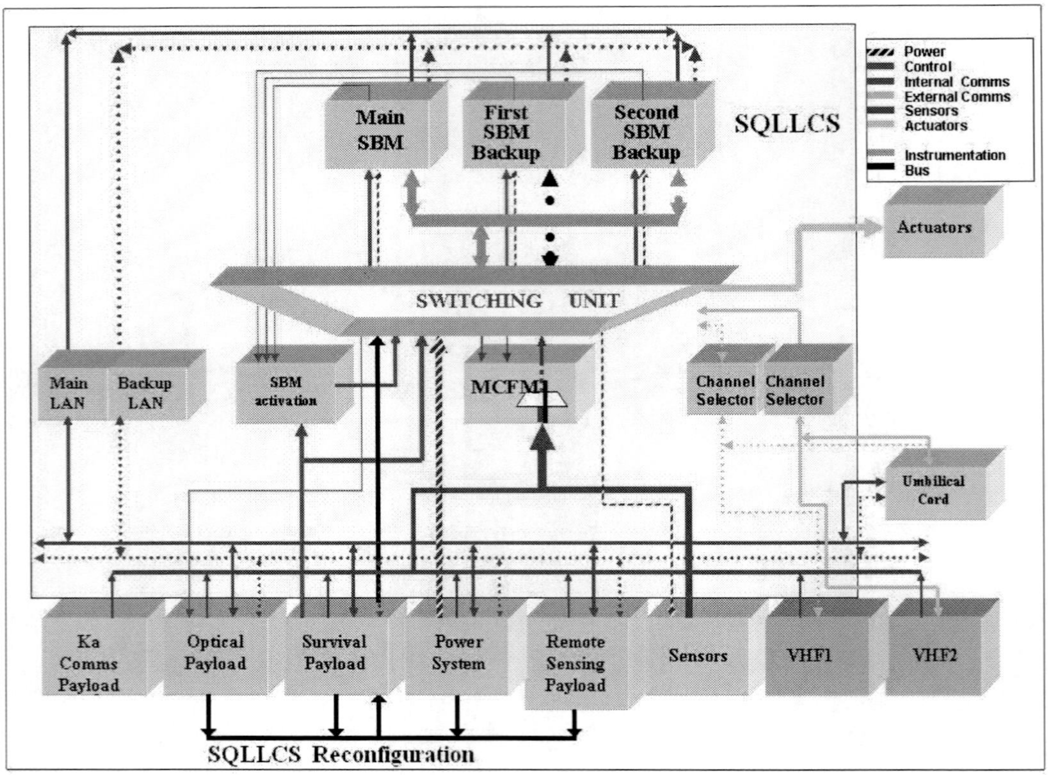

Figure 3. Flight computer architecture developed for SQLLCS.

In order to generate a cost-effective computer, the architecture employs three SBMs with identical printed circuit boards (PCB). This led to the design of a single PCB that employs jumpers to program the SBM identity as well as to provide separate energization paths for three different processor configurations: main, first backup and second backup (see figure 3).

The SQLLCS architecture also demanded the design of a compact digital switching unit to interconnect any one of the SBMs to the satellite instrumentation. Moreover, quad digital arrays of switches (IRFF130) were employed to provide energization for every SBM. By these means, SQLLCS reconfiguration (maintenance) can be commanded from an external source, in this case from a microcomputer payload. The last function can be controlled either in automated fashion from a microcomputer payload or by remote means from the ES

[Vicente, 2006]. This electronic feature also allowed the definition of fault containment regions in the SQLLCS architecture composed by every SBM, to avoid damage propagation when failures take place. Regarding the switching unit, it is based on both analog (CD4053BMJ) and digital (SN54HC157) multiplexers.

In addition, the following peripheral interfaces were completed for the SQLLCS:

- A multiplexing, conditioning and filtering module to allow the acquisition of up to 48 signals from satellite sensors (fine sun sensors, magnetometers, current, voltage, light presence, etc.). The module is based on analog multiplexers CD4051BMJ, TLC2201 amplifiers and discrete logic.
- Electronics for both main and backup LAN onboard the satellite
- Connectors to interface the SQLLCS I/O signals to the satellite instrumentation as well as the required line drivers for signal coupling

The hardware architecture developed for the flight computer server is shown in figure 3. As can be seen the microsatellite contains four payloads, three of them with dedicated microcomputers interfaced to the main and backup onboard LANs.

2. Single Board Microcomputers

In order to increase satellite availability the SQLLCS architecture was conceived and developed to admit automatic maintenance through hardware reconfiguration facilities. However strong limitations in weight, volume and available power forced the automatic maintenance process to take advantage of onboard microcomputers to implant an N modular redundancy computer architecture. The detailed architecture along with schemes for fault diagnosis, fault detection and automatic maintenance are described in [Vicente, 2004].

The SQLLCS core can be assembled with up to three SBMs, figures 4 and 5, each one capable of being activated through digital signals coming from an external source.

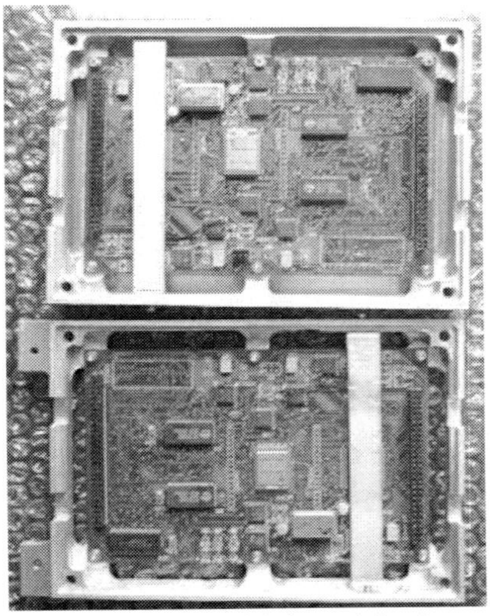

Figure 4. SBMs fabricated for the space qualified computer server.

Figure 5. SQLLCS boards being assembled.

Under these circumstances, one SBM can be activated at a time and remaining SBMs comprise cold standby spares that can be employed when failures are detected at the main SBM. The exposed scheme enhances the relation among availability and redundancies projecting a computer with power consumption demands very likely to that one of a simplex system. Each SBM contains:

- A 40 Mhz 16 bit Siemens RISC microcontroller (RISCM) with 76 I/O lines, 16-priority-level interrupt controller, ten-channel 10-bit A/D converter, seven 16-bit timers and a programmable watchdog timer
- 64 Kb PROM containing the basic satellite software
- 1.2 Mb SRAM for data management, for upgraded software uploading, as well as for payload data storage
- Military qualified 16-bit flow-through EDAC protection (29C516E) for whole static RAM in order to prevent single-event-upsets generated by cosmic radiation
- 1.2 Mb SRAM for EDAC syndrome storage
- Three serial channels, one assigned for communications with ES and the two other for FTRLAN operation
- Local temperature sensor (LM135AH) with associated conditioning electronics
- Lateral connectors that allow the interconnection of up to three identical SBMs in a piggyback manner, each one rotated 180 degrees against each other. In this case every SBM is jumper programmed for digital identification purposes as well as to provide independent energization means.

For weight- and space-saving purposes the flight computer PCBs were designed to hold electronic components on both of its faces. This was achieved with the use of surface mount technology for both active and passive electronic parts, and by avoiding the use of electrolytic capacitors which are not recommended for use in space.

The physical switching capabilities of SBMs offers redundancy support for the RISCM and its external peripherals, such as: program memory, data memory, I/O lines, A/D converter, network ports, timers, interrupt controller, oscillator, etc. In addition the basic satellite software stored in PROM is also replaced whenever a SBM reconfiguration takes place. In this way, the substitution of SBMs allows the extension of life for the flight computer. Consequently, the satellite availability is therefore increased as well as its useful life even when permanent faults arise.

3. Fault Protections For Single Board Microcomputers

The semiconductor chips from satellite equipment are exposed to space radiation whose effects are usually a function of the total radiation dose seen by each component. The radiation is originated by anyone of the following phenomena:

- Particles associated with magnetic planetary fields, as in the case of Van Allen rings that affect satellites orbiting the Earth
- Cosmic rays from deep space, such as Gamma rays, X rays, etc.
- High energy protons generated during solar explosions

Charged particles can pass through the electronic devices and generate a cloud of electrical charge. This charge can induce single-event-upsets (SEU) that may generate serious consequences to the integrated circuit, for example, the lost of stored bits from important software variables. The permanent exposition to radiation generates charge accumulation, which modify the semiconductors properties related with transistor beta, threshold voltages, leakage currents, and so on [Pisacane, 1994]. This means that digital logic slows down, operational amplifiers offset voltages change, power dissipation goes up, etc.

Some times SEUs will be followed by the latch-up phenomenon characterized by very large currents flowing in the device. The large currents will usually destroy the electronic part within a few tens of milliseconds. Although SEU events can occur in various integrated circuits technologies, such as CMOS and TTL bipolar devices, latch-up phenomenon has almost exclusively been observed in CMOS. It is important to highlight that the SQLLCS was designed and assembled employing mostly CMOS devices for power saving reasons, therefore special protections had to be aggregated for safety reasons.

3.1 Protections Against Seu Events

SRAM from SQLLCS will always be exposed to radiation and consequently will present random data errors, in average 1 error every 15 hours according with experimental data reported by [Shaneyfelt, 1994; Laurence, 1993] when using COTS memory with EDAC protection. In addition, published reports show that errors are more often generated in the South America area where relatively much radiation was indirectly detected [Underwood, 2003; Lee, 1992].

To counteract error generation in memories each SBM integrates a 100 pins military qualified 16-bit flow-trough EDAC device placed between the microcontroller and SRAM memory. The 29C516E EDAC allows the monitoring and correction of data values coming from data memory. It detects and corrects 100% of all single-bit errors and detects all double bit-errors. However, EDAC utilization introduces a penalization because additional syndrome SRAM has to be included with the same capacity as that from the memory to be protected. In our case 1.2 Mb of SRAM were added for this purpose.

In addition, dedicated software is required for "memory washing" purposes which in this case takes place every ten minutes within satellite operations. In this way when single errors are detected and corrected an interrupt request is sent to the RISCM in order to calculate the number of errors. This information is attached to telemetry to notify the ES personnel about the operative state of the flight computer memory. Besides, when multiple errors are detected the hardware automatically forces the SQLLCS to be temporally switched off taking into consideration that onboard operations software can be seriously damaged. Moreover, as it is considered that multiple errors might be related with the starting of a latch-up phenomenon, the only safe process to overcome a latch-up event is to remove energy from the affected component [Pisacane, 1994], however, for safety reasons, the whole SQLLCS is switched off.

3.2 Latch-Up Protection

The integrated circuits employed at every SBM, excluding few devices and the RISCM, are all military qualified components according with MIL-STD-883. The military components contain electronic protection to avoid the latch-up phenomenon. However COTS parts as the RISCM are not latch-up free. To overcome this risk two different types of protections were implemented. The first one consists of an EDAC based latch-up protection for SBMs, which was described in above paragraph. The second one is a dedicated protection for the RISCM integrated at each SBM board. This is an analog sensor that permanently measures the microcontroller current looking for measurements whose value falls above microcontroller nominal current. When this event is detected by the sensor an electrical pulse of 16 seconds is generated to automatically interrupt power application to the whole SQLLCS. After the programmed time gets over energy is applied again to the SQLLCS. However if the latch-up process resumes its operation the sensor protection keeps working over and over again.

3.3 Other Protections

COTS critical components from SQLLCS will also be protected for space flight with thin sheet shields either from Tungsten or Tantalum, as recommended by [Winokour, 1994] to prevent charged particles from interacting with COTS electronic devices.

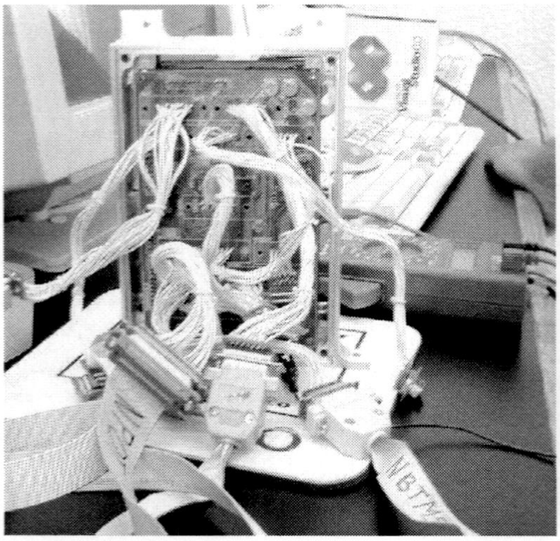

Figure 6. SQLLCS under laboratory testing.

Figure 7. SQLLCS being tested in satellite structure

4. Sqllcs Integration And Validation

The SQLLCS hardware was progressively validated in several development stages. At the beginning of the project must of the electronic designs were validated in laboratory with the help of breadboards and COTS components. While waiting for the arrival of military qualified components some PCBs were designed and sent for production. Later the PCBs were carefully checked and then parts assembling took place with the application of continuous testing to guarantee PCB operation. The overall process conducted to the identification and elimination of errors as well as to the development of updated versions of PCB for every one of the five electronic cards that comprise the SQLLCS architecture. Afterwards the validation among adjacent boards as well as connector`s pins testing between SQLLCS and validation tools was performed (figure 6). This process took to the installation of every single board into space qualified aluminium containers built at university workshops (figures 5 and 7).

4.1 Hardware And Software Tools Developed For Sqllcs Validation

To reach the validation of SQLLCS hardware as well as its progressive software development most of the onboard satellite equipments were needed in continuous working stages. However, the required satellite subsystems were also developed at the same time by other institutions and therefore were unable for testing purposes. For this reason, provisions were made since the starting of the project to develop hardware and software tools for SQLLCS validation purposes. The following hardware and software tools were developed: a satellite simulator (SIMSAT), an emulation software for satellite payloads called SOFDEVO and the Earth station software (ESS). These tools allowed the progressive development of both hardware and software for the SQLLCS. It is important to highlight that computer server validation and evaluation was possible thanks to the help of those special tools developed for the project (figure 8).

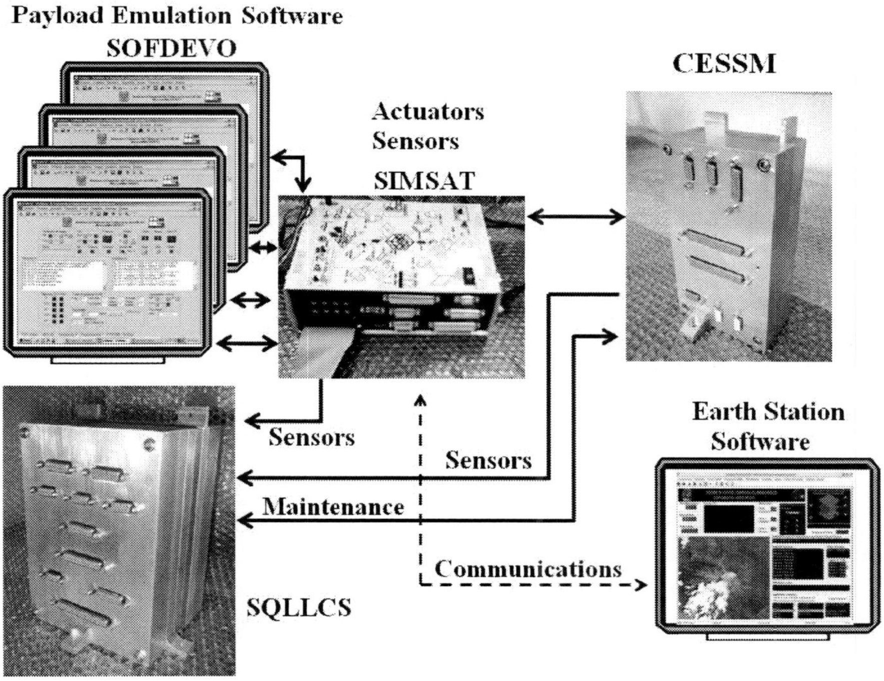

Figure 8. Hardware and software tools developed for SQLLCS validation.

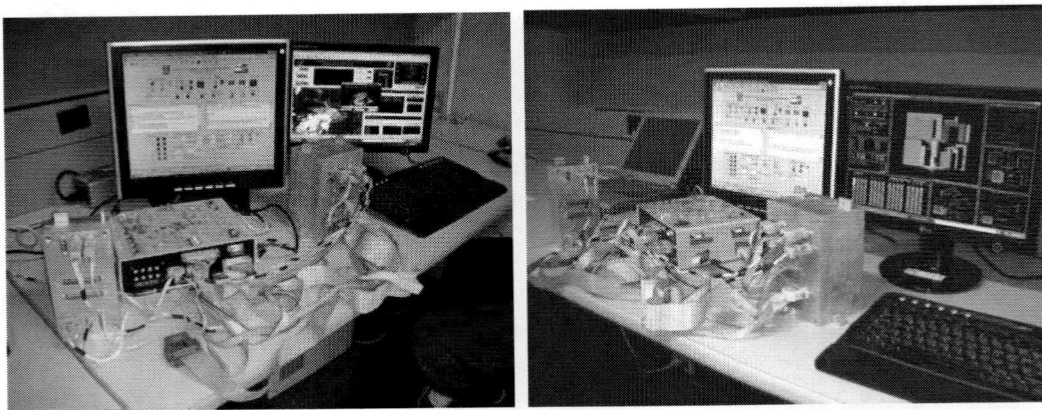

Figure 9. SQLLCS validation testing undergone in laboratory, SOFDEVO software at the left and ESS software at the right.

In this way, SQLLCS preliminary software testing procedures were achieved through the direct connection of a single SBM either to SOFDEVO or to ESS. Later on testing was carried out among ESS, SBM and SOFDEVO. The progressive development and testing phases of referred subsystems took to the validation approach seen in figures 8 and 9, where emulation of satellite sensors was performed with the help of two modules. The first one called "conditioning electronics for satellite sensors and maintenance" (CESSM) and the second constituted by SIMSAT. In this way the SQLLCS was continuously validated with the CESSM flight model and updated versions of ESS and SOFDEVO. Successfully testing involved following actions:

- Transmission of new programs from ESS to SQLLCS
- Programming and transmission of satellite missions from ESS to SQLLCS
- Execution of either new programs or new missions at SQLLCS with the support of both SIMSAT and SOFDEVO testing tools
- Telemetry transmission from SQLLCS to ESS
- Telemetry downloading and consequently telemetry displaying at ESS

4.2 Satellite Simulator

The satellite simulator shown in figures 8 and 9 was constructed to allow SQLLCS to interact with satellite subsystems such as sensors, actuators, payloads and other subsystems. SIMSAT contains: 48 passive sensors to provide real telemetry collection; electronic representation for UHF radios and magnetic torquer coils; emulation of gravity gradient release and 10 I/O connectors of different size for interconnection with SQLLCS. Furthermore contains several on-off switches to manually control the SQLLCS reconfiguration as well as to command the execution of programs in the computer server either from PROM or from SRAM memory. The switches are used in SQLLCS manual operation mode and this was mainly employed in the first validation stage of computer server reconfiguration processes. In addition, SIMSAT allocates LAN ports to interconnect up to 4 personal computers, each one running a copy of SOFDEVO software for payload emulation purposes and consequently for validation of computer server operations (figure 8).

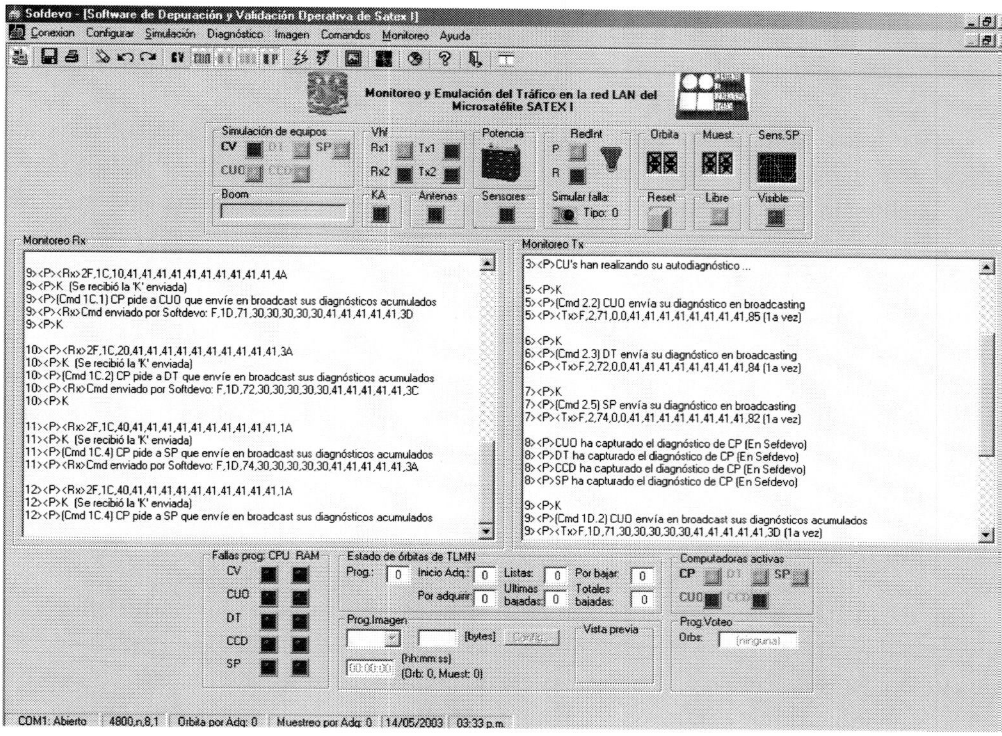

Figure 10. SOFDEVO software created to debug both SQLLCS hardware and SQLLCS software.

4.3 SOFDEVO Software

SOFDEVO resulted in an outstanding tool that allowed the depuration, emulation and operative validation of software not only for SQLLCS but also for satellite payload computers. Therefore, the software allowed the depuration and operative validation of SQLLCS hardware and software. SOFDEVO operation is based on the fact that all computers on board the space vehicle are connected through a redundant LAN where digital communications traffic takes place. In this way SOFDEVO runs dedicated software for every emulated computer, generating automatic payload operations. In terms of network traffic those operations performs as the real payload computer subsystem. The software manages network protocols, fault-tolerant network attributes [Vicente, 2004], and also contains all kind of programmed answers for network queries related to the emulated computer subsystem, figure 10.

Besides, some software features are allowed to be programmed in advance by the user through the help of menus to generate automated answers according with specific needs (e.g. image file prepared in advance to an image acquisition and transference process related to the remote sensing payload). Whenever a likely process is required it has to be programmed before starting the specific emulation process in such a way that SOFDEVO real time performance is achieved.

Moreover SOFDEVO contains two data windows with scroll capabilities, figure 10, one shows decoded messages collected from the network, allowing the user to be informed about real time network operations. The second window shows the automatic answers generated by SOFDEVO when network enquires are directed to anyone of the emulated computer subsystems.

On the other hand, when computer emulation takes place SOFDEVO software allows fault programming related to the emulated computing nodes. The emulated fault is delivered to the satellite instrumentation whenever computer diagnostic results are collected by the SQLLCS. This software feature was exhaustively employed to validate SQLLCS and satellite software behaviour under the presence of injected failures. By these reasons SOFDEVO software was invaluable for hardware validation as well as for software debugging.

The referred operative scheme was used as fault injection mechanism to validate fault detection, fault identification and alarm activation tools at the Earth station software. Whenever the fault injection mechanism is employed it has to be programmed before starting the specific emulation process.

4.4 Earth Station Software

The Earth station software was also created in parallel fashion along with other satellite subsystems according with mission needs and satellite flight software requirements. The software contains reception and transmission functions related to satellite operating activities such as communication commands, communications protocols, digital data packaging, etc. The software is friendly and keeps the user informed about satellite status and operations.

From initial stages of the satellite project the progresses made in satellite automation software allowed the validation of operations among satellite instrumentation and ESS. In this way the continuous verification of automated processes for the satellite took to the progressive development and validation of the whole SQLLCS hardware and system software. Besides, the project offered enough time to generate three updated versions of ESS from which the last one represents a highly visual environment where satellite telemetry is exhibited with digital animations and virtual instruments (figures 9, 11, 12 and 13).

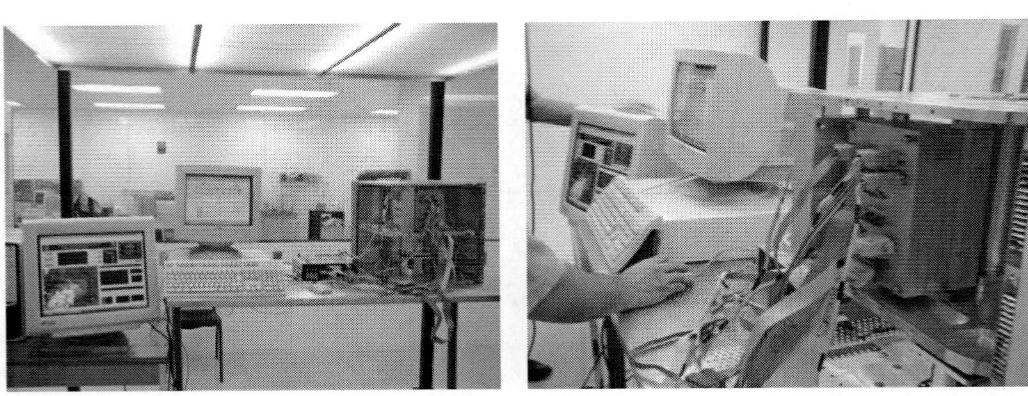

Figure 11. SQLLCS testing undergone at clean room facilities.

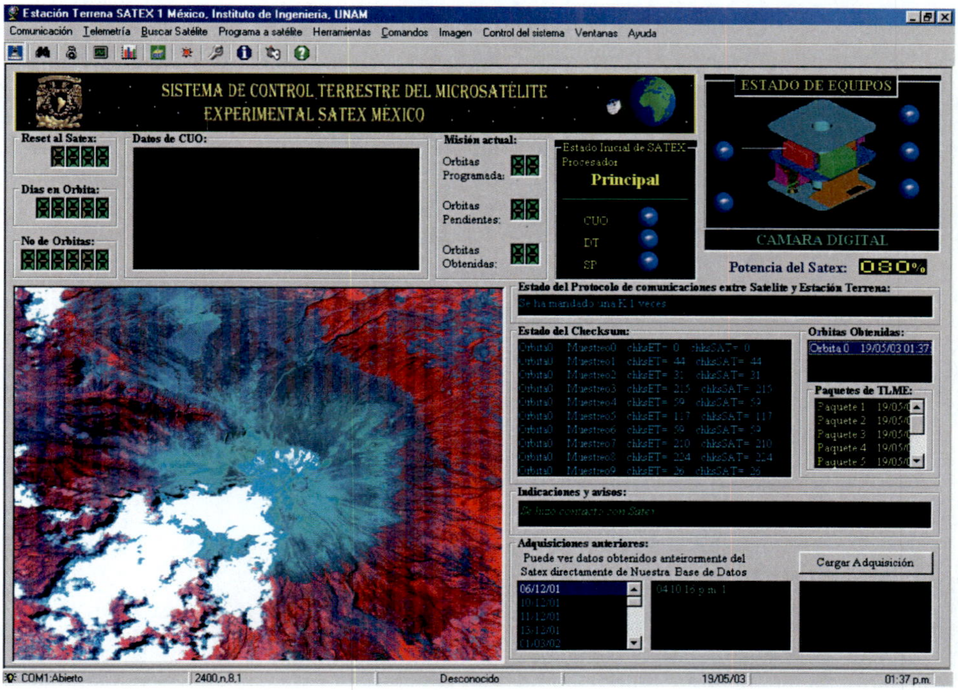

Figure 12. Earth station software downloading telemetry from SQLLCS.

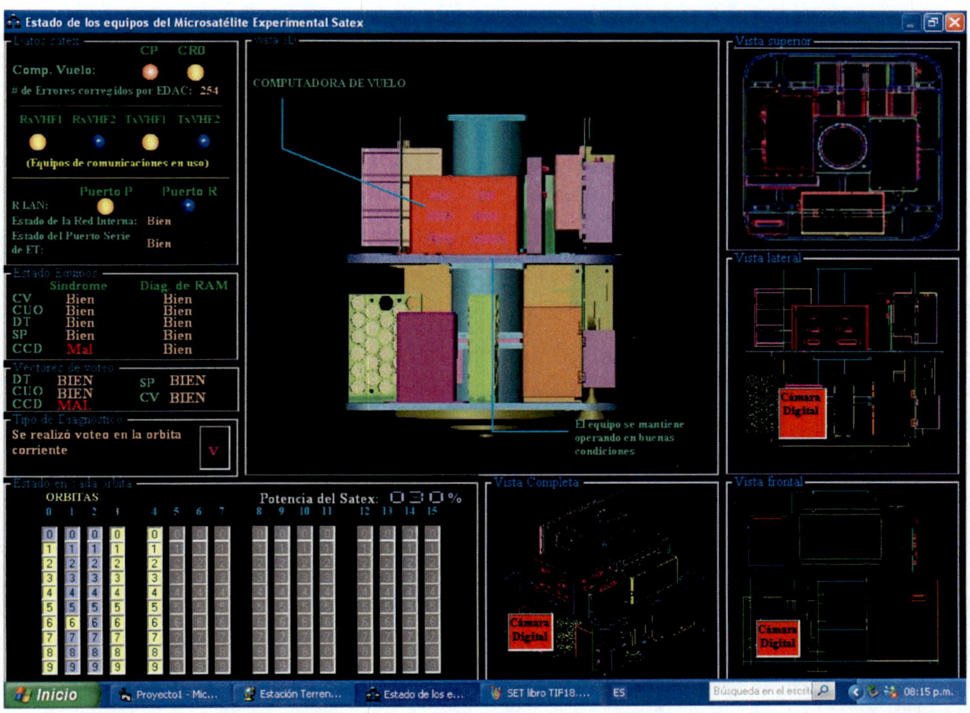

Figure 13. Earth station software displaying status of computer server as well as a fault associated to the digital camera injected with SOFDEVO software.

In addition, ESS contains a set of menus which allow following actions to be controlled:

- Communication speed among satellite and Earth station
- Automatic satellite search
- Programming of alarm thresholds for anyone of the satellite sensors
- Transmission of new programs to the SQLLCS
- Definition of satellite telemetry missions (Normal or Special)
- Telemetry downloading
- Statistics of initialization procedures performed by SQLLCS
- Telemetry displaying with virtual instruments for up to 64 satellite sensors, 48 of them attached to SQLLCS
- Animated displaying of satellite equipment status
- Numerical telemetry displaying
- Image acquisition programming
- Image downloading and fast local view

5. Reliability Study Of Sqllcs Hardware

Considering that reliability data of an electronic system depends on the reliability of every one of its electronic devices, the first reliability model generated for the SQLLCS was formed with interconnected modules according with the server architecture. Those modules where connected in series and parallel fashion according with employed redundancies. The critical points of the server architecture are those series modules which lacks of redundancy support (single points of failure). Its failure may take the whole operative line they belong to an incorrect operation. Meanwhile, the parallel modules tolerate operative failures according with the number of redundancies included.

In order to know the SQLLCS reliability gains introduced by cold standby SBM´s and redundant LAN under space environment, an analysis based on the exponential failure law, the military standard MIL-HBK217f notice 2 and Markov models, was elaborated.

To obtain the failure rates for each one of the electronic components contained in SQLLCS, see table 1, the exponential failure law was employed. According with this law the reliability for an electronic device in its useful life region is given by:

$$R(t) = e^{-\lambda t},$$

where λ (failures per hour) is the constant value of the failure rate function. While in its useful life period, an electronic device offers a safety and predictable service.

In addition, the calculation procedures for failure rates suggested in the military norm "MIL-HDBK-217f Notice 2" where employed. Full information about the procedures to calculate reliability rates for every SQLLCS electronic component was directly obtained from the military norm MIL-HDBK-217f Notice 2, provided by the USA Department of Defence. The norm takes into account several factors such as: components quality, operation environment and thermal issues. From those factors the norm specifies equations to calculate failure rates for specific electronic devices such as: microprocessors, memories, analog devices, digital devices, transistors and so on. In particular, the part stress analysis method was utilized, which requires detailed information regarding the hardware under study. Therefore, this method can be applied in systems under the final design stage or in those systems already designed like in the SQLLCS case. The manual contains 23 sections, from them the sections 5 to 23 contain failure rates models for a wide variety of electronic devices.

Once the failure rates for every electronic device from the SQLLCS architecture were obtained, see table 1, the architecture was divided into modules: SBMs, switching unit for SBMs, main LAN, redundant LAN and so on. Afterwards each module was analyzed to check for serial or parallel arrangements. In this case, every module was represented with a failure rate whose value depends on the determination of the failure rates for every one of the grouped electronic components. Particularly, the failure rates for modules were obtained by applying combinatorial techniques [Johnson, 1989]. In this case the reliability for N series systems was obtained with:

$R_{Sist\_Serie}(t) = e^{-\sum_{i=1}^{N} \lambda_i t}$, while the reliability for N parallel identical systems was calculated with:

$$R_{Sist\_//}(t) = 1 - (1 - R(t))^N \quad \text{or} \quad R_{Sist\_//}(t) = 1 - \left(1 - e^{-\lambda_{Sist} t}\right)^N$$

Table 1. Calculated failure rates (failures/hr) for electronic components of SQLLCS.

Description	Acronym	Failure rate
16 bit RISCM	SAB80	8.36297E-07
16 bit EDAC	EDAC	9.12533E-08
SRAM memory (128K x 8)	RAM128	1.01393E-07
SRAM memory (512K x 8)	RAM512	3.11022E-07
EPROM memory (32K por 8)	PROM	1.30852E-08
Comparator LM139FK	COMP	3.08985E-08
Nand gate SN54HC00FK	NAND	1.0771E-08
Or gate SN54HC32FK	OR	1.0771E-08
And gate SN54HC08FK	AND	1.0771E-08
Operational amplifier TLC2201FK	AMP2201	3.54674E-08
Operational amplifier TLC271FK	AMP271	3.64674E-08
Digital multiplexer SN54HC157FK	MUXDIG	1.1571E-08
Line driver SN55189FK	LD189	3.86674E-08
Line driver SN55188FK	LD188	3.86674E-08
Analog multiplexer CD4053BM	MUX4053	2.52931E-08
Monostable MM54HC221A	MONO	3.78951E-08
40MHz oscilator	XTAL	0.000000015
Hexfet (MOSFET) IRFF130	IRFF	1.16E-08

Table 2. Reliability values given by Hoffman for electronic space equipment.

Mission	Success Reliability
1 year	0.99
3 years	0.97
5 years	0.95
10 years	0.9

For the later case, following modules were calculated:

lamda_2_MONO_en_P = -log(1-(1-exp(-(1/2)*MONO))^2);
lamda_2_HEXFET_en_P = -log(1-(1-exp(-IRFF))^2);
lamda_2_MUX4053_en_P = -log(1-(1-exp(-1/3)*MUX4053))^2);

Then, according with the electronic components integrated at every module several equations for several modules were obtained. In those equations SBM was denoted as F, the switching unit for SBMs was indicated as C, main LAN was designated as R, and redundant LAN as RR.

Equation obtained for Single Board Microcomputers:

F = XTAL + SAB80 + 2*PROM + 2*RAM128 + 2*RAM512 + RAM512 + RAM128 + EDAC + (1/2)*AND + (3/4)*OR + (1/2)*NAND + (1/2)*OR + NAND + (3/4)*OR + (3/4)*AND + AMP2201 + AMP2201 + AMP271 + (1/4)*LM139 + LM136 + (1/4)*OR + (1/4)*OR + lamda_2_MONO_en_P + (1/4)*OR + (1/4)*OR + (1/4)*LD188 + 2*lamda_2_HEXFET_en_P;

Equation obtained for the switching unit of single board microcomputers:

C = (1/4)*MUXDIG + (2/3)*LD188 + (3/4)*MUXDIG + (2/3)*MUX4053 + MUXDIG + (2/3)*MUX4053 + (1/3)*MUX4053 + 2*lamda_2_MUX4053_en_P + (2/3)*MUX4053 + (1/4)*OR + (1/2)*LD188 + 2*lamda_2_HEXFET_en_P + 2*lamda_2_HEXFET_en_P + (1/4)*OR + (1/4)*AND + AND + (1/2)*OR + MUX4053;

Equations obtained for local area networks:

R = LD189 + AND +(3/4)*AND + LD188 + (1/4)*OR + (1/4)*AND;
RR = LD189 + AND + (1/2)*NAND + (1/4)*AND + LD188 + (1/4)*OR;

Equations obtained for parallel hardware:

lamda_2_MONO_en_P = -log(1-(1-exp(-(1/2)*MONO))^2);
lamda_2_HEXFET_en_P = -log(1-(1-exp(-IRFF))^2);
lamda_2_MUX4053_en_P =-log(1-(1-exp(-1/3)*MUX4053))^2);

Previous equations describe with detail the SQLLCS architecture content for every one of its modules.

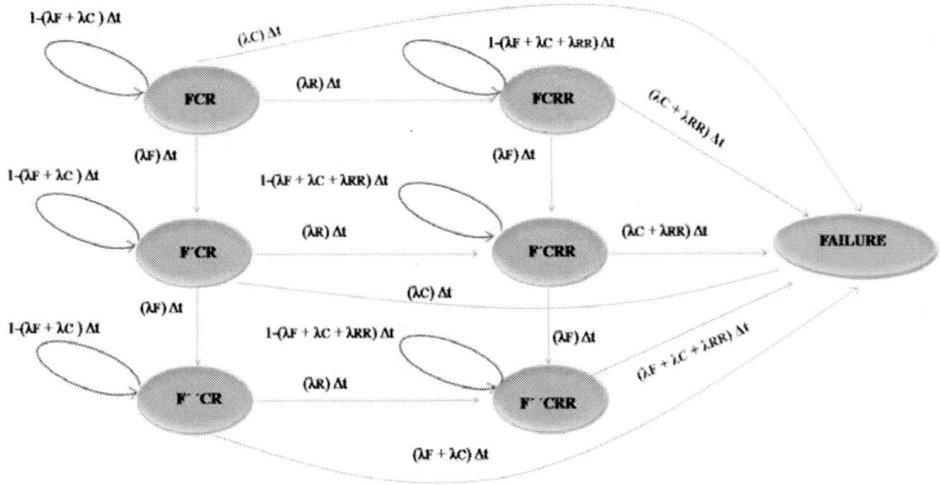

Figure 14. Markov model describing SQLLCS operating states.

In order to describe the SQLLCS architecture operation in terms of its redundancies a Markov model was then generated (figure 14). Every state model represents a different operative state of the server architecture, in such a way that it has a state probability to preserve the state (indicated by a loop vector) given by the associated expression to the vector. In case the server architecture experiences an operative failure, a Markov state change will be generated, signaled in the Markov model by a vector which connects a previous state with a future state (figure 14). Again, the probability for state switching is given by the expression associated to the vector.

In the Markov model shown in figure 14, λ_F stands for SBM failure rate, λc is the failure rate for the switching unit, λ_R is the failure rate for main LAN, and λ_{RR} is the failure rate for redundant LAN. In figure 14 the state titled as FCR denotes that SQLLCS operates with: main SBM symbolised with letter F (from three alternatives, F, F´ and F´´), the switching unit designated with letter C, and main LAN indicated with letter R (from two alternatives, R and RR). Once generated the Markov probabilistic model, shown in figure 14, the following Markov equations are obtained from each model state.

$$p_{FCR}(t+\Delta t) = p_{FCR}(t)(1-(\lambda_F+\lambda_C)\Delta t)$$
$$p_{F'CR}(t+\Delta t) = p_{F'CR}(t)(1-(\lambda_F+\lambda_C)\Delta t) + p_{FCR}(t)(\lambda_F\Delta t)$$
$$p_{F''CR}(t+\Delta t) = p_{F''CR}(t)(1-(\lambda_F+\lambda_C)\Delta t) + p_{F'CR}(t)(\lambda_F\Delta t)$$
$$p_{FCRR}(t+\Delta t) = p_{FCRR}(t)(1-(\lambda_F+\lambda_C+\lambda_{RR})\Delta t) + p_{FCR}(t)(\lambda_R\Delta t)$$
$$p_{F'CRR}(t+\Delta t) = p_{F'CRR}(t)(1-(\lambda_F+\lambda_C+\lambda_{RR})\Delta t) + p_{FCRR}(t)(\lambda_F\Delta t) + p_{F'CR}(t)(\lambda_R\Delta t)$$
$$p_{F''CRR}(t+\Delta t) = p_{F''CRR}(t)(1-(\lambda_F+\lambda_C+\lambda_{RR})\Delta t) + p_{F'CRR}(t)(\lambda_F\Delta t) + p_{F''CR}(t)(\lambda_R\Delta t)$$
$$p_{FAILURE}(t+\Delta t) = p_{FAILURE}(t)(\Delta t) + p_{FCR}(t)(\lambda_C\Delta t) + p_{F'CR}(t)(\lambda_C\Delta t) + p_{F''CR}(t)((\lambda_F+\lambda_C)\Delta t)$$
$$+ p_{FCRR}(t)((\lambda_C+\lambda_{RR})\Delta t) + p_{F'CRR}(t)((\lambda_C+\lambda_{RR})\Delta t) + p_{F''CRR}(t)((\lambda_F+\lambda_C+\lambda_{RR})\Delta t)$$

The previous equations can be written in Matrix form as P(t+Δt) = AP(t), where:

$$P(t+\Delta t = \begin{bmatrix} \textbf{PFCR (t + (t)} \\ \textbf{PF'CR (t + (t)} \\ \textbf{PF''CR (t + (t)} \\ \textbf{PFCRR (t + (t)} \\ \textbf{PF'CRR (t + (t)} \\ \textbf{PF''CRR (t + (t)} \\ \textbf{PFAILURE (t + (t)} \end{bmatrix} \qquad P(t) = \begin{bmatrix} \textbf{PFCR (t)} \\ \textbf{PF'CR (t)} \\ \textbf{PF''CR (t)} \\ \textbf{PFCRR (t)} \\ \textbf{PF'CRR (t)} \\ \textbf{PF''CRR (t)} \\ \textbf{PFAILURE (t)} \end{bmatrix}$$

While Matrix **A** has the following form:

$$\mathbf{A=} \begin{bmatrix} \mathbf{a1} & \mathbf{0} & \mathbf{0} & \mathbf{0} & \mathbf{0} & \mathbf{0} & \mathbf{0} \\ \mathbf{b1} & \mathbf{b2} & \mathbf{0} & \mathbf{0} & \mathbf{0} & \mathbf{0} & \mathbf{0} \\ \mathbf{0} & \mathbf{c2} & \mathbf{c3} & \mathbf{0} & \mathbf{0} & \mathbf{0} & \mathbf{0} \\ \mathbf{d1} & \mathbf{0} & \mathbf{0} & \mathbf{d4} & \mathbf{0} & \mathbf{0} & \mathbf{0} \\ \mathbf{0} & \mathbf{e2} & \mathbf{0} & \mathbf{e4} & \mathbf{e5} & \mathbf{0} & \mathbf{0} \\ \mathbf{0} & \mathbf{0} & \mathbf{f3} & \mathbf{0} & \mathbf{f5} & \mathbf{f6} & \mathbf{0} \\ \mathbf{m1} & \mathbf{m2} & \mathbf{m3} & \mathbf{m4} & \mathbf{m5} & \mathbf{m6} & \mathbf{m7} \end{bmatrix}$$

And the Matrix A terms different from zero are:

$$a1 = (1-(\lambda_F+\lambda_C)\Delta t)$$
$$b1 = (\lambda_F \Delta t)$$
$$b2 = (1-(\lambda_F+\lambda_C)\Delta t)$$
$$c2 = (\lambda_F \Delta t)$$
$$c3 = (1-(\lambda_F+\lambda_C)\Delta t)$$
$$d1 = (\lambda_R \Delta t)$$
$$d4 = (1-(\lambda_F+\lambda_C+\lambda_{RR})\Delta t)$$
$$e2 = (\lambda_R \Delta t)$$
$$e4 = (\lambda_F \Delta t)$$
$$e5 = (1-(\lambda_F+\lambda_C+\lambda_{RR})\Delta t)$$
$$f3 = (\lambda_R \Delta t)$$
$$f5 = (\lambda_F \Delta t)$$
$$f6 = (1-(\lambda_F+\lambda_C+\lambda_{RR})\Delta t)$$
$$m1 = (\lambda_C \Delta t)$$
$$m2 = (\lambda_C \Delta t)$$
$$m3 = ((\lambda_F+\lambda_C)\Delta t)$$
$$m4 = ((\lambda_C+\lambda_{RR})\Delta t)$$
$$m5 = ((\lambda_C+\lambda_{RR})\Delta t)$$
$$m6 = ((\lambda_F+\lambda_C+\lambda_{RR})\Delta t)$$
$$m7 = 1$$

Afterwards, the vector of probabilities to stay at each one of the model states was obtained, in such a way that the survival reliability function is determined by grouping the SQLLCS success states. To generate the calculation of such a vector an interactive software in Matlab was created to plot reliability curves for variable times.

On the other hand, in the publication from [Hoffman, 1990] some reliability values that space equipments have to accomplish are given (see table 2). Besides, electronic equipment utilized in long-life space applications must achieve a minimal reliability of 0.95 for a 10-year time period [Johnson, 1989]. These indexes are employed later to elaborate about the reliability results obtained when considering few SQLLCS assembled with different number of SBMs.

5.1 SQLLCS Maintenance

Regarding the associated operations with fault-detection, fault-diagnosis and SQLLCS reconfiguration, full information is presented in [Vicente, 2004] which describes a fault-tolerant semivirtual computer architecture applied to a low Earth orbit microsatellite. The architecture takes advantage of on-board intelligent subsystems and payload microcomputers to conform an otherwise difficult to implant N-modular redundancy computer. Communications among processors (physical and virtual parts) are accomplished over a fault-tolerant LAN, allowing a versatile operating behaviour in terms of data communication as well as in terms of distributed fault tolerance (figure 15).

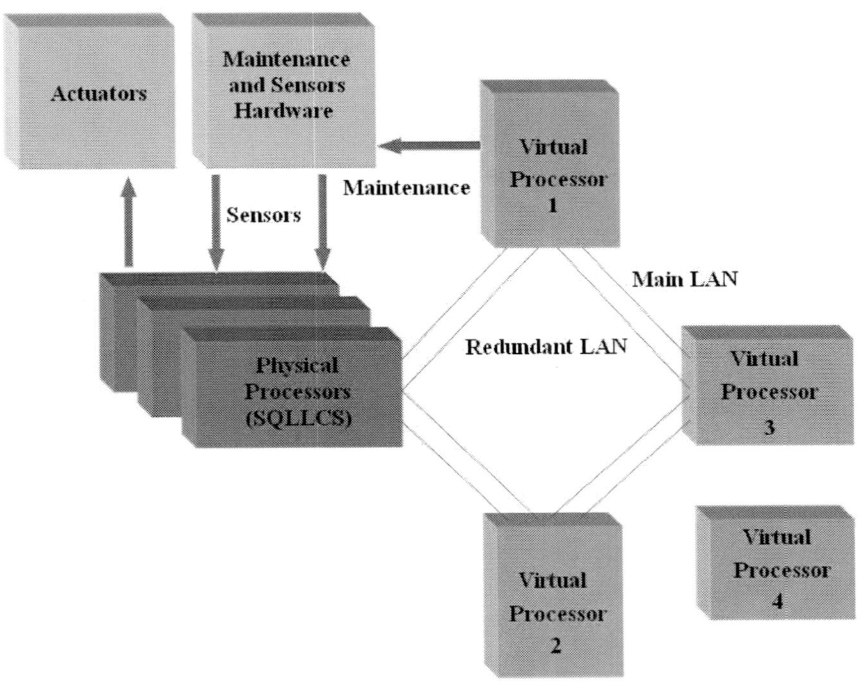

Figure 15. Semivirtual computer architecture.

The network was developed with both hardware and software diversity to avoid generic failures. Fault-tolerant operation for the semivirtual architecture was achieved by software algorithms performing processors diagnosis, fault detection and hardware reconfiguration to allow SQLLCS continuous operations. The diagnosis process for a processor covers the type of failures that can be present in hardware resources from processors. It does not include RAM memory because fault-tolerance for this resource is provided by EDAC operation and interrupt software. Fault detection is based on the Bizantine resilient theory. This was implemented with a majority voting process of results at every computing node. The amount of connectivity paths among networked nodes requested by the Bizantine technique was modified in order to generate a practical computer architecture implementation for a 55 Kg satellite. To fulfill the required connectivity resources among processors a fault-tolerant LAN was developed. In this way the semivirtual computer architecture accomplished an otherwise difficult to implement "N" modular redundancy computer tightly coupled by software to carry out fault-tolerance operations. During semivirtual computer architecture operation all diagnosis results from each node are broadcasted, one by one, followed by a second broadcasting of global diagnosis results from every node. The process enabled a consensual algorithm to establish a majority voting mechanism to detect semivirtual computer architecture failures, with special focus on the SQLLCS. In addition, in [Vicente, 2006] a detailed Markov model depicting the semivirtual computer architecture operations in terms of maintenance for both the physical and virtual part of the architecture is presented.

In this chapter, a reduced Markov Model is analized. The main difference from that one presented in [Vicente, 2006] is that no virtual processors are included in the SQLLCS study to highlight the reliability participation of physical processors (SQLLCS). This assumption reduced by half the size of the Markov model shown in figure 14.

5.2 Reliability of an SQLLCS Assembled with Three SBMs

To highlight the overall benefits of SQLLCS hardware assembled with three SBMs for a LEO microsatellite, a reliability comparison among three different implementations was performed, employing the Markov model shown in figure 14. One assembled with COTS electronic components, another assembled with military electronic devices and a last one assembled with space qualified electronic parts. Figure 16.a shows the reliability results for three different SQLLCS server versions. The highest reliability is obtained when the computer is assembled with space qualified electronic components and the lowest reliability is achieved with COTS parts.

As seen in figure 16.a, the three server versions assembled respectively with space, military and COTS parts fulfill the reliability requirements stated by [Hoffman, 1990] (see table 2). Nevertheless, the COTS assembled computer does not meet the terms of long-life reliability criterion (0.95 reliability for a 10-year time period).

Considering that space qualified components are expensive and require longer arrival times, the best reliability-cost relation for a microsatellite application is obtained when the computer server is assembled with military qualified components.

In figure 16.a the reliability obtained for a SQLLCS assembled with military components in a one year time operation, is important because it represents the minimal expected life for Satex experimental satellite. A reliability of 99.94 was obtained for this time, showing a very good forecast because of the spares included for the single points of failure in the architecture. The reliability analysis shows the advantages of using good quality electronic parts as well as the improvement obtained from using spares to increase the expected operational life for the SQLLCS and therefore for the satellite.

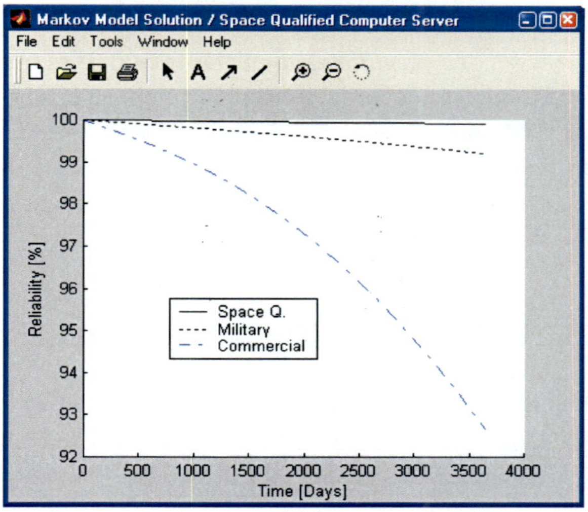

(a)

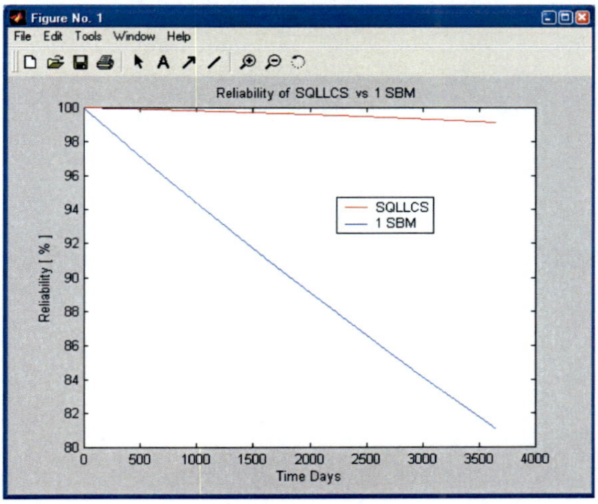

(b)

Figure 16. Reliability results calculated in a 10 year time period: (a) For three SQLLCS versions (each one assembled with 3 SBM); (b) For a SQLLCS (assembled with 3 SBM) and a single SBM all of them assembled with military parts.

5.3 Reliability for a Single SBM

In figure 16.b the reliabilities for a single SBM assembled with military parts are as follows: 97.9, 93.8, 89.9 and 81.2 for 1, 3, 5 and 10 years of operation, respectively. Comparing these reliability values with reliability data given by [Hoffman, 1990], we may conclude that a non redundant SBM assembled with military qualified parts does not accomplishes the reliability indexes required by electronic space equipment. This is the quantitative justification to develop a SQLLCS with cold standby spares in order to increase satellite survival possibilities.

As also shows figure 16.b, an expected reliability of 0.95 for a single SBM is reached in a time period of 2.44 years. On the other hand, a reliability of 0.95 is important in aeronautical applications [Hills, 1988], because once an electronic equipment reaches this value preventive maintenance has to be programmed on it in order to avoid unexpected failures. Then, if this principle is employed in a satellite mission, the single SBM assembled with military parts must undergo maintenance before 2.44 years. Hence it does not qualify as long-life space equipment either.

5.4 Reliability for a SQLLCS Assembled with Two SBMs

In addition, a reliability study was made for a SQLLCS architecture assembled with just two SBMs. For this architecture a new Markov model was obtained, similar to that shown in figure 14. For this model, the states F´´CR and F´´CRR were eliminated (see figure 17).

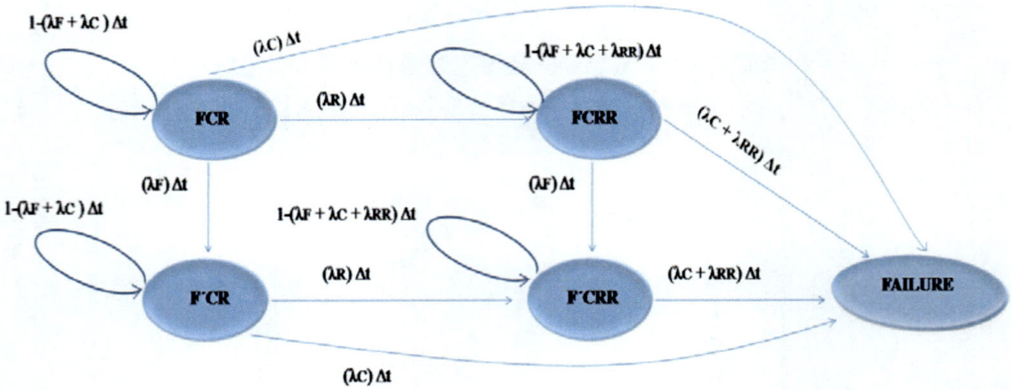

Figure 17. Markov model for a SQLLCS assembled with 2 SBMs.

Then the respective Markov equations were obtained for each state shown in the Markov model,

$$p_{FCR}(t+\Delta t) = p_{FCR}(t)(1-(\lambda_F+\lambda_C)\Delta t)$$
$$p_{F'CR}(t+\Delta t) = p_{F'CR}(t)(1-(\lambda_F+\lambda_C)\Delta t) + p_{FCR}(t)(\lambda_F \Delta t)$$
$$p_{FCRR}(t+\Delta t) = p_{FCRR}(t)(1-(\lambda_F+\lambda_C+\lambda_{RR})\Delta t) + p_{FCR}(t)(\lambda_R \Delta t)$$
$$p_{F'CRR}(t+\Delta t) = p_{F'CRR}(t)(1-(\lambda_F+\lambda_C+\lambda_{RR})\Delta t) + p_{FCRR}(t)(\lambda_F \Delta t) + p_{F'CR}(t)(\lambda_R \Delta t)$$

$$\begin{aligned} p_{FAILURE}(t+\Delta t) = {} & p_{FAILURE}(t)(\Delta t) + p_{FCR}(t)(\lambda_C \Delta t) + p_{F'CR}(t)(\lambda_C \Delta t) \\ & + p_{FCRR}(t)((\lambda_C+\lambda_{RR})\Delta t) + p_{F'CRR}(t)((\lambda_C+\lambda_{RR})\Delta t) \end{aligned}$$

The previous equations can also be written in Matrix form as P(t+Δt) = AP(t), generating a Matrix [A] of 5x5 elements, where:

$$P(t+\Delta t = \begin{bmatrix} \mathbf{PFCR\ (t + (t)} \\ \mathbf{PF'CR\ (t + (t)} \\ \mathbf{PFCRR\ (t + (t)} \\ \mathbf{PF'CRR\ (t + (t)} \\ \mathbf{PFAILURE(t +.(t)} \end{bmatrix} \qquad P(t) = \begin{bmatrix} \mathbf{PFCR\ (t)} \\ \mathbf{PF'CR\ (t)} \\ \mathbf{PFCRR\ (t)} \\ \mathbf{PF'CRR\ (t)} \\ \mathbf{PFAILURE\ (t)} \end{bmatrix}$$

Afterwards, the vector of probabilities to stay at each one of the model states was extracted. Then grouping the SQLLCS success states the survival reliability was determined. Again, to make the calculation of such a vector it was elaborated another interactive software in Matlab to generate reliability curves.

Figure 18.a shows the reliability results for three different types of SQLLCS, each one assembled with two SBMs. The first one was assembled with commercial parts, the second with military components and a third one with space qualified electronic parts. As can be seen, reliability results for both the space and the military versions are above 0.95 for 1, 3, 5 and 10 years of operation. This means that both equipments accomplish Hoffman´s principle as well as the requirements for long-life space equipment. In the case of the commercial version, reliability data are: 0.992, 0.966, 0.927 and 0.798 for 1, 3, 5 and 10 years of operation, respectively. In this case only the reliability for one year of operation satisfies the Hoffman´s criteria. In other words, a commercial SQLLCS containing two SBM may be used in space, but only in a satellite mission lasting less than one year, if and only if the commercial parts are protected for latch-up phenomena.

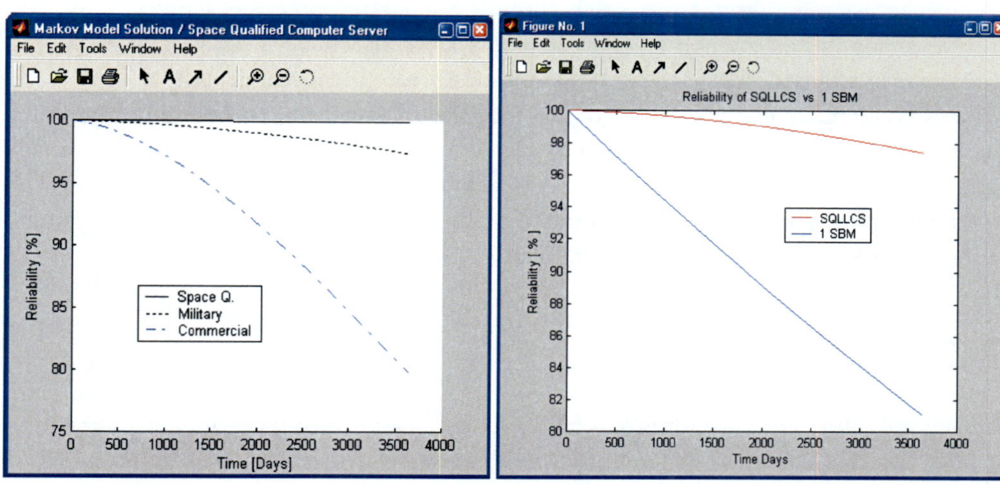

(a) (b)

Figure 18. (a) Reliability data for three SQLLCS versions each one assembled with two SBM; (b) Reliability of a SQLLCS; assembled with two SBMs and a single SBM, both of them assembled with military parts.

In figure 18.b reliability results are given for a SQLLCS (containing two SBMs) as well as for a single SBM, both of them assembled with military parts. The SQLLCS reliability values for 1, 3, 5 and 10 years are all above the Hoffman´s criterion. In addition, it accomplishes the reliability indexes for long-life space equipment. On the other hand reliability values for the single SBM assembled with military qualified parts does not accomplishes the reliability indexes required by electronic space equipment and neither satisfies the requirements for a long-life space equipment.

Equipment	Qualification	Hoffman´s reliability criterion	Long life reliability criterion	Best reliability-cost criterion	Early maintenance according with Hill´s criterion (years)
Single SBM	Militar	Not passed	Not passed	No	2.4
SQLLCS assembled with 2 SBMs	COTS	Not passed	Not passed	No	3.9
	Militar	Passed	Passed	Yes	No
	Space	Passed	Passed	No	No
SQLLCS assembled with 3 SBMs	COTS	Passed	Not passed	No	7.8
	Militar	Passed	Passed	Yes	No
	Space	Passed	Passed	No	No

Table 3. Reliability results obtained for different versions of SQLLCS.

Table 3 summarizes the preceding results. With them can be affirmed that SQLLCS reliability forecast results are encouraging and better than those obtained from centralized computer systems without redundancies. The summarizing results also show that the SQLLCS assembled with military parts, either with 3 or 2 SBMs, fulfils the Hoffman´s criterion of success reliability as well as the long life reliability criterion. However, the best reliability-cost solution is the SQLLCS assembled with 2 SBMs, because of savings in cost, physical space and weight, which are important issues in satellite instrumentation.

Reliability results obtained and exposed in this chapter are encouraging for Satex microsatellite computer instrumentation. They have to be taken in consideration to evaluate the behaviour of satellite missions. However, reliability data need to be evaluated with experimental performance of space missions to reach better estimation consensus regarding the equipment expectations in space applications.

6. Conclusion

The space qualified architecture for a long-life computer server specially developed for a 50 Kg microsatellite has been depicted. Taking into account that SQLLCS is in charge of critical microsatellite operations, it deserved an architecture design with cold standby spares to avoid single points of failures. Besides, the architecture takes into account the strong limitations associated with small space vehicles in terms of cost, weight and available physical space. It also considers electronic and physical protections to cope the harsh environment expected at low Earth orbit, in particular that related with the effects of cosmic radiation in electronic parts.

The SQLLCS consists of a maximum of three single board microcomputers (main computer and cold spares), each one with enough resources to fully control the microsatellite operations. Besides, SQLLCS contains three further boards that comprise an electronic switching unit that allows any one of the SBM to be energized and connected to the satellite instrumentation, a multiplexing and conditioning unit to read up to 48 electronic signals from satellite sensors, as well as hardware for a redundant LAN to interconnect the satellite microcomputers and protections against latch-up phenomenon.

Validation of the space computer server required the availability of complementary satellite subsystems for hardware and software debugging. However, all of the satellite subsystems were designed and fabricated in parallel fashion; by this reason the full validation of the computer server flight model required the development of special tools. In this sense a satellite simulator, emulation software for payloads, as well as the Earth station software were created to enable SQLLCS validation without the need of complementary hardware satellite subsystems. The ESS provides virtual animation facilities to monitor the computer server behavior at a glance. With referred resources, the SQLLCS was debugged in hardware and software, conducting to a successful system validation.

To know the reliability gains obtained from SQLLCS hardware spares, a reliability prediction study was performed. For this purpose, combinatorial techniques, the exponential failure law and the military standard MIL-HDBK217f notice 2 were employed to reduce the SQLLCS hardware architecture into modules. From these modules the participation of hardware spares were represented by Markov models, and then the Markov equations were obtained. To obtain reliability forecast data, dedicated Matlab software was also created. The obtained reliability curves show that aggregated redundancies in critical parts of the architecture generate good reliability expectations for the SQLLCS and therefore for the microsatellite.

In particular, SQLLCS´s reliability results for devices assembled either with three or two SBMs and military components fulfill the reliability requirements established for long-life satellite equipment. At the same time these computer servers accomplish a good reliability-cost relation for a satellite mission. However, the best reliability-cost solution is the SQLLCS assembled with two SBMs, because of savings in cost, physical space and weight, which are important issues in the satellite field.

Acknowledgments

The authors would like to acknowledge the federal agencies from México, TELECOM and COFETEL, which supported the work described in this paper. In addition, acknowledgments are given to the Mexican Research Institutions that collaborated in the development of the microsatellite mission, in alphabetical order: CICESE, CITEDI, CIMAT, INAOE, IPN and UNAM. Special thanks are also given to Miguel Ángel Pérez, Andrés Sánchez, Luis Ramón Guitérrez, José Luis Gutiérrez, Carlos Pineda, Adán Espinoza, Juan Reza, Juan Ramón Tórres, Miguel Falcón, Alberto Rangel, Hugo Ortíz, Iris Mejía, Galdino Gutierrez, Martín Juárez and Victor Melo, for their participation at different stages of this project.

References

Bretschneider T., Tan S. H., Goh C. H., Arichandran K., Koh W. E., & Gill E., (2005), "X-SAT Mission Progress", In: Röser H.-P., Sandau R., Valenzuela A. (Eds.), *Small*

Satellites for Earth Observation; Selected Proceedings of the 5th International Academy of Astronautics (pp. 145-152), Berlin, Germany.

Cardarilli G.C., Leandri A., Marinucci P. & Ottavi M., (2003), "Design of a Fault Tolerant Solid State Mass Memory", *IEEE Transactions On Reliability*, Vol. 52, No. 4, pp. 476-491.

Elfving A., Stagnaro L. & Winton A., (2003), "SMART-1: key technologies and autonomy implementations", *Acta Astronautica Journal, Vol. 52, Issues 2-6*, pp. 475-486.

Guldager P.B., Thuesen G. G. & Jørgensen J. L., (2005), "Quality Assurance for Space Instruments Built with COTS", *Acta Astronautica Journal, Vol. 56, Issues 1-2*, pp. 279-283.

Caldwell D. and Rennels D., (2000) "A Minimalist Fault-Tolerant Microcontroller Design for Embedded Spacecraft Computing", *The Journal of Supercomputing, Volume 16, numbers 1-2*, pp. 7-25.

Chau S. N. and Lang M., "*A Multi-Mission Testbed for Advanced Technologies*", Forum on Innovative Approaches to Outer Planetary Exploration 2001-2020, Jet Propulsion Laboratory, Houston, Texas, USA, February (2001).

http://www.lpi.usra.edu/meetings/outerplanets2001/pdf/4030.pdf .

"*The Cibola Satellite Tests a Reconfigurable Computer*", 1663 Los Alamos Science and Technology Magazine, Los Alamos National laboratory, May (2007).

http://www.lanl.gov/news/index.php/fuseaction/1663.article/d/20075/id/10389

Hills A.D., and Mirsa N.A., "*Fault Tolerant Avionics*", IEEE/AIAA 8th Digital Avionics Systems Conference, October (1988).

Hoffman E.J. & Moore R.C., (1990), *"Spacecraft Tracking, Telemetry and Control, Course Notes"*, Baltimore, Maryland, The Johns Hopkins University, Applied Physics Laboratory Space Department.

Johnson B. W., (1988), *"Design & Analysis of Fault Tolerant Digital Systems"*, Boston, MA, Addison-Wesley Longman Publishing Co., Inc.

Laurence J.F., Hersman, C. B., Williams, S. P., & Conde, R. F., *" A General-Purpose MIL-STD-1750A Spacecraft Computer System*", 9th AIAA Computing in Aerospace Conference, pages 1421-1429, San Diego, CA, USA, October (1993).

Lee I., Sung D.K. & Choi S.D., *"Experimental Multimission Microsatellites-Kitsat Series"*, Proceedings of 6th Annual USU/AIAA Conference on Small Satellites, Logan, Utah, September, (1992).

Pisacane V. L. and Moore R. C., (1994), "Fundamentals of Space Systems", New York, Oxford University Press.

Renaudie C., Markgraf M., Montenbruck O. & Garcia M., "*Radiation Testing of Commercial-off-the-Shelf GPS Technology for Use on Low Earth Orbit Satellites*", RADECS 2007, Proceedings of the 9th European Conference Radiation and Its Effects on Components and Systems, Deauville, France, September (2007).

Shaneyfelt M.R., Winokur P.S., Meisenheimer T.L., Sexton F.W., Roeske S.B. & Knoll M.G., (1994), "Hardness Variability in Commercial Technologies", *IEEE Transactions on Nuclear Science, Vol. 41, No. 6*, pp. 2536-2543.

Surrey Satellite Technology Limited MicroSat-70 Website, (2008), "SSTL MicroSat-70 : The Original Modular Microsatellite", http://zenit.sstl.co.uk/index.php?loc=49.

Surrey Satellite Technology Limited MiniSat Website, (2008), "MiniSat-400, Flight Proven Minisatellite", http://zenit.sstl.co.uk/index.php?loc=49.

Sperber F., *"Amsat-Phase 3-D a 400 Kgs International Communication and Experimental Satellite in a High-Elliptical Orbit"*, Proceedings of the 3rd International Symposium on Small Satellite Systems and Services, Annecy, France, (1996).

Sweeting M. N., (2001), "25 Years of Space at Surrey—Pioneering Modern Microsatellites", *Acta Astronautica Journal,* Volume 49, Issue 12, pp. 681-691.

Underwood C., (2003), "Observations of Radiation in the Space Radiation Environment and its Effects on Commercial of-the-shelf Electronics in Low-Earth Orbit", *Philosophical Transactions: Mathematical, Physical and Engineering Sciences, Science and Applications of the Space Environment: New Results and Interdisciplinary Connections, Vol. 361, No. 1802*, pp. 193-197.

Vicente Vivas Esaú, *"Fault-Tolerant Semivirtual Computer Architecture with automatic maintenance Capabilities Applied to a LEO Microsatellite"*, PhD thesis in Electrical Engineering (in Spanish), México City, Universidad Nacional Autónoma de México, 2004.

Vicente-Vivas E., García-Nocetti D.F. & Mendieta-Jiménez F. J., (2006), "Automatic Maintenance Payload on board of a Mexican LEO Microsatellite", *Acta Astronautica Journal,* Volume 58, Issue 3, pp. 149-167.

Wimmer W., (1997), "Lessons to Be Learned From European Science and Applications Space Missions", *Control Engineering Practice*, Volume 5, Issue 2, pp. 155-165.

Winokur P.S., Fleetwood D.M., & Sexton F.W., (1994), "Radiation-Hardened Microelectronics for Space Applications", *International Journal on Radiation Phys. Chem. Vol 43,* pp. 175-190.

In: VLSI and Computer Architecture
Editor: Kenzo Watanable

ISBN 978-1-60692-075-6

Chapter 4

NUMERICAL SIMULATION OF QUANTUM WAVEGUIDES

***Anton Arnold*[1], *Matthias Ehrhardt*[2] *and Maike Schulte*[3]**
[1] Institut für Analysis und Scientific Computing,
TU Wien, Wiedner Hauptstr. 8, 1040 Wien, Austria
[2] Weierstraß–Institut für Angewandte Analysis und Stochastik,
Mohrenstr. 39, 10117 Berlin, Germany
[3] Institut für Numerische und Angewandte Mathematik,
Universität Münster, Einsteinstr. 62,
48149 Münster, Germany

Abstract

This chapter is a review of the research of the authors from the last decade and focuses on the mathematical analysis of the Schrödinger model for nano–scale semiconductor devices. We discuss *transparent boundary conditions* (TBCs) for the time–dependent Schrödinger equation on a two dimensional domain.

First we derive the two dimensional discrete TBCs in conjunction with a conservative Crank–Nicolson–type finite difference scheme and a compact nine–point scheme. For this difference equations we derive *discrete transparent boundary conditions* (DTBCs) in order to get highly accurate solutions for open boundary problems. The presented discrete boundary–valued problem is unconditionally stable and completely reflection–free at the boundary.

Then, since the DTBCs for the Schrödinger equation include a convolution w.r.t. time with a weakly decaying kernel, we construct *approximate* DTBCs with a kernel having the form of a finite sum of exponentials, which can be efficiently evaluated by recursion.

In several numerical tests we illustrate the perfect absorption of outgoing waves independent of their impact angle at the boundary, the stability, and efficiency of the proposed method. Finally, we apply inhomogeneous DTBCs to the transient simulation of quantum waveguides with a prescribed electron inflow.

1. Introduction

Today's semiconductor devices like transistors and nanoscale split-gate devices are rapidly shrinking in their size. In this context, modeling and numerical simulations play an important role in the development and design of new devices. We focus on devices with ballistic

[Schematic view of a double gate metal oxide semiconductor field-effect transistor (DG-MOS). The electron transport takes place from source to drain in x-direction. An external potential is applied at the gates.]

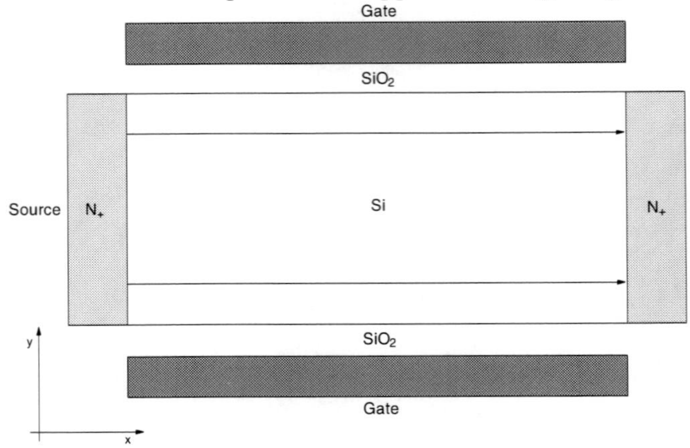

[Simplified model of a DG-MOS]

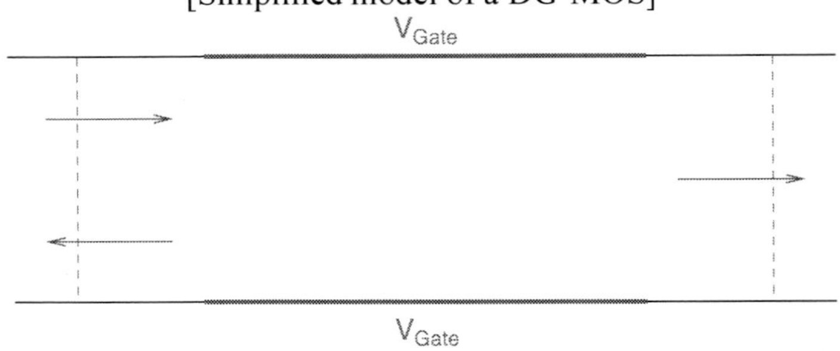

Figure 1. Schematic view and simplified model of a DG-MOS.

electron transport, such as electron quantum waveguide devices. Their functionality depends on the formation of a 2D electron gas and on wave interference effects (cf. [23]).

Speaking of ballistic transport means that electrons are assumed to not suffer any collision during their transit through the device (e.g. high–purity materials, very small time or length scales, and at low temperatures). A schematic view of such a device, a *double gate metal oxide semiconductor field–effect transistor* (DG–MOS), is shown in Figure 1.. At the gates there is an applied external potential and the electron transport takes place from source to drain.

We consider the *effective mass approximation*, where the mass m_* is assumed to be constant in homogenized parts of the device. The different materials (e.g. Si, SiO_2) have different effective masses. We simplify this model like it is presented in Figure 1., where only one effective mass is used and external potentials $V_{\rm Gate}$ could be applied at the gates. But regarding different materials and therefore different effective masses won't change our proposed model in principle.

Quantum waveguides are novel electronic switches of nanoscale dimensions. They are made of several different semiconductor materials such that the electron flow is confined to small channels or waveguides. Due to their sandwiched structure the relevant geometry for the electron current is roughly two dimensional.

Using external electrostatic potentials the "allowed region" for the electrons, and hence the geometry can be modified. This fact allows to control the current flow through such an electronic device. It makes it a switch, which resembles a transistor, but on a nanoscale (cf. §2.1 of [2], e.g.).

Being quantum particles, the electron transport through a quantum waveguide can be modeled in good approximation by the following two dimensional, *transient Schrödinger equation*

$$\begin{aligned} i\hbar\frac{\partial}{\partial t}\psi(\mathbf{x},t) &= -\frac{\hbar^2}{2m^*}\Delta_{\mathbf{x}}\psi(\mathbf{x},t)+V(\mathbf{x},t)\psi(\mathbf{x},t), \\ &\qquad\qquad \mathbf{x}=(x,y)\in\Omega(t),\ t>0, \\ \psi(\mathbf{x},0) &= \psi^I(\mathbf{x}), \quad \mathbf{x}\in\Omega(0), \\ \psi(\mathbf{x},t) &= 0, \quad \mathbf{x}\in\partial\Omega(t), \end{aligned} \tag{1}$$

on a time–dependent geometry $\Omega(t)\subset\mathbb{R}^2$ with initial data $\psi^I\in L^2(\Omega(0))$ and homogeneous Dirichlet boundary conditions. Here, $\hbar$ and m^* denote the Planck constant and effective mass, respectively. The external real valued potential satisfies $V(.,t)\in L^\infty(\Omega(t))$ and $V(\mathbf{x},.)$ is piecewise continuous. The solution ψ to (1) is a time–dependent complex valued wave function with $\psi(.,t)\in L^2(\Omega(t))$.

The spatial domain consists of (very long) leads and the active switching region, which sometimes has the shape of a stub. The structure can be realized as an etched layer $\Omega(t)$ of GaAs between two layers of doped AlGaAs. Here, we shall only consider domains $\Omega(t)$ that are piecewise constant in t and monotonously growing in time. At the jump discontinuities of the domain we shall extent the solution ψ by zero, as a new initialization.

In typical applications, electrons are continuously fed into the leads. Depending on the size and shape of the stub, the electron current is either reflected (*off–state* of the device) or it can pass through the device (*on–state*). Since the applied external potential can modify the stub size, it hence allows to switch the device. Important device data for practitioners are the ratio between the on- and the (residual) off–current as well as the switching time between these two stationary states. These data can be obtained from numerical simulations of the described Schrödinger equation model (1).

The leads are very long compared to the typical size of the active region and they usually only carry (linear combinations of) plane wave solutions. For the efficiency of numerical simulations it is therefore desirable to restrict the simulation model to a small computational region $\tilde{\Omega}(t)$ close to the stub (see Fig. 2). Hence, the leads should be cut off by using artificial boundary conditions. This is possible without changing the solution of the Schrödinger equation by introducing *open boundary conditions* [24], which are non–local in time (convolution type) and in space. Open boundary conditions are called *transparent*, if they yield identical solutions both on the original large domain $\Omega(t)$ and the reduced (computational) domain $\tilde{\Omega}(t)$.

This chapter is organized as follows: In Section 2. we introduce the concept of transparent boundary conditions (TBCs). In Section 3. we derive and analyze a discrete analogue of the analytic TBCs in conjunction with a fourth order compact finite difference scheme of the Schrödinger equation. We present some numerical simulations to illustrate the ef-

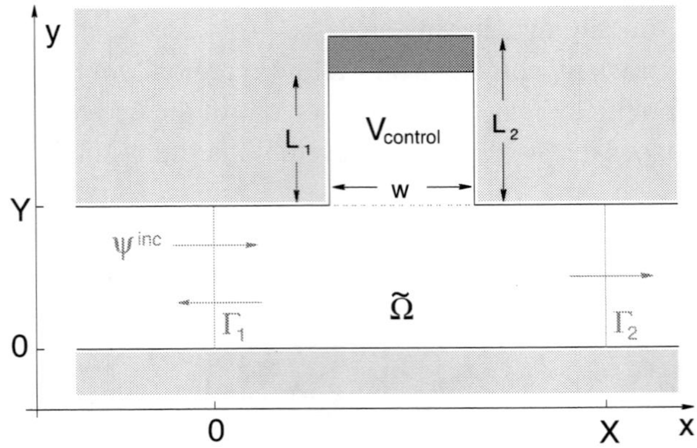

Figure 2. T-shaped structure Ω with the length X, a channel width Y, and a stub width w. It is possible to enlarge the stub length from L_1 to L_2. Inhomogeneous TBCs have to be proposed at $x = 0$, homogeneous TBCs at $x = X$. The inflow at $x = 0$ is modeled by an incoming function ψ^{inc} given by linear combination of plane waves.

fectiveness and accuracy of our DTBCs in Section 4.. Finally, we give an application of inhomogeneous DTBCs to a 2D waveguide simulation with a T-shaped quantum transistor.

2. Transparent Boundary Conditions for the Two Dimensional Schrödinger Equation

To illustrate the idea of deriving transparent boundary conditions (TBCs) we first consider the one dimensional, time–dependent Schrödinger equation (1) with a potential that satisfies for simplicity the following assumptions: $V(x,t) \equiv V_l$ for $x \le 0$ and $V(x,t) \equiv V_r$ for $x \ge X$ and all $t \ge 0$. For the treatment of nonconstant exterior potentials we refer the reader to [18], [20], [22].

The first step of the derivation is to cut the original *whole–space problem* into three subproblems, the interior problem on the bounded domain $0 < x < X$, and a left and right exterior problem. These problems are coupled by the assumption that the wave function ψ and its spatial derivative ψ_x are continuous across the artificial boundaries at $x = 0, x = X$. Hence, the *interior problem* reads

$$\begin{aligned}
i\hbar\psi_t &= -\frac{\hbar^2}{2m^*}\psi_{xx} + V(x,t)\psi, \qquad 0 < x < X, \quad t > 0,\\
\psi(x,0) &= \psi^I(x), \quad 0 < x < X,\\
\psi_x(0,t) &= (T_l\psi)(0,t), \quad t > 0,\\
\psi_x(X,t) &= (T_r\psi)(X,t), \quad t > 0.
\end{aligned} \tag{2}$$

$T_{l,r}$ denote the *Dirichlet–to–Neumann (DtN) maps* at the left/right boundaries, and they are obtained by solving the two *exterior problems*:

$$\begin{aligned} i\hbar v_t &= -\frac{\hbar^2}{2m^*} v_{xx} + V_r v, \quad x > X,\ t > 0, \\ v(x,0) &= 0, \quad x > X, \\ v(\infty,t) &= 0 \quad \text{and} \quad v(X,t) = \Phi(t), \quad t > 0,\ \Phi(0) = 0, \end{aligned} \tag{3}$$

which yields $(T_r\Phi)(t) = v_x(X,t)$ and analogously for the left mapping T_l at $x = 0$. Since the potential is constant in the exterior problems, we can solve them explicitly by the Laplace method and thus obtain the two boundary operators $T_{l,r}$ needed in (2) (cf. Fig. 3).

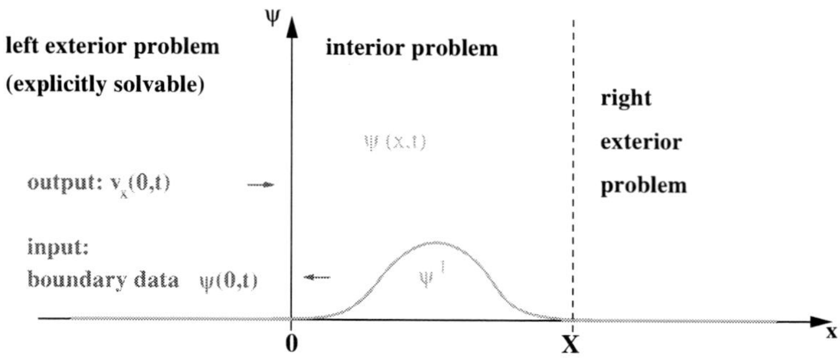

Figure 3. Construction idea for transparent boundary conditions in 1D.

The Laplace transformation of v is given by

$$\hat{v}(x,s) = \int_0^\infty v(x,t)\, e^{-st}\, dt, \quad s = \eta + i\xi,\ \xi \in \mathbb{R},\ \eta > 0 \text{ fixed}.$$

Now the right exterior problem (3) is transformed to

$$\begin{aligned} \hat{v}_{xx} + i\frac{2m^*}{\hbar}\Big(s + i\frac{V_r}{\hbar}\Big)\hat{v} &= 0, \quad x > X, \\ \hat{v}(X,s) &= \hat{\Phi}(s). \end{aligned} \tag{4}$$

Since its solutions have to decrease as $x \to \infty$ (since we have $\psi(.,t) \in L^2(\mathbb{R})$), we obtain

$$\hat{v}(x,s) = e^{-\sqrt[+]{-i\frac{2m^*}{\hbar}(s+i\frac{V_r}{\hbar})}(x-X)}\, \hat{\Phi}(s).$$

Hence, the Laplace–transformed Dirichlet–to–Neumann operator T_r reads

$$\widehat{T_r\Phi}(s) = \hat{v}_x(X,s) = -\sqrt{\frac{2m^*}{\hbar}}\, e^{-i\frac{\pi}{4}} \sqrt[+]{s + i\frac{V_r}{\hbar}}\, \hat{\Phi}(s), \tag{5}$$

and T_l is calculated analogously. Here, $\sqrt[+]{\ }$ denotes the branch of the square root with nonnegative real part.

An inverse Laplace transformation of (5) yields the *right TBC* at $x = X$:

$$\psi_x(X,t) = -\sqrt{\frac{2m^*}{\hbar\pi}}\, e^{-i\frac{\pi}{4}} e^{-i\frac{V_r}{\hbar}t} \frac{d}{dt} \int_0^t \frac{\psi(X,\tau)\, e^{i\frac{V_r}{\hbar}\tau}}{\sqrt{t-\tau}}\, d\tau \tag{6}$$

and analogously for the left artificial boundary at $x = 0$. These BCs are *non–local in* t and of memory–type, thus requiring the storage of all previous time levels at the boundary in a numerical discretization. A second difficulty in numerically implementing the continuous TBC (6) is the discretization of the singular convolution kernel. A simple calculation shows that (6) is equivalent to the *impedance boundary condition*:

$$\psi(X,t) = -\sqrt{\frac{\hbar}{2m^*\pi}}\, e^{i\frac{\pi}{4}} \int_0^t \frac{\psi_x(X,t-\tau)\, e^{-i\frac{V_r}{\hbar}\tau}}{\sqrt{\tau}}\, d\tau. \tag{7}$$

Integrating by parts in (6) and carrying out the t-derivative, one sees that the resulting kernel behaves like $\mathcal{O}(t^{-3/2})$ for $t \to \infty$. We remark that the TBC (7) was first derived in 1982 by Papadakis [31] in the context of underwater acoustics. Since the Schrödinger equation (2) has (formally) a similar structure as the heat equation, analogous DtN maps for the heat equation were already given by Carslaw and Jaeger [14] in 1959.

It is possible to extend the one dimensional TBCs (6) to rectangular geometries $(0,X)\times(0,Y)$ in 2D (cf. [8] for details). Now we consider the two dimensional time–dependent Schrödinger equation (1) on the infinite stripe $\Omega = \mathbb{R}\times(0,Y)$. The derivation of two dimensional TBCs is based on taking the partial Fourier series of ψ w.r.t. y:

$$\psi(x,y,t) = \sum_{m\in\mathbb{N}} \hat{\psi}_m(x,t) \sin\left(\frac{m\pi y}{Y}\right). \tag{8}$$

Again we assume that the potential V is constant in each of the two exterior domains: $V(x,y,t) = V_{\text{ext}}$ for $(x,y) \in \Omega\backslash\tilde{\Omega}$, $t \geq 0$. The time evolution of the modes $\hat{\psi}_m(x,t)$, $m \in \mathbb{N}$ is decoupled there. Hence, each mode satisfies at $x = 0$ and $x = X$ a one dimensional TBC

$$\frac{\partial}{\partial\eta}\hat{\psi}_m(x,t) = -\sqrt{2}e^{-i\pi/4}e^{-iV_m t}\sqrt{\partial_t}\left(e^{iV_m t}\hat{\psi}_m(x,t)\right), \quad m \in \mathbb{N}, \tag{9}$$

with the potentials $V_m := V_{\text{ext}} + \frac{1}{2}\left(\frac{m\pi}{Y}\right)^2$, and the unit outward normal vector η. Here, $\sqrt{\partial_t}$ denotes the fractional time derivative of order $1/2$ with the Fourier symbol $\sqrt{-i\omega}$.

Remark 2.1 *Note that the exterior potential V_{ext} in 2D may depend on y. Then, the orthogonal mode decomposition of* (8) *has to use the eigenfunctions of the stationary Schrödinger equation in y with* non–constant *potential $V(y)$ (cf. [12]).*

3. Discrete Transparent Boundary Conditions for the Two Dimensional Schrödinger Equation

The numerical discretizations of the artificial boundary conditions (6), (7) and (9) is delicate, as it may easily render the initial–boundary value problem only conditionally stable (e.g. [28]). DTBCs for a Crank-Nicolson finite difference discretization of the Schrödinger equation were first given in [6], [17] and [16] (cf. also [4] for a recent review of the various alternative approaches and [32], [33] for enhancements of the discrete TBCs for the Schrödinger equation). In this section we shall follow the "philosophy" of [6], [7], [17], [20] and *derive DTBCs*, instead of *discretizing the continuous TBC* (9).

For the derivation of the DTBCs we will now mimic the derivation of the continuous TBCs presented in Section 2. on a discrete level. First, we will use the unconditionally stable Crank–Nicolson time–discretization scheme combined with a compact nine–point discretization in space. The DTBCs will be constructed directly for the resulting difference equations.

3.1. The Difference Equations

We first consider the scaled time–dependent Schrödinger equation (TDSE)

$$\begin{aligned} i\frac{\partial}{\partial t}\psi(x,y,t) &= -\frac{1}{2}\Delta\psi(x,y,t) + V(x,y,t)\psi(x,y,t), \quad (x,y)\in\mathbb{R}^2,\, t>0, \\ \psi(x,y,0) &= \psi^I(x,y), \qquad (x,y)\in\mathbb{R}^2, \end{aligned} \tag{10}$$

on the whole space $\mathbb{R}^2$. For the derivation of the associated difference equations we introduce the equidistant grid $\Omega_{\Delta x,\Delta y}$ with the spatial grid points $x_j = j\Delta x$ and $y_k = k\Delta y$ for $j,\, k \in \mathbb{Z}$. We use the uniform time discretization $t_n = n\Delta t$, $n \in \mathbb{N}_0$. Hence, $\psi_{j,k}^n \sim \psi(x_j, y_k, t_n)$ denotes an approximation of the solution $\psi(x,y,t)$ of the Schrödinger equation (10) on the space–time–grid. Using the compact nine–point finite difference scheme in space combined with a Crank-Nicolson time–stepping, the discretized two dimensional scaled Schrödinger equation reads

$$\widetilde{D}^2\,\psi_{j,k}^{n+\frac{1}{2}} = \left(I + \frac{\Delta x^2}{12}D_x^2 + \frac{\Delta y^2}{12}D_y^2\right)\left[2V_{j,k}^{n+\frac{1}{2}}\psi_{j,k}^{n+\frac{1}{2}} - 2iD_t^+\psi_{j,k}^n\right], \tag{11}$$

with $j,\, k \in \mathbb{Z}$, $n \geq 0$. Here, we make use of the following difference operators

$$\begin{aligned} D_t^+\psi_{j,k}^n &:= \frac{\psi_{j,k}^{n+1} - \psi_{j,k}^n}{\Delta t}, \\ D_x^2\,\psi_{j,k}^n &:= \frac{\psi_{j-1,k}^n - 2\psi_{j,k}^n + \psi_{j+1,k}^n}{\Delta x^2}, \\ D_y^2\,\psi_{j,k}^n &:= \frac{\psi_{j,k-1}^n - 2\psi_{j,k}^n + \psi_{j,k+1}^n}{\Delta y^2}, \\ \widetilde{D}^2 &:= D_x^2 + D_y^2 + \frac{\Delta x^2 + \Delta y^2}{12}D_x^2\,D_y^2, \end{aligned}$$

the identity operator I, and the abbreviations

$$\psi_{j,k}^{n+\frac{1}{2}} := \frac{1}{2}\left(\psi_{j,k}^{n+1} + \psi_{j,k}^n\right), \qquad V_{j,k}^{n+\frac{1}{2}} := V\left(x_j, y_k, t_{n+\frac{1}{2}}\right).$$

It can be shown by Taylor series that the compact difference scheme (11) approximates the scaled Schrödinger equation (10) with the order $\mathcal{O}(\Delta x^4 + \Delta y^4 + \Delta t^2)$.

Theorem 3.1 (preservation of ℓ^2-norm, [33]) *Let the grid function $V^{n+\frac{1}{2}}$ be bounded for all $n \in \mathbb{N}_0$. For the whole space problems of the 2D time-dependent Schrödinger equation the scheme (11) then preserves the ℓ^2-norm*

$$\|\psi^n\|_{\ell^2(\mathbb{Z}^2)} := \sqrt{\Delta x\Delta y\sum_{j,k\in\mathbb{Z}}|\psi_{j,k}^n|^2} \tag{12}$$

in time.

3.2. Derivation of DTBCs for the Two Dimensional Schrödinger Equation

Here, we review the results obtained in [33]. First, we consider the scaled Schrödinger equation

$$\begin{aligned} i\frac{\partial}{\partial t}\psi(x,y,t) &= -\frac{1}{2}\Delta\psi(x,y,t) + V(x,y,t)\psi(x,y,t), \quad (x,y)\in\Omega,\, t>0,\\ \psi(x,y,0) &= \psi^I(x,y), \qquad (x,y)\in\Omega,\\ \psi(x,0,t) &= \psi(x,Y,t) = 0, \qquad x\in\mathbb{R},\, t>0, \end{aligned} \tag{13}$$

on the infinite stripe $\Omega = \mathbb{R}\times(0,Y)$ with some $Y>0$ (cf. Fig. 1). Let the initial function $\psi^I \in L^2(\Omega)$ be compactly supported on the computational domain $\tilde{\Omega}$:

$$\text{supp } \psi^I(x,y) \subset (0,X)\times(0,Y) =: \tilde{\Omega}.$$

Remark 3.2 *For the case that the initial data $\psi^I(x,y)$ is not compactly supported inside the computational domain $\tilde{\Omega}$ we refer the reader to [21].*

The potential $V(x,y,t)$ is assumed to be an $L^\infty(\tilde{\Omega}\times\mathbb{R}^+)$ function in space and time, and constant on each of the two exterior domains $\Omega^C := \Omega\backslash\tilde{\Omega}$. DTBCs will now be derived at the boundaries $x=0$ and $x=X$ of the computational domain $\tilde{\Omega}$.

We introduce the uniform grid $\Omega_{\Delta x,\Delta y} := \{(j\Delta x, k\Delta y)\mid j\in\mathbb{Z};\, k=0,\ldots,K\in\mathbb{N}\}$ with $x_J = X$, $y_K = Y$ and use the time steps $t_n = n\Delta t$, $n\in\mathbb{N}$. We approximate the TDSE (13) by the difference equation (11). Adapting the continuous strategy and the idea from [8] we take the explicit discrete solution on the exterior domain to eliminate the exterior problem. This is done using first a *discrete sine–transformation*

$$\widehat{\psi}^n_{j,m} := \frac{2}{K}\sum_{k=1}^{K-1}\psi^n_{j,k}\,\sin\left(\frac{\pi k m}{K}\right), \qquad m=1,\ldots,K-1, \tag{14}$$

in y-direction and then a $\mathcal{Z}$-transformation

$$\mathcal{Z}\left(\widehat{\psi}^n_{j,m}\right) := \Phi_{j,m}(z) := \sum_{n=0}^{\infty}\widehat{\psi}^n_{j,m}z^{-n} \quad \text{with } z\in\mathbb{C},\ |z|>1, \tag{15}$$

in the discrete time variable. The sine–transformed scheme (11) for the modes $\widehat{\psi}^n_{j,m}$, $m = 1,\ldots,K-1$, $j\le 0$ and $j\ge J$ reads

$$\begin{aligned} &\gamma_m\,\widehat{\psi}^{n+1}_{j+1,m} + \gamma_m\,\widehat{\psi}^{n+1}_{j-1,m} + \rho_m\,\widehat{\psi}^{n+1}_{j,m}\\ &\qquad = (2W-\gamma_m)\,\widehat{\psi}^n_{j+1,m} + (2W-\gamma_m)\,\widehat{\psi}^n_{j-1,m} + (\kappa_m-\rho_m)\,\widehat{\psi}^n_{j,m}, \end{aligned} \tag{16}$$

where we use the abbreviations

$$\begin{aligned}
D &:= \frac{\Delta x^2}{\Delta y^2}, \qquad C := \frac{\Delta x^2 + \Delta y^2}{12\Delta y^2}, \qquad W := \frac{i\Delta x^2}{3\Delta t},\\
\gamma_{j,m} &:= 1 + 2C\left(\cos\left(\frac{\pi m}{K}\right) - 1\right) + W - \frac{\Delta x^2}{6}V_j,\\
\kappa_m &:= 4\left(\cos\left(\frac{\pi m}{K}\right) + 4\right)W,\\
\rho_{j,m} &:= -2 - 2D + 4C + 8W - \frac{4\Delta x^2}{3}V_j\\
&\quad + \left(2D - 4C + 2W - \frac{\Delta x^2}{3}V_j\right)\cos\left(\frac{\pi m}{K}\right),
\end{aligned} \tag{17}$$

$m = 1, \ldots, K-1$, and V_j denotes the constant potential on Ω^C, which may take different values (i.e. V_0, V_J) on each outer domain. Performing the $\mathcal{Z}$–transformation of (16) we obtain

$$\Phi_{j+1,m}(z) + \left[\frac{\rho_{j,m}(z+1) - \kappa_m}{\gamma_{j,m}(z+1) - 2W}\right]\Phi_{j,m}(z) + \Phi_{j-1,m}(z) = 0, \tag{18}$$

$j \leq 0$, $j \geq J$, $m = 1, \ldots, K-1$. Note that the coefficients are constant in j on each part of the outer domain. For the derivation of (18) we have used the fact that the initial function has compact support on the computational domain, hence

$$\widehat{\psi}^0_{j+1,m} = \widehat{\psi}^0_{j-1,m} = \widehat{\psi}^0_{j,m} = 0, \qquad m = 1, \ldots, K-1,\ j = 0, J.$$

With the physical constraint that the solution $\psi^n_{j,k}$ of (11) decays for $|j| \to \infty$ we calculate the unique solution $\Phi_{j,m}(z) = (\nu_{J,m}(z))^j$ of equation (18). $\nu_{J,m}(z)$ denotes that solution of the characteristic equation

$$(\nu_{J,m}(z))^2 + \left[\frac{\rho_{J,m}(z+1) - \kappa_m}{\gamma_{J,m}(z+1) - 2W}\right]\nu_{J,m}(z) + 1 = 0,$$

which satisfies $|\nu_{J,m}(z)| < 1$. We note that this is always possible for $|z| > 1$. $\Phi_{j,m}(z)$ then fulfills the $\mathcal{Z}$-transformed DTBCs at $j = 0,\ J$ for each mode:

$$\Phi_{1,m}(z) = \frac{1}{\nu_{0,m}(z)}\Phi_{0,m}(z), \tag{19a}$$

$$\Phi_{J-1,m}(z) = \frac{1}{\nu_{J,m}(z)}\Phi_{J,m}(z), \tag{19b}$$

with

$$\nu_{j,m}(z) = \frac{-\rho_{j,m}(z+1) + \kappa_m + \sqrt{\zeta_{j,m}z^2 - 2\xi_{j,m}z + \theta_{j,m}}}{2\gamma_{j,m}(z - \eta_{j,m})}, \qquad j = 0, J. \tag{20}$$

Here we use that branch of the square root, which yields $|\nu_{j,m}(z)| < 1$ and we introduce the abbreviations

$$\begin{aligned}
\eta_{j,m} &:= \frac{2W}{\gamma_{j,m}} - 1,\\
\zeta_{j,m} &:= (\rho_{j,m})^2 - 4(\gamma_{j,m})^2,\\
\theta_{j,m} &:= (\kappa_m - \rho_{j,m})^2 - 4(\gamma_{j,m}\eta_{j,m})^2,\\
\xi_{j,m} &:= -(\rho_{j,m})^2 - 4(\gamma_{j,m})^2\eta_{j,m} + \rho_{j,m}\kappa_m,
\end{aligned} \tag{21}$$

$m = 1, \dots, K-1$. With some tedious work one can calculate analytically the $\mathcal{Z}$-inverse of (20): $\mathcal{Z}^{-1}(\nu_{j,m}(z))^{(n)} =: \ell_{j,m}^{(n)}$. We use the auxiliary function

$$F(z, \mu_{j,m}) := \frac{z}{\sqrt{z^2 - 2\mu_{j,m} z + 1}},$$

with

$$\mu_{j,m} := \frac{\xi_{j,m}}{\sqrt{\zeta_{j,m}}\sqrt{\theta_{j,m}}}, \qquad m = 1, \dots, K-1. \tag{22}$$

Using the abbreviations

$$\begin{aligned} \lambda_{j,m} &:= \frac{\sqrt{\zeta_{j,m}}}{\sqrt{\theta_{j,m}}}, \\ \tau_{j,m} &:= \frac{\theta_{j,m}}{\eta_{j,m}} - \zeta_{j,m}\eta_{j,m} - 2\xi_{j,m}, \qquad m = 1, \dots, K-1, \end{aligned} \tag{23}$$

we obtain

$$\frac{\sqrt{\zeta_{j,m} z^2 - 2\xi_{j,m} z + \theta_{j,m}}}{z - \eta_{j,m}} = \frac{1}{\sqrt{\zeta_{j,m}}}\Big(\zeta_{j,m} - \frac{\theta_{j,m}}{z\eta_{j,m}} + \frac{\tau_{j,m}}{z - \eta_{j,m}}\Big) F(z, \mu_{j,m})$$

by comparison of coefficients. Hence, we have

$$\begin{aligned} \nu_{j,m}(z) = &-\frac{\rho_{j,m}}{2\gamma_{j,m}} \frac{z}{z - \eta_{j,m}} - \frac{\rho_{j,m} - \kappa_m}{2\gamma_{j,m}} \frac{1}{z - \eta_{j,m}} \\ &+ \frac{1}{2\gamma_{j,m}} \frac{1}{\sqrt{\zeta_{j,m}}}\Big(\zeta_{j,m} - \frac{\theta_{j,m}}{z\eta_{j,m}} + \frac{\tau_{j,m}}{z - \eta_{j,m}}\Big) F(z, \mu_{j,m}), \end{aligned}$$

and its inverse $\mathcal{Z}$–transform $\ell_{j,m}^{(n)} = \Big(\mathcal{Z}^{-1}\Big[\frac{1}{\nu_{j,m}(z)}\Big]\Big)_n$ reads

$$\begin{aligned} \ell_{j,m}^{(n)} = &-\frac{\rho_{j,m}}{2\gamma_{j,m}}\eta_{j,m}^n - \frac{\rho_{j,m} - \kappa_m}{2\gamma_{j,m}}\Big(\eta_{j,m}^{n-1} - \frac{1}{\eta_{j,m}}\delta_n^0\Big) + \frac{\sqrt{\theta_{j,m}}}{2\gamma_{j,m}}\Big[\lambda_{j,m}^{1-n} P_n(\mu_{j,m}) \\ &- \frac{1}{\eta_{j,m}}\lambda_{j,m}^{-n} P_{n-1}(\mu_{j,m}) + \frac{\tau_{j,m}}{\eta_{j,m}\sqrt{\theta_{j,m}\zeta_{j,m}}} \sum_{k=0}^{n-1} (\lambda_{j,m}\eta_{j,m})^n P_k(\mu_{j,m})\Big], \ n \in \mathbb{N}_0, \end{aligned} \tag{24}$$

with the Legendre polynomials P_n ($P_{-1} \equiv 0$), the Kronecker symbol δ_n^0, and the abbreviations used in (17), (21), (22) and (23). The sine-transformed DTBCs at $j = 0$ and $j = J$ for the 2D discrete Schrödinger (11) follow with the inverse $\mathcal{Z}$-transformation of (19):

$$\widehat{\psi}_{1,m}^n - \ell_{0,m}^{(0)} \widehat{\psi}_{0,m}^n = \sum_{p=1}^{n-1} \ell_{0,m}^{(n-p)} \widehat{\psi}_{0,m}^p, \quad n \geq 1, \tag{25a}$$

$$\widehat{\psi}_{J-1,m}^n - \ell_{J,m}^{(0)} \widehat{\psi}_{J,m}^n = \sum_{p=1}^{n-1} \ell_{J,m}^{(n-p)} \widehat{\psi}_{J,m}^p, \quad n \geq 1. \tag{25b}$$

Remark 3.3 *The convolution coefficients $\ell_{j,m}^{(n)}$ are highly oscillatory as a function of the time step n. In [33] it is shown that the convolution coefficients $\ell_{j,m}^{(n)}$ given in* (24) *have the asymptotic behaviour*

$$\ell_{j,m}^{(n)} \sim \sigma_{j,m}\, e^{i\vartheta_{j,m} n} \tag{26}$$

as $n \to \infty$, with

$$\sigma_{j,m} := -\frac{\rho_{j,m}}{2\gamma_{j,m}} + \frac{\kappa_m - \rho_{j,m}}{2\gamma_{j,m}\eta_{j,m}} + \frac{\sqrt{\tau_{j,m}}}{2\gamma_{j,m}\sqrt{\eta_{j,m}}}, \qquad \vartheta_{j,m} = \arg(\eta_{j,m})$$

for $j = 0,\ J$ and $m = 1, \ldots, K-1$. This behaviour deviates from the $\mathcal{O}(t^{-3/2})$–decay of the continuous convolution kernel in (7). *Hence, it may lead to numerical cancellations in the calculation of the convolution sums* (25). *As an alternative we shall derive coefficients that decay like $\mathcal{O}(n^{-3/2})$. For the left DTBCs we therefore add equation* (25a) *for n and $n+1$ with the corresponding weighting factor* 1 *and $-e^{i\vartheta_{1,m}} = -\eta_{1,m}$ (the case $j = J$ is analogous) and proceed like in [17]. We define the* summed coefficients

$$s_{j,m}^{(n)} := \begin{cases} \ell_{j,m}^{(n)} - \eta_{j,m}\,\ell_{j,m}^{(n-1)}, & n \geq 1, \\ \ell_{j,m}^{(0)}, & n = 0, \end{cases} \tag{27}$$

for $m = 1, \ldots, K-1$; $j = 0$ and $j = J$. In Fig. 4 we give an example on the asymptotic behaviour of the convolution coefficients. The free Schrödinger equation is discretized with $J = K = 50$, $\Delta x = \Delta y = 0.02$ and $\Delta t = 2 \cdot 10^{-5}$. A solution ψ is calculated for $n = 1, \ldots, 250$ time steps. In Fig. 3.2. we present the real part of $\ell_{J,m}^{(n)}$ and the absolute value $|\ell_{J,m}^{(n)}|$ in Fig. 3.2. for all modes $m = 1, \ldots, K-1$. The errors $\mathrm{Re}\,(\sigma_{J,m}e^{i\vartheta n} - \ell_{J,m}^{(n)})$ *and $|\sigma_{J,m}e^{i\vartheta n} - \ell_{J,m}^{(n)}|$ between the convolution coefficients and the asymptotic expression* (26) *– which are converging to zero – are shown in Fig. 3.2. and Fig. 3.2..*

Real part (a) and absolute value (b) of the convolution coefficients $\ell_{J,m}^{(n)}$ and real part (c) and absolute value (d) of the error $\sigma_{J,m}e^{i\vartheta n} - \ell_{J,m}^{(n)}$ between the asymptotic expansion (26) and the convolution coefficients for the modes $m = 1, \ldots, K-1$ as a function of the time steps $n = 1, \ldots, 250$. We consider the computational domain $\tilde{\Omega} = (0,1) \times (0,1)$ and choose the discretization parameters $J = K = 50$, $\Delta x = \Delta y = 0.02$ and $\Delta t = 2 \cdot 10^{-5}$.

Theorem 3.4 (DTBCs for the 2D Schrödinger equation, [33]) *The sine-transformed DTBCs at $j = 0$ and $j = J$ for the discrete Schrödinger equation* (11) *read*

$$\widehat{\psi}_{1,m}^{n} - s_{0,m}^{(0)}\widehat{\psi}_{0,m}^{n} = \sum_{p=1}^{n-1} s_{0,m}^{(n-p)}\widehat{\psi}_{0,m}^{p} + \eta_{1,m}\widehat{\psi}_{1,m}^{n-1}, \tag{28a}$$

$$\widehat{\psi}_{J-1,m}^{n} - s_{J,m}^{(0)}\widehat{\psi}_{J,m}^{n} = \sum_{p=1}^{n-1} s_{J,m}^{(n-p)}\widehat{\psi}_{J,m}^{p} + \eta_{J-1,m}\widehat{\psi}_{J-1,m}^{n-1}. \tag{28b}$$

The coefficients $s_{j,m}^{(n)}$ for $j = 0, J$, $m = 1, \ldots, K-1$ are given by equation (27). *For $n \geq 2$, they can be calculated by the formula*

$$s_{j,m}^{(n)} = -\frac{\sqrt{\theta_{j,m}}}{2\gamma_{j,m}}\lambda_{j,m}^{1-n}\,\frac{P_n(\mu_{j,m}) - P_{n-2}(\mu_{j,m})}{2n-1}, \tag{29}$$

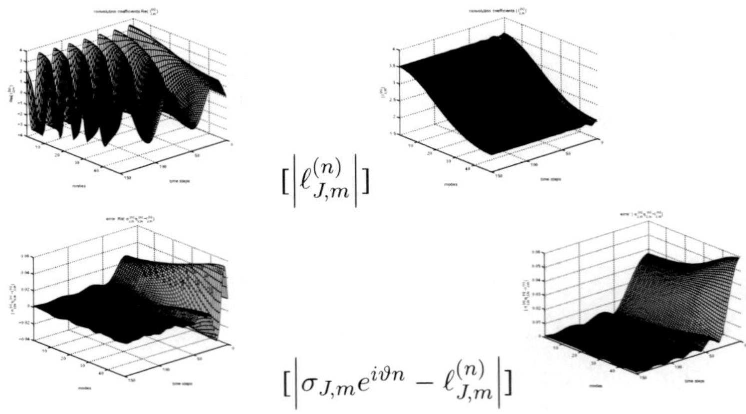

[Re $\left(\ell_{J,m}^{(n)}\right)$] [$\left|\ell_{J,m}^{(n)}\right|$]

[Re $\left(\sigma_{J,m}e^{i\vartheta n} - \ell_{J,m}^{(n)}\right)$] [$\left|\sigma_{J,m}e^{i\vartheta n} - \ell_{J,m}^{(n)}\right|$]

Figure 4. Real part (a) and absolute value (b) of the convolution coefficients $\ell_{J,m}^{(n)}$ and real part (c) and absolute value (d) of the error $\sigma_{J,m}e^{i\vartheta n} - \ell_{J,m}^{(n)}$ between the asymptotic expansion (26) and the convolution coefficients for the modes $m = 1, \ldots, K-1$ as a function of the time steps $n = 1, \ldots, 250$. We consider the computational domain $\tilde{\Omega} = (0,1) \times (0,1)$ and choose the discretization parameters $J = K = 50$, $\Delta x = \Delta y = 0.02$ and $\Delta t = 2 \cdot 10^{-5}$.

or by the recursion

$$s_{j,m}^{(n+1)} = \frac{2n-1}{n+1}\frac{\mu_{j,m}}{\lambda_{j,m}} s_{j,m}^{(n)} - \frac{n-2}{n+1}(\lambda_{j,m})^{-2} s_{j,m}^{(n-1)}, \tag{30}$$

for $j = 0,\ J$ and $m = 1, \ldots, K-1$. These new coefficients have the asymptotic behaviour

$$s_{j,m}^{(n)} \sim \mathcal{O}(n^{-3/2}). \tag{31}$$

Remark 3.5 *The idea of deriving DTBCs is to eliminate the exterior problem by using the explicit solution on the outer domain Ω^C. This is the reason for assuming a uniform grid on the exterior domain Ω^C. On the computational domain $\tilde{\Omega}$, however, the grid can be non–uniform, or even adaptive in time.*

Remark 3.6 *Recently, it was discovered by the authors that a more convenient formulation of* (19) *is given by*

$$\nu_{0,m}(z)\Phi_{1,m}(z) = \Phi_{0,m}(z), \quad \nu_{J,m}(z)\Phi_{J-1,m}(z) = \Phi_{J,m}(z).$$

Here, the inverse $\mathcal{Z}$-transformation of $\nu_{j,m}(z)$, $j = 0, J$ already decays like $\mathcal{O}(n^{-3/2})$. Instead of (28)*, this approach yields DTBCs with discrete convolutions at the 'interior' grid points $j = 1,\ J-1$ (cf. [9] for details).*

3.3. Approximation of the DTBCs by Sums of Exponentials

An ad-hoc implementation of the discrete convolution at the right boundary $x_J = X$

$$\sum_{p=1}^{n-1} s_{J,m}^{(n-p)} \widehat{\psi}_{J,m}^{p}, \qquad m = 1, \ldots, K-1$$

in (28b) with convolution coefficients $s_{J,m}^{(n)}$ from (27) has still one disadvantage. The boundary conditions of this kind are non–local both in time and space (w.r.t. the y-direction) and therefore computations are too expensive. As a remedy, to get rid of this time non–locality, we proposed already in [8] the *sum of exponentials ansatz*, i.e. to approximate the kernel (27) by a finite sum (say L terms) of exponentials that *decay* with respect to time. This approach allows for a fast (approximate) evaluation of the discrete convolution in (28b) since the convolution can now be evaluated with a simple recurrence formula for L auxiliary terms and the numerical effort now remains constant in time. On the Laplace–transformed level this approximation amounts to replace the symbol $\hat{s}_{J,m}(z) = \frac{z-\eta_{J,m}}{z}\nu_{J,m}(z)$ (cf. (27)) of the convolution by a rational approximation.

In the sequel we will briefly review this ansatz [8]. In order to derive a fast numerical method to calculate the discrete convolution in (28b), we approximate the coefficients $s_{J,m}^{(n)}$ for each mode m by the following ansatz (*sum of exponentials*):

$$s_{J,m}^{(n)} \approx \tilde{s}_{J,m}^{(n)} := \begin{cases} s_{J,m}^{(n)}, & n = 0, 1, \ldots, \upsilon_m - 1 \\ \sum_{l=1}^{L_m} b_{m,l} q_{m,l}^{-n}, & n = \upsilon_m, \upsilon_m + 1, \ldots, \end{cases} \tag{32}$$

where $L_m, \upsilon_m \in \mathbb{N}$ are a fixed numbers. Evidently, the approximation properties of $\tilde{s}_{J,m}^{(n)}$ depend on L_m, υ_m, and the corresponding set $\{b_{m,l}, q_{m,l}\}$. Thus, the choice of an (in some sense) optimal approximation would be a difficult nonlinear problem, which we do not pursue here. Instead, we propose below a deterministic method of finding $\{b_{m,l}, q_{m,l}\}$ for fixed L_m, υ_m and for each mode m.

The "split" definition of $\{\tilde{s}_{J,m}^{(n)}\}$ in (32) is motivated by the fact that the implementation of the right discrete TBC (28b) involves a convolution sum with p ranging only from 1 to $p = n - 1$. Since the first coefficient $s_{J,m}^{(0)}$ does not appear in this convolution, it makes no sense to include it in our sum of exponential approximation, which aims at simplifying the evaluation of the convolution. Hence, one may choose $\upsilon_m = 1$ in (32). The "special form" of $\ell_{J,m}^{(0)}$ and $\ell_{J,m}^{(1)}$ given in [8] suggests even to exclude $s_{J,m}^{(1)}$ from this approximation and to choose $\upsilon_m = 2$ in (32). We use this latter choice in our numerical implementation in the Example in the following §4..

Also, there is an additional motivation for choosing $\upsilon_m = 2$: With the choice $\upsilon_m = 0$ (or $\upsilon_m = 1$) we typically obtain (for each mode) two (or, resp., one) coefficient pairs $(b_{m,l}, q_{m,l})$ of big magnitude. These "outlier" values reflect the different nature of the first two coefficients. Including them into our discrete sum of exponentials would then yield less accurate approximation results.

Let us fix L_m and consider the formal power series:

$$g_m(x) := s_{J,m}^{(\upsilon_m)} + s_{J,m}^{(\upsilon_m+1)} x + s_{J,m}^{(\upsilon_m+2)} x^2 + \ldots, \quad |x| \le 1. \tag{33}$$

If there exists the $[L_m - 1|L_m]$ *Padé approximation*

$$\tilde{g}_m(x) := \frac{P_{L_m-1}(x)}{Q_{L_m}(x)}$$

of (33), then its Taylor series

$$\tilde{g}_m(x) = \tilde{s}_{J,m}^{(\upsilon_m)} + \tilde{s}_{J,m}^{(\upsilon_m+1)} x + \tilde{s}_{J,m}^{(\upsilon_m+2)} x^2 + \dots$$

satisfies the conditions

$$\tilde{s}_{J,m}^{(n)} = s_{J,m}^{(n)}, \qquad n = \upsilon_m, \upsilon_m + 1, \dots, 2L_m + \upsilon_m - 1, \tag{34}$$

due to the definition of the Padé approximation rule.

Theorem 3.7 ([8]) *Let $Q_{L_m}(x)$ have L_m simple roots $q_{m,l}$ with $|q_{m,l}| > 1$, $l = 1, \dots, L_m$. Then*

$$\tilde{s}_{J,m}^{(n)} = \sum_{l=1}^{L_m} b_{m,l} q_{m,l}^{-n}, \qquad n = \upsilon_m, \upsilon_m + 1, \dots, \tag{35}$$

where

$$b_{m,l} := -\frac{P_{L_m-1}(q_{m,l})}{Q'_{L_m}(q_{m,l})} q_{m,l} \neq 0, \qquad l = 1, \dots, L_m. \tag{36}$$

Remark 3.8 *Let us note that the assumption in Theorem 3.7 on the roots of $Q_{L_m}(x)$ to be simple is not essential. For multiple roots one only has to reformulate Theorem 3.7. All our practical calculations confirm that this assumption holds for any desired L_m, although we cannot prove this.*

Evidently, the approximation to the convolution coefficients $s_{J,m}^{(n)}$ by the representation (32) using a $[L_m - 1|L_m]$ Padé approximant to (33) behaves as follows: the first $2L_m$ coefficients are reproduced exactly, see (34). However, the asymptotic behaviour of $s_{J,m}^{(n)}$ and $\tilde{s}_{J,m}^{(n)}$ (as $n \to \infty$) differ strongly – algebraic versus exponential decay.

3.4. Fast Evaluation of the Discrete Convolution

Let us consider the approximation (32) of the discrete convolution kernel appearing in the right discrete TBC (28b). With these "exponential" coefficients the *approximated convolution*

$$\tilde{C}_{J,m}^{(n-1)} := \sum_{p=1}^{n-1} \tilde{s}_{J,m}^{(n-p)} \widehat{\psi}_{J,m}^{p}, \qquad \tilde{s}_{J,m}^{(n)} = \sum_{l=1}^{L_m} b_{m,l} q_{m,l}^{-n}, \tag{37}$$

$|q_{m,l}| > 1$, of a discrete function $\widehat{\psi}_{J,m}^{p}$, $p = 1, 2, \dots$, with the kernel coefficients $\tilde{s}_{J,m}^{(n)}$, can be calculated by recurrence formulas, and this will reduce the numerical effort significantly.

A straightforward calculation (cf. [8]) yields: The value $\tilde{C}_{J,m}^{(n-1)}$ from (37) for $n \geq 2$ is represented by

$$\tilde{C}_{J,m}^{(n-1)} = \sum_{l=1}^{L_m} \tilde{c}_{J,m,l}^{(n-1)}, \tag{38}$$

where

$$\tilde{c}_{J,l,m}^{(0)} \equiv 0,$$

$$\tilde{c}^{(n-1)}_{J,m,l} = q^{-1}_{m,l}\tilde{c}^{(n-2)}_{J,m,l} + b_{m,l}q^{-\nu_m}_{m,l}\psi^{n-2}_{J+1,m}, \tag{39}$$

$n = 2, 3, \ldots, l = 1, \ldots, L_m$.

Remark 3.9 (Transformation of approximated convolution coefficients, [8]) *Let $v_m = 2$. Let the approximated convolution coefficients $\tilde{s}^{(n)}_{J,m}$ and the coefficient pairs $\{b_{m,l}, q_{m,l}\}$ from Theorem 3.7 be given for a set $\{\Delta x, \Delta y, \Delta t, V\}$ for all modes $m = 1, \ldots, K-1$. Then, define for another parameter set $\{\Delta x^*, \Delta y^*, \Delta t^*, V^*\}$ the coefficients $\{b^*_{m,l}, q^*_{m,l}\}$ given by*

$$\begin{aligned}
q^*_{m,l} &:= \frac{q_{m,l}\bar{a}_m - \bar{b}_m}{a_m - q_{m,l}b_m}, \\
b^*_{m,l} &:= b_{l,m}q_{m,l}\frac{a_m\bar{a}_m - b_m\bar{b}_m}{(a_m - q_{m,l}b_m)(q_{m,l}\bar{a}_m - \bar{b}_m)}\frac{1+q^*_{m,l}}{1+q_{m,l}}, \\
a_m &:= 2\frac{\Delta x^2}{\Delta t} + 2\frac{(\Delta x^*)^2}{\Delta t^*} + i(\Delta x^2 V - (\Delta x^*)^2 V^*), \\
b_m &:= 2\frac{\Delta x^2}{\Delta t} - 2\frac{(\Delta x^*)^2}{\Delta t^*} - i(\Delta x^2 V - (\Delta x^*)^2 V^*).
\end{aligned}$$

The resulting convolution coefficients $\left(\tilde{s}^{(n)}_{J,m}\right)^*$ (obtained via (37)) are in practise good approximations for $\left(s^{(n)}_{J,m}\right)^*$, the exact convolution coefficients for the parameters $\{\Delta x^*, \Delta y^*, \Delta t^*, V^*\}$ (cf. (29) or (30)).

Finally we summarize the sum-of-exponentials approach by the following algorithm. For each mode $m = 1, \ldots, K-1$:

1. Prescribe L_m, ν_m, take $\Delta x = \Delta y = \Delta t = 1$ and calculate the coefficients $s^{(n)}_{J,m}$, $n = \nu_m, \nu_m + 1, \ldots, 2L_m + \nu_m - 1$ with (29) or (30).
2. Calculate $\{b_{m,l}, q_{m,l}\}$ and $\tilde{s}^{(n)}_{J,m}$ via Padé–algorithm.
3. For given Δx^*, Δy^*, Δt^*, V^* use Remark 3.9 with $\Delta x = \Delta y = \Delta t = 1$ and $\{b_{m,l}, q_{m,l}\}$ for the computation of $\{b^*_{m,l}, q^*_{m,l}\}$.
4. The corresponding coefficients $b_{m,l}$, $q_{m,l}$ are used for the computation of $\tilde{s}^{(n)}_{J,m}$ and for the efficient calculation of the discrete convolutions.

Steps 1 and 2 are made once and for all, see [8] for tables of coefficient pairs $\{b_{m,l}, q_{m,l}\}$ or `http://www.dtbc.de.vu/` for the implemented Padé algorithm (Maple code).

3.5. Implementation of the DTBCs

In (28) the DTBCs are written in sine–transformed space. A direct implementation in position space would necessitate tremendous numerical costs, hence they are implemented in y sine-transformed space (cf. [10]). The discrete convolution

$$\widehat{C}^{(n-1)}_{J,m} := \sum_{p=1}^{n-1} s^{(n-p)}_{J,m}\widehat{\psi}^p_{J,m}, \quad m = 1, \ldots, K-1 \tag{40}$$

for the right boundary $x_J = J\Delta x$ is calculated in sine-transformed space and inverse transformed by

$$C_{J,k}^{(n-1)} = 2\sum_{m=1}^{K-1} \sin\left(\frac{\pi mk}{K}\right)\Big(\sum_{p=1}^{n-1} s_{J,m}^{(n-p)}\widehat{\psi}_{J,m}^{p}\Big), \quad k = 1,\dots,K-1.$$

Since the convolution (40) only involves the solution at the boundary at past time levels (i.e. for $p \le n-1$), one can directly store the sine-transformed boundary data $\widehat{\psi}_{J,m}^{p}$. Moreover, this part of the DTBCs only enters the inhomogeneity of the linear system to be solved at each time level.

The part $s_{J,m}^{(0)}\widehat{\psi}_{J,m}^{n}$ of the left hand side of the DTBCs (28b) has to be inverse transformed to physical space and we get the couplings

$$\begin{aligned}\left(s_{J,m}^{(0)}\widehat{\psi}_{J,m}^{n}\right)^{\vee}_{J,k,l} &= 2\sum_{m=1}^{K-1} \sin\left(\frac{\pi mk}{K}\right) s_{J,m}^{(0)}\widehat{\psi}_{J,m}^{n} \\ &= \frac{2}{K}\sum_{m=1}^{K-1}\sum_{l=1}^{K-1} s_{J,m}^{(0)} \sin\left(\frac{\pi mk}{K}\right)\sin\left(\frac{\pi kl}{K}\right)\psi_{J,l}^{n}\end{aligned}$$

for $k,\, l = 1,\dots,K-1$. Hence, the 9-diagonal system of the discrete 2D Schrödinger equation (11) obtains additional entries due to the DTBCs.

In order to model the electron influx from the left lead, we shall prescribe an incoming plane wave $\varphi(x,y,t)$ at the left boundary. Hence, inhomogeneous DTBCs have to be used at $x_0 = 0$:

$$\begin{aligned}&\widehat{\psi}_{1,m}^{n} - \widehat{\varphi}_{1,m}^{n} - s_{0,m}^{(0)}\left(\widehat{\psi}_{0,m}^{n} - \widehat{\varphi}_{0,m}^{n}\right) \\ &\quad = \sum_{p=1}^{n-1} s_{0,m}^{(n-p)}\left(\widehat{\psi}_{0,m}^{p} - \widehat{\varphi}_{0,m}^{p}\right) - \eta_{1,m}\left(\widehat{\psi}_{1,m}^{n-1} - \widehat{\varphi}_{1,m}^{n-1}\right), \quad n \ge 1, \qquad (41)\end{aligned}$$

with the discrete, sine-transformed incoming wave $\varphi_{j,k}^{n}$, $0 \le k \le K, 0 \le j \le J$. These boundary conditions are implemented analogously to the right DTBCs at $x_J = X$.

4. Numerical Results

In this section we first present some rather mathematical examples on DTBCs for the Schrödinger equation in two dimensions. We verify numerically the accuracy of the DTBCs for the free Schrödinger equation. Then we apply the DTBCs to the simulation of quantum waveguides.

4.1. Travelling Gaussian Wave Functions

In this first example we solve the two dimensional, transient Schrödinger equation (1) discretized with the compact nine-point scheme (11) on the time-constant domain $\Omega = (0,1)^2$.

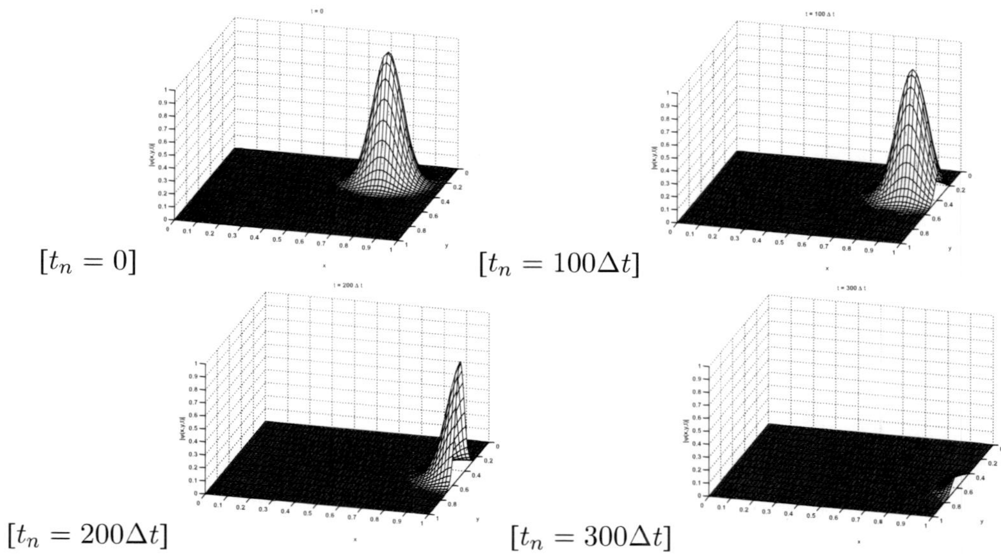

Figure 5. Absolute value of the initial function (42) and the absolute value of the solution to the Schrödinger equation at some time steps t_n calculated with exact DTBCs at $x = 0$ and $x = 1$. The wave impinges on the boundary at a non–orthogonal angle. The discretization parameters are $\Delta x = \Delta y = 1/120$ and $\Delta t = 2 \cdot 10^{-5}$.

As an initial function we choose the y-periodic Gaussian wave function

$$\psi^I(x,y) = \sum_{\ell \in \mathbb{Z}} (-1)^\ell e^{-\frac{\alpha}{2}\left[(x-x_0)^2+(y-y_0+\ell)^2\right]+ik_x x+ik_y y}, \quad (x,y) \in \Omega \tag{42}$$

with the parameters $\alpha = 240$, $x_0 = 3/4$, and $y_0 = 1/4$. As specified by the wave numbers $k_x = 140$, $k_y = 120$ the resulting wave has a non–orthogonal impact on the boundary (cf. Fig. 5 (b-d)). This is typically a "rough test" for TBCs, as high orthogonal solution modes then become significally coupled into the system. Exact DTBCs according to (28) are implemented at $x = 0$, $x = 1$. We consider the discretization parameters $\Delta x = \Delta y = 1/120$, $\Delta t = 2 \cdot 10^{-5}$. In Figure 4.1. we show the absolute value of the initial function (42). The evolution of this initial function according to the Schrödinger equation is presented in Figure 5 (b), (c), (d) for some times t_n. The Gaussian beam leaves the computational domain through the artificial boundary $x = 1$ without being reflected back. For the determination of the error due to the artificial boundary conditions we compare the numerical solution ψ with a numerical reference solution $\tilde{\psi}$ on $\tilde{\Omega} = (0,2) \times (0,1)$. The reference solution is calculated with the same discretization scheme, and with DTBCs at $x = 0$, $x = 2$. We obtain the relative L^2-error

$$L(t) = \frac{||\psi(.,.,t) - \tilde{\psi}(.,.,t)||_{\ell^2(\Omega)}}{||\psi^I(.,.)||_{\ell^2(\Omega)}}. \tag{43}$$

Within this test the error due to the cut–off of the initial function is also included. The effects of the artificial boundary at $x = 2$ should be negligible here, because ψ essentially

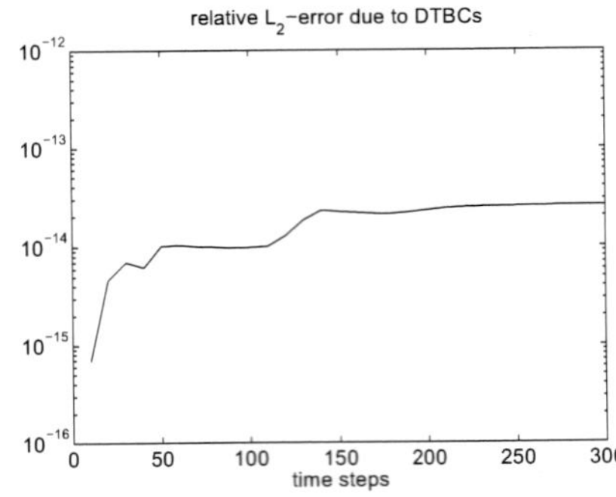

Figure 6. Relative error $L(\psi, \tilde{\psi}, t_n)$ due to the boundary conditions.

does not cross this boundary during the simulation period. In Figure 6 this error $L(t)$ is plotted. We remark that the magnitude of this error is about the rounding error of Matlab.

4.2. Quantum Waveguide Simulation

Here we will present a physical application of DTBCs. Artificial boundary conditions play an essential role in Schrödinger based simulations of the electron transport through quantum semiconductor devices. Typical examples of practical relevance include the ballistic transport along the channel of MOSFETs (cf. [27], [37]) or quantum waveguides (cf. [13]) for an analysis of T-shaped *quantum interference transistors*. These are novel electronic switches of nano–scale dimensions. They are made of several different layers of semiconductor materials such that the electron flow is confined to small channels or waveguides. Due to their sandwiched structure the relevant geometry for the electron current is essentially two dimensional.

Following the simulation of a GaAs–waveguide in [13], we choose the T–shaped geometry shown in Fig. 2 to simulate a quantum waveguide transistor. In x-direction the channel has a length of $X = 60$nm; the channel width Y and the stub width w are 20nm. In order to control the current through the channel, the stub length can be changed from $L_1 = 32$nm to $L_2 = 40.5$nm. Homogeneous DTBCs are implemented at $x = X$. An inhomogeneous DTBC at $x = 0$ (cf. (41)) models the prescribed influx of electrons. All other boundaries are considered as hard walls, i. e. we use homogeneous Dirichlet boundary conditions for ψ. A (discrete) time harmonic incoming wave function

$$\varphi_{j,k}^{n} = \sin\left(\frac{\pi k}{K}\right) e^{ik_x j\Delta x} e^{-\frac{iE_x n\Delta t}{\hbar}}, \qquad k = 0, \ldots, K \tag{44}$$

is modeling the mono-energetic constant incoming current at $x = 0$. Here, φ includes only the lowest transversal mode. But any linear combination of higher modes would work equally well, which is a great advantage compared to other artificial boundary conditions (e.g. [13]). In our example the energy E of the incoming wave equals 29.9meV and the effective electron mass has the value $m^* = 0.067m_0$, which corresponds to GaAs. In the subsequent simulations we are mostly interested in the switching and the large time behaviour of this waveguide. Therefore we first need to compute a stationary state corresponding to a given incoming plane wave function ψ^{Inc}. For this initialization process we choose the following (somewhat arbitrary) initial function

$$\psi^I(x,y) = \begin{cases} \sin\left(\frac{y\pi}{Y_1}\right) e^{ik_x x} & 0 \le x < x_1 \\ \frac{1}{2}\sin\left(\frac{y\pi}{Y_1}\right) e^{ik_x x}\Big[1+\cos\left(\pi\frac{x-x_1}{x_2-x_1}\right)\Big] & x_1 \le x < x_2 \\ 0 & x \ge x_2 \end{cases} \tag{45}$$

with $x_1 = 5$nm and $x_2 = 15$nm, which is consistent with the incoming wave. Then we solve the TDSE until stationarity is reached. The value of k_x can be derived from the discrete dispersion relation.

In the analytic case the dispersion relation for the free Schrödinger equation

$$i\hbar\frac{\partial}{\partial t}\psi = -\frac{\hbar^2}{2m_*}\Delta\psi + V(x,y,t)\psi, \quad (x,y)\in\Omega,\, t>0\,, \tag{46}$$

on a domain $\mathbb{R}\times(0,Y_1)$ with a plane wave solution in the first orthogonal mode (cf. (44)) reads

$$\epsilon(k_x) = \frac{\hbar^2 k_x^2}{2m_*} + \frac{\hbar^2\pi^2}{2m_* Y_1^2}, \tag{47}$$

which needs to be modified for the discretized Schrödinger equation. For a given inflow energy E, the value of k_x appearing in (45) can be derived from the following *discrete dispersion relation*. To derive it, we first put the ansatz $\psi_{j,1} = e^{ik_x j\Delta x}\sin\left(\frac{\pi\Delta y}{Y_1}\right)$, $j\in\mathbb{Z}$ into the spatial semi-discretization (by the compact nine-point scheme) analogous to (11):

$$\begin{aligned} E_{\mathrm{space}}(k_x) = \Big[&-\frac{\hbar^2}{m_*\Delta x^2}\left(\cos(k_x\Delta x)-1\right) - \frac{\hbar^2}{m_*\Delta y^2}\left(\cos\left(\frac{\pi\Delta y}{Y_1}\right)-1\right) \\ &-\frac{\hbar^2(\Delta x^2+\Delta y^2)}{6m_*\Delta x^2\Delta y^2}\left(\cos(k_x\Delta x)-1\right)\left(\cos\left(\frac{\pi\Delta y}{Y_1}\right)-1\right)\Big] \\ &\times\Big[1+\frac{1}{6}\left(\cos(k_x\Delta x)-1\right)+\frac{1}{6}\left(\cos\left(\frac{\pi\Delta y}{Y_1}\right)-1\right)\Big]^{-1}. \end{aligned} \tag{48}$$

This is the dispersion relation modified due to the spatial discretization. Adding now the correction due to the Crank-Nicolson time discretization yields the dispersion relation

$$E(k_x) = \frac{\hbar}{i\Delta t}\ln\left(\frac{2i\hbar - \Delta t E_{\mathrm{space}}(k_x)}{2i\hbar + \Delta t E_{\mathrm{space}}(k_x)}\right) \tag{49}$$

for the discrete Schrödinger equation (analogous to (11)) with a time-harmonic plane wave solution. For a detailed analysis of the discrete dispersion relation we refer to [33, 32].

For the following simulation we solve the Schrödinger equation (1) by the difference equation (11) without external potential, i.e. $V = 0$. For realistic simulations of MOSFET–channels, (1) should be coupled to the self–consistent Coulomb potential inside the channel. Since we focus on DTBCs, we shall not include this here. But a coupling to the Poisson

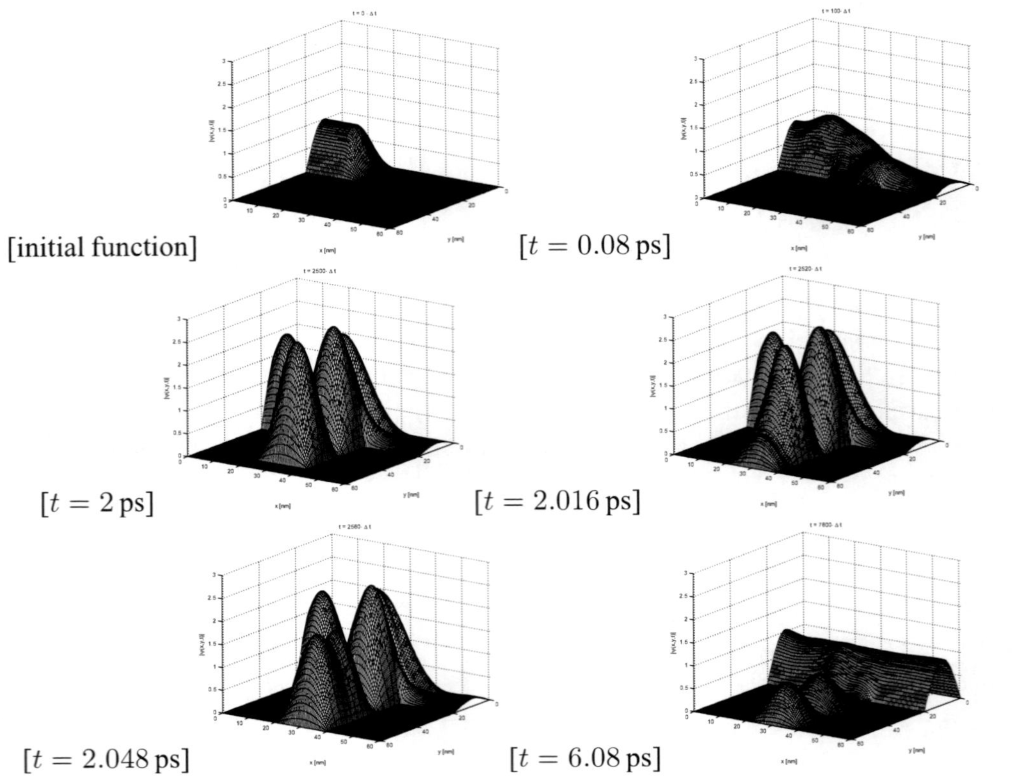

Figure 7. Absolute value of the solution $\psi(x, y, t)$ of the time-dependent Schrödinger equation (46) on the T–shaped structure from Figure 2. The discretization parameters are $\Delta x = \Delta y = 0.25$nm, $\Delta t = 0.8$fs, $V = -E = -29.9$meV, $m_* = 0.067 m_0$. (c) shows the steady state corresponding to the short stub with $L_1 = 32$nm. (f) is the steady state for the long stub with $L = 40.5$nm.

equation *inside* the computational domain does *not* change the derivation or discretization of our open BC (cf. [32]).

Fig. 7 shows the temporal evolution of the solution $|\psi(x, y, t)|$. In this simulation the stub length is first fixed to $L_1 = 32$nm. After 1.68ps the solution reaches (essentially) a steady state (*off-state* of the waveguide). Phenomenologically speaking, in this case only $1\frac{1}{2}$ wave packets "fit" into the stub (cf. Fig. 4.2.). Hence, they block the current flow through the waveguide. Then, at $t = 1.68$ps the stub is enlarged at once to $L_2 = 40.5$nm. After some transient phase, the solution converges to a new steady state (*on-state* of the waveguide, cf. Fig. 4.2.). Here, two wave packets "fit" into the stub, and the current can flow almost unblocked through the device.

Conclusion

In this chapter we have reviewed *discrete transparent boundary conditions* for the transient two dimensional Schrödinger equation. In particular, we discussed them for a fourth order

Numerov finite difference scheme. Their numerical efficiency is demonstrated in numerical tests on a rectangular geometry as well as for quantum waveguide simulations (for details cf. [33, 8, 10]).

Acknowledgements

The first author (A.A.) was partly supported by the DFG–project AR 277/3–3 and the Wissenschaftskolleg *Differentialgleichungen* of the FWF.

References

[1] K. Aihara, M. Yamamoto and T. Mizutani, Three–terminal conductance modulation of a quantum interference device using a quantum wire with a stub structure, *Appl. Phys. Lett.* **63** (1993), 3595–3597.

[2] G. Allaire, A. Arnold, P. Degond, T.Y. Hou, *Quantum Transport — Modelling, Analysis and Asymptotics*, Lecture Notes in Mathematics 1946, Springer, Berlin (2008).

[3] B. Alpert, L. Greengard and T. Hagstrom, Rapid evaluation of nonreflecting boundary kernels for time–domain wave propagation, *SIAM J. Numer. Anal.* **37** (2000), 1138–1164.

[4] X. Antoine, A. Arnold, C. Besse, M. Ehrhardt and A. Schädle, A Review of Transparent and Artificial Boundary Conditions Techniques for Linear and Nonlinear Schrödinger Equations, *Commun. Comput. Phys.* **4** (2008), 729–796. (open–access article)

[5] J. Appenzeller, C. Schroer, T. Schäpers, A. v.d. Hart, A. Förster, B. Lengler and H. Lüth, Electron interference in a T-shaped quantum transistor based on Schottky-gate technology, *Phys. Rev. B* **53** (1996), 9959–9963.

[6] A. Arnold, Numerically absorbing boundary conditions for quantum evolution equations, *VLSI Design* **6** (1998), 313–319.

[7] A. Arnold, Mathematical concepts of open quantum boundary conditions, *Transp. Theory Stat. Phys.* **30** (2001), 561–584.

[8] A. Arnold, M. Ehrhardt and I. Sofronov, Discrete transparent boundary conditions for the Schrödinger equation: Fast calculation, approximation, and stability, *Commun. Math. Sci.* **1** (2003), 501–556.

[9] A. Arnold, M. Ehrhardt, M. Schulte and I. Sofronov, *Discrete transparent boundary conditions for the Schrödinger equation on circular domains* , submitted to: Commun. Math. Sci., 2008.

[10] A. Arnold and M. Schulte, *Transparent boundary conditions for quantum-waveguide simulations*, to appear in Mathematics and Computers in Simulation (2008).

[11] V.A. Baskakov and A.V. Popov, Implementation of transparent boundaries for numerical solution of the Schrödinger equation, *Wave Motion* **14** (1991), 123–128.

[12] N. Ben Abdallah, F. Méhats and O. Pinaud, On an open transient Schrödinger-Poisson system, *Math. Models Methods Appl. Sci.* **15** (2005), 667–688.

[13] L. Burgnies, *Mécanismes de conduction en régime ballistique dans les dispositifs électroniques quantiques*, Ph. D. thesis, Université des Sciences et Technologies de Lille, 1997.

[14] H.S. Carslaw and J.C. Jaeger, *Conduction of heat in solids*, Clarendon Press, Oxford, UK, 1959.

[15] A. Dedner, D. Kröner, I. Sofronov and M. Wesenberg, Transparent Boundary Conditions for MHD Simulations in Stratified Atmospheres, *J. Comput. Phys.* **171** (2001), 448–478.

[16] M. Ehrhardt, *Discrete Artificial Boundary Conditions*, Ph.D. Thesis, Technische Universität Berlin, 2001.

[17] M. Ehrhardt and A. Arnold, Discrete Transparent Boundary Conditions for the Schrödinger Equation, *Riv. Mat. Univ. Parma* **6** (2001), 57–108.

[18] M. Ehrhardt and R.E. Mickens, Solutions to the discrete Airy equation: Application to parabolic equation calculations, *J. Comput. Appl. Math.* **172** (2004), 183–206.

[19] M. Ehrhardt and A. Zisowsky, Fast calculation of energy and mass preserving solutions of Schrödinger–Poisson systems on unbounded domains, *J. Comput. Appl. Math.* **187** (2006), 1–28.

[20] M. Ehrhardt and A. Zisowsky, Discrete non–local boundary conditions for Split–Step Padé approximations of the one–way Helmholtz equation, *J. Comput. Appl. Math.* **200** (2007), 471–490.

[21] M. Ehrhardt, Discrete transparent boundary conditions for Schrödinger–type equations for non–compactly supported initial data, *Appl. Numer. Math.* **58** (2008), 660–673.

[22] M. Ehrhardt and C. Zheng, Exact artificial boundary conditions for problems with periodic structures, *J. Comput. Phys.* **227** (2008), 6877–6894.

[23] D.K. Ferry and S.M. Goodnick: *Transport in Nanostructures*, Cambridge University Press, Cambridge (1997).

[24] W.R. Frensley Boundary conditions for open quantum systems driven far from equilibrium, *Rev. Mod. Phys.* **62** (1990), 745–791.

[25] L. Greengard and J. Strain, A fast algorithm for the evaluation of heat potentials, *Comm. Pure Appl. Math.* **43** (1990), 949–963.

[26] T. Hagstrom, Radiation boundary conditions for the numerical simulation of waves, *Acta Numerica* **8** (1999), 47–106.

[27] G. Jing and M. S. Lundstrom, A computational study of thin–body, double–gate, Schottky barrier MOSFETs, *IEEE Tran. on Elec. Dev.* **49** (2002), 1897–1902.

[28] B. Mayfield, *Non-local boundary conditions for the Schrödinger equation*, Ph. D. thesis, University Rhode Island, Providence, RI (1989).

[29] C.A. Moyer, Numerov extension of transparent boundary conditions for the Schrödinger equation in one dimension, *Amer. J. Phys.* **72** (2004), 351–358.

[30] C.A. Moyer, Numerical solution of the stationary state Schrödinger equation using discrete transparent boundary conditions, *Computing in Science and Engineering* **8** (2006), 32–40.

[31] J.S. Papadakis, Impedance formulation of the bottom boundary condition for the parabolic equation model in underwater acoustics, *NORDA Parabolic Equation Workshop, NORDA Tech. Note* **143** (1982).

[32] M. Schulte, *Numerical Solution of the Schrödinger Equation on Unbounded Domains* , Ph.D. Thesis, Universität Münster, 2007.

[33] M. Schulte and A. Arnold, Discrete transparent boundary conditions for the Schrödinger equation – a higher compact order scheme, *Kinetic and Related Models* **1** (2008), 101–125.

[34] I.L. Sofronov, Conditions for Complete Transparency on the Sphere for the Three–Dimensional Wave Equation, *Russian Acad. Sci. Dokl. Math.* **46** (1993), 397–401.

[35] I.L. Sofronov, Artificial Boundary Conditions of Absolute Transparency for Two- and Threedimensional External Time–Dependent Scattering Problems, *Euro. J. Appl. Math.* **9** (1998), 561–588.

[36] I.L. Sofronov, Non–reflecting inflow and outflow in wind tunnel for transonic time–accurate simulation, *J. Math. Anal. Appl.* **221** (1998), 92–115.

[37] J. Wang, E. Polizzi and M. Lundstrom, A three-dimensional quantum simulation of silicon nanowire transistors with the effective–mass approximation, *J. Appl. Phys.* **96** (2004), 2192–2203.

In: VLSI and Computer Architecture
Editor: Kenzo Watanable
ISBN 978-1-60692-075-6

Chapter 5

REFERENCE ARCHITECTURE MODEL AND TOOLS FOR MULTI-MODAL PERCEPTUAL SYSTEMS

***Jan Kleindienst*[*] *and Jan Cuřín*[†]**
IBM Research, Prague, Voice Technologies and Systems
V Parku 2294/4, Praha 4 - Chodov, 14800, Czech Republic

Abstract

Designing multi-modal perceptual systems is a non-trivial endeavor that requires interdisciplinary effort, solid architectural infrastructure, and efficient development tools. We introduce an architecture for building intelligent services in smart rooms – i.e., environments equipped with a collection of sensors ranging from low-level fire detectors to complex perceptual components processing signals from an array of cameras and microphones. The architecture presents a unified approach towards designing, implementing and testing multi-modal perceptual systems. We describe the reference architecture we co-authored for such applications, and present a set of tools we have developed to facilitate the collaborative building of such context-aware applications and services. We also show the benefits of the tools for application authoring in the context of the presented architecture. Specifically, we introduce a modeling tool that captures the use cases in a form of virtual animated 3D scenarios, facilitates the controlled triggering of sensor events at proper time intervals as reactions to scenario activities, and provides means to aggregate such events into higher-level information – i.e., the situation model – utilized by the contextual services. Our tool supports direct replacement of virtualized sensors by real sensors and perceptual components once they become available in the development cycle, without the necessity to change the API contracts. We have used the architectural principles and the tooling successfully in two large international projects dealing with multi-modal perception. In addition, the reference architecture has passed a process of formal and functional evaluation.

1. Introduction

The vision providing the bounding box for the work introduced herein, aims at developing next-generation services that assist humans in a natural and unobtrusive way by utilizing

[*]E-mail address: jankle@cz.ibm.com
[†]E-mail address: jan_curin@cz.ibm.com

the state-of-the-art in machine perception and context modeling.

Smart spaces are physical environments equipped with perceptual sensors and devices that provide contextual information to applications and services helping them to react to people in a natural manner [1]. Applications in smart spaces rely on highly distributed and heterogeneous components consisting of a wide range of hardware and software elements. Building and deploying services over such infrastructures complicates the task for the developers, which have to allocate considerable effort to integration.

The key task of the work was to come up with an architecture description and tools supporting development of applications in the smart space environment, where the distributed perceptual components (sensors, cameras, microphones) capture and process information in the form of a contextual knowledge utilized by the application logic. Several key design principles govern the design of the architecture for multi-modal perceptual systems, such as:

- Architecture that facilitates development, integration, and debugging, while minimizing reengineering and dependencies on specific technologies and platforms.
- Standardized and disciplined approach, so that partners can draw on contributions from the other partners, exchanging and sharing components between sites.
- Support for integration and common maintenance of perceptual components, context-aware modules, communications middleware, and other software elements into intelligent services.
- Good metaphors for modeling intelligent sensors and context-awareness, as these are the key prerequisite for building intelligent services in multi-modal way.
- Approaching the architecture design is an iterative process, hence the architecture evaluations and continuous developer feedback are inherent parts of the design process.

The design principles formulated questions along various design dimensions, including functional and non-functional requirements, e.g.:

- Interoperability, e.g., How to achieve structured component integration? What is the component model of such integration? What APIs? What is the meaning of architecture compliance and how it helps in supporting the exchange of components?
- Ubiquity & Distribution, e.g., What is the right middleware model? How to efficiently support high bandwidth data transfers from sensors to applications (e.g., showing a camera stream on a user mobile device)? What is the most suitable paradigm for the event passing mechanism in such a system? What is the approach for fault-tolerance, auto-configuration, etc.?
- Information Control and Data Flows, e.g., How does information travel through the system? What are data and control paths? Where is the context captured and processed?

- Data sources and data sinks, e.g., What are the proper abstractions for sensors and actuators? Is it possible to find a common metaphor to define a basic lifecycle on the plethora of sensors and actuators?
- Tooling, e.g., What are the add-value tools and utilities for component integration? How to maximize the impact of these tools to streamlining the process of application design across sites and partners?
- Meta-architecture, e.g., What is the process of developing and cultivating the architecture blueprints in a multi-partner environment including both industry and academia? To what extent should we reuse existing solutions and to what extend should we develop from scratch? What are the prospects of extensibility given the about decisions?

Many projects have developed architectural frameworks that facilitate integration in complex context-acquisition environments, e.g., UbiREAL [2], Gaia [3], or Context Toolkit [4]. But in contrast to these systems, which integrate contextual information directly from various sensors, we needed a system which relies on information provided by more complex perceptual components, i.e., more complex context-acquisition components such as person trackers and speech recognizers. A body tracker might be at once capable of detecting location, heading and posture of persons, identifying them, and tracking subjects of their interest. One particular design decision we emphasize is related to the situation modeling aspects of the architecture design introduced in Section 2. Such scenarios led us to separate the perceptual components layer from the layer that deals with higher abstraction – the situation modeling. Defining and modeling situations based on a wide range of context-acquisition components was not supported by other environments, so we decided to implement a new framework, called SitCom, which is described in Section 3. Our system is used as an integrator for the perceptual component providers and is capable of dynamically exchanging perceptual components of the same kind, as well as enabling the seamless replacement of these components by simulators, a requirement from service developers. Section 4. illustrates the application of the principles and the tooling in two projects dealing with multi-modal perception. Section 5. summarizes the features of the presented system.

2. The Reference Architecture Model

The construction of contextual services demands an interdisciplinary effort because it has to integrate at minimum three related parts: environment sensing, situation modeling, and application logic. Thus, even a simple system built for an intelligent room requires at least three different developer roles, each with a different set of skills:

- **Perceptual Technology Providers** supply sensing components such as person trackers, sound and speech recognizers, activity detectors, etc., to see, hear, and feel the environment. The skills needed here include signal processing, pattern recognition, statistical modeling, etc.
- **Context Model Builders** make models that synthesize the flood of information acquired by the sensing layer into semantically higher-level information suitable for

user services. Here, the skills are skewed towards probabilistic modeling, inferencing, logic, etc.

- **Service Developers** construct the context-aware services by using the abstracted information from the context model (the situation modeling layer). Needed skills are application logic and user interface design; multi-modal user interfaces typically require yet an additional set of roles and skills.

To illustrate how these roles are exercised during the process of service construction, we use an example of a service developer creating a new context-aware service that determines if people gathered in a meeting room are having a presentation or if they are taking a coffee-break. This would for example let staff enter the room only when they do not disturb.

The service developer does not, typically, have access to a fully-equipped room with sensors, nor does she have the technology for detecting people and meetings. But for an initial prototype of such a service, pre-recorded input data may suffice.

A service developer then talks to a context-model builder to define what new contextual abstractions will be needed, and which can be reused from an existing catalogue. The context-model builder in turn talks to technology providers to verify that capturing the requested environment state is within state-of-the art of the sensing technology. If not, there is still a possibility to synthesize the requested information from other sensors under a given quality of service; for example, the number of people in a room may be inferred through sound and speech analysis in case no video analysis technology is available.

One of the key features of our approach is an effective separation of the effort of different development groups facilitating the possibility of easy integration. This resulted in the layered architecture design—*Reference Architecture Model* [5].

The *Reference Architecture Model* (see schema in Figure 1) provides a collection of structuring principles, specifications and *Application Programming Interfaces* (APIs) that govern the assemblage of components into highly distributed and heterogeneous interoperable systems.

Each level derives new information from the abstraction of the information processed and data flow characteristics such as latency and bandwidth, based on the functional requirements from the system design phase.

The vertical columns – utilities and ontology – provide support across layers. The utilities provide global timing and other basic services that are relevant to all layers. The *Ontology* provides a definition of modeled concepts and a directory service that gives access to information at any layer. It is worth to emphasize that the layers are not strictly isolated. For example, the multi-modal service residing at the user front-end can render a video stream originating from a logical sensor at the bottom of the model's schema. Now, we shortly describe the individual layers of the *Reference Architecture Model*.

2.1. Low-level Distributed Data Transfer

The control and metadata layers provide mechanisms for data annotation, synchronous and asynchronous system control, effective storing and searching multi-media content and metadata generated by data sources. The important capability of this layer is a synchronization of various data streams.

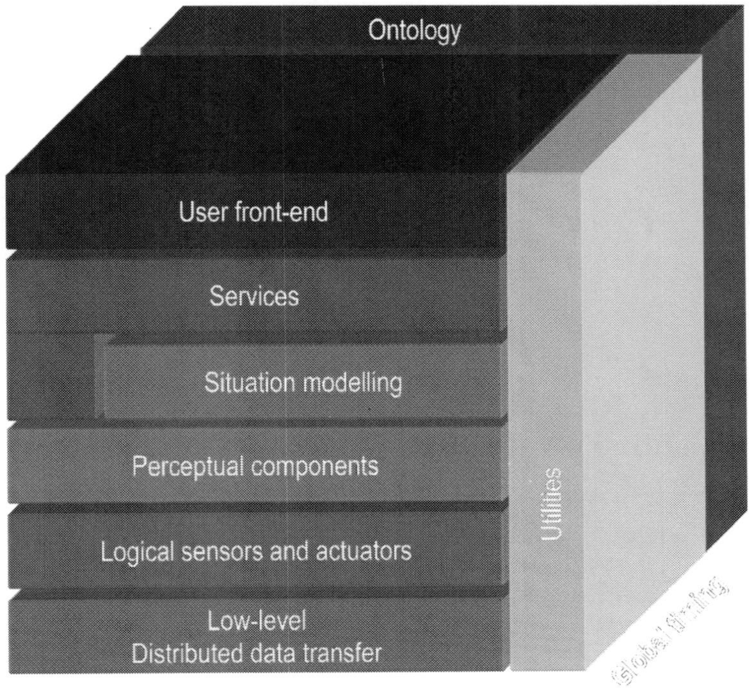

Figure 1. Schema of the *Reference Architecture Model*.

For transferring various low and high bandwidth data streams from producers (sensors, cameras, etc.) to consumers (perceptual components, services, etc.) in real-time with a desired quality, the system may use for example the NIST SmartFlow middleware [6] or the *CHIL Flow* [7].

2.2. Logical Sensors and Actuators

The components at the *Logical Sensors and Actuators* layer are either sensors feeding their data to the *Perceptual Component* layer, or various actuators, such as output devices receiving data or steering mechanisms receiving control instructions. The aim of this layer is to organize the sensors and actuators into classes with well-defined and well-described functionality, to make them available remotely over a network and to standardize their data access. The sensors will be classified mainly based on the kind of output stream they produce and the actuators will be classified mainly based on the input stream they consume.

2.3. Perceptual Components

The *Perceptual Components* extract meaningful events from continuous streams of video, audio and other sensor signals. All kinds and variations of perceptual components live at this layer to perceive the user actions in a smart environment. In one project, we counted a total of 64 such components provided by a dozen of vendors. These components process the sensor signals from one modality (body trackers, face recognizers) or modality combinations (audio-visual speech-recognition) and deliver events to upper layers. The *Reference*

Architecture defines the API contract (Access, Subscriber, Control, Introspection, and Admin APIs) as well as the lifecycle (Unregistered, Registered, Launched, Running) to which the perceptual components must adhere to be considered compliant.

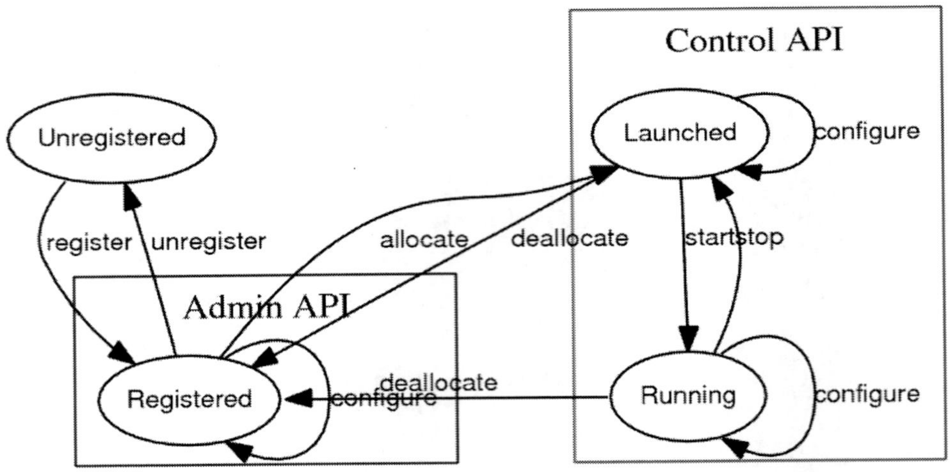

Figure 2. The Perceptual Component's APIs and lifecycle.

A key aspect of perceptual components is that they implement a specific API and a well-defined lifecycle (see Figure 2). The API is defined along *functional* aspects of the perceptual components, e.g., *body location tracker API*, so that a body tracker may do its tracking based on video signals, audio signals, or even be a combination of other monomodal trackers. The important part is that the API represents the type of information that we can get from the component, not how it got computed.

Such an API also presents the advantage that certain components, which may deliver multiple types of information (e.g., speaker ID and audio transcription), will implement multiple simple APIs (e.g., a speaker ID API and a transcription API) instead of one single-purpose combined API. This simplifies the situation model, as it can then request information from only one aspect of such a component. It also reduces the set of APIs we have to deal with, directly benefiting the integration and deployment of the system.

The component lifecycle contains several states that help preserve resources for components that may be used but are not needed at a particular point in time. When a component is registered, it can be remotely activated, but does not allocate any resource before that point. Some components have non-immediate initialization times, like speech recognizers or body trackers, so that it makes sense for them to decouple resource allocation from initialization.

2.4. Situation Modeling

The *situation modeling* layer is the place where the situation context received from audio and video sensors is processed and modeled. The context information acquired by the components at this layer helps services to respond better to varying user activities and environment changes. For example, the Situating Modeling answers questions such as *Is there a meeting going on in the smart room? Who is the person speaking at the whiteboard? Has this person been in the room before?* The situation modeling layer is implemented, in this *Reference Architecture*, by the SITCOM framework.

This layer is also a collection of abstractions representing the environment context in which the user interacts with the application. Ideally, it should act as a database that maintains up-to-date state of objects (people, artifacts, situations) and their relationships. The situation model itself acts as a kind of directed inference engine, which will watch for the occurrence of certain *situations* in the current environment. The searched-for situations are dictated by the needs of active services, as for the detection of the level of interruptibility of a particular person in a room for whom a service has to monitor incoming calls.

In the *Reference Architecture*, the situation model takes its information from the perceptual components and produces a set of situation events that are provided to services. There is no particular mandatory implementation of such a model, and therefore we have provided both a framework for handling situation models and an implementation based on *situation machines*, described in Section 3.

2.5. Services and User Front-end

The *user services* are implemented as software agents and manage interactions with humans by means of user services hosted on various interaction devices. The components at this level are responsible for communicating with the user and presenting the appropriate information at appropriate time-spatial interaction spots. The User Services utilize the contextual information available from the situation modeling layer.

Moreover, the *services* layer provides reusable components that help users solve tasks quickly and easily. These reusable components constitute elementary services (constructs/entities) that can be aggregated to form services representing non-obtrusive services.

The *user front-end* layer comprises service agents, and their management, by encapsulating/wrapping the underlying agent platform. This layer performs the user specialized requirements of the *user front-end-layer* to the underlying *services-layer* by agent-based mechanisms. This layer also provides access to the knowledge-base, so that a common ontology can be used for communication and interpretation of contents.

The user interface layer contains the *interface agents* [8], which act as personal assistants of users. *interface agents* take care of the demands of users by subscribing them to the *Services* they want to use. To do this, the *interface agent* interprets the current role profile of a user and its specific settings in the user profile or reacts to a direct input of a user.

3. SitCom: Tool for Perceptual Systems Developers

Since situation modeling is the layer where perceptual information meets the application logic, it is the natural place from which to start application development.

A closer look at the layered architecture in Figure 1 reveals that the layer that primarily deals with context acquisition and processing – the *situation modeling* layer – isolates (and at the same time glues) two fundamental infrastructure parts: the perceptual technologies, as the providers of perceptual and context information; and the context-aware services, as the primary consumers of context and perceptual data.

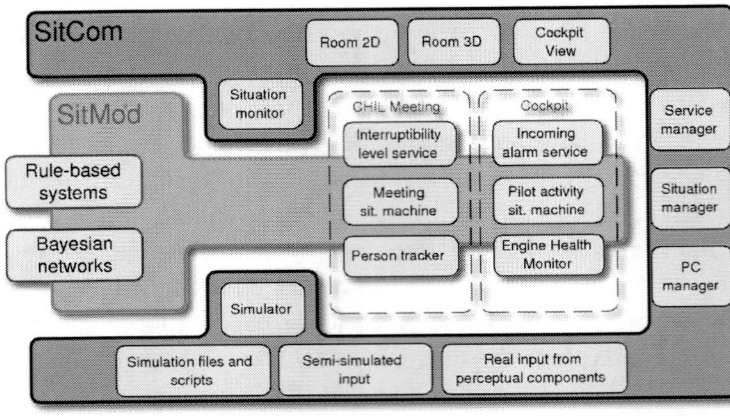

Figure 3. Schema of SITCOM Pluggable Modules.

Due to their different roles and functions in the system, services and technologies typically have different development life-cycles, and they show a tendency to evolve independently, especially during the early stages of the development cycle. By the specification and cultivation of the conceptual abstractions and components that link these two subsystems at the *situation modeling* layer, the software architects ensure that both services and technologies converge and seamlessly “plug in” during the later stages of the cycle.

To accommodate this task, the need for a conceptually new software utility emerged. Such a utility performs the role of a “Swiss-army knife” acting as a multi-function tool utilized by different developer roles at different stages of the development cycle. To deal with the development and integration of heterogeneous context-aware applications, we have built a tooling platform called SITCOM.

SITCOM (Situation Composer) [9] is a 3D simulation tool and runtime for the development of context-aware applications and services. As stated previously, context-aware applications draw data from the surrounding environment (such as an ongoing meeting in a room, a person location, body posture, etc.) and their behavior depends on the respective situation (e.g., while in meeting, make the phone silent). In SITCOM, the environment characteristics are captured by *situation machines* (introduced later in the text) that receive input events from real sensor inputs (cameras, microphones, proximity sensors, etc.), simulated data, or a combination of real and simulated inputs. SITCOM allows for the composition of *situation machines* into hierarchies to provide event filtering, aggregation, and combination to construct higher-level meaning. Through the IDE controls, SITCOM also facilitates the capture and creation of situations (e.g., a 10-minute sequence of several people meeting in a conference room) and their subsequent realistic rendering as 3D scenarios. These scenarios can be re-played to invoke situations relevant for the application behavior, and thus provide a mechanism for a systematic testing of the context-aware application under different environment conditions.

Figure 3 shows SITCOM’s main building blocks: the Situation Composer, and *SitMod*, the Situation Model framework.

3.1. Controlling SitCom's Input

SITCOM operates with both real and simulated input. Simulated input includes recorded data, which can be either synthetic or real. Simulated input is particularly important at the beginning of a pervasive system project, when very sparse data is usually available for most parts of a complex system. In such cases, working with synthetic data is the best way to bootstrap the development of situation modeling and service design. S ITCOM supports the creation of synthetic data, while also allowing service testing and simulation with such data. Thus, simulated input can be split into two aspects: input data generation and input data use.

Along with generating data, SITCOM can also record and/or store data using the same data formats. Data recordings can be useful in recording services in operation, and subsequently replay or post-process them. Using SITCOM, recorded events and services can be viewed multiple times based on different views (e.g., cameras) and situation models. SITCOM is oblivious of the origin of the events that it receives. This provides support for interactive simulated input, i.e., input which mixes real-time live information with recorded information. Interactive simulated input is handy in cases where some part of the system has to be tightly controlled (e.g., for recording or demonstration purposes). Information mixing in SITCOM can be based on the following methods:

- Manual triggering: This involves the lower level of interactivity, where a button is used to start a particular scenario. Manual triggering is most useful in a demonstration mode, as one can simulate any event that might be of interest for a scenario.

- Manually inserting entities: This allows SITCOM users to interact with the world model as perceived by the situation model. For example, in a meeting simulation, one could manually add a person or two, and observe how the situation model or service reacts to this new event stream.

- Interacting with a mock-up: This provides the highest level of interactivity through having the models use a mock-up of a scenario. The mock-up will make entities behave in some programmed way. An example would be to simulate an incoming call to a (live) participant, and having the participant decide if she would answer. The mock-up would then have to deal with that fact, and may be calling back later if the call has not been answered, or playing a voice and not calling back if the participant has answered.

All three methods have been proved to be very valuable tools for the verification of situation models, demonstrations, as well as service development.

3.2. Internal Representation Structures

SITCOM relies on data representation structures, which are specially designed to bridge context information and pervasive services in smart spaces. The basic unit visible at the context-acquisition level is an entity. An *entity* is characterized by its type (e.g., person, whiteboard, room) and access to a set of property streams. A *property stream* represents a set of events for a particular aspect of the entity. As an example, properties for an entity of the type person are location (indicating the physical location of the person), heading (tracking direction the person is looking in), and identity (i.e., a stream sending events about the identification of the person). At the *Situation Modeling* layer, entities are stored in a container called entity repository. The *Entity repository* is capable of triggering events upon creation and registration of new entities. Accordingly, situation models and services can be notified about changes for a particular *entity* or *stream*, based on an appropriately designed subscription mechanisms.

For processing contextual information, the concept of a situation network [10] is used – we introduce the abstraction of *situation machines*. *Situation machines* (SM) interpret information inferred from observing entities in the entity repository and current states of other situations. As SMs may use different techniques, such as rule-based or statistical approach, SITCOM can accommodate a mix of models at the same time. More details about the actual implementation are given in Section 4.1.

3.3. SitCom's Graphical Presentation

SITCOM users, typically developers of context-acquisition components or services, are provided with a rich GUI. The user can allocate, start and stop real or simulated perceptual components, load services and *situation machines*, and observe the current state of entities and streams in various SITCOM views (including 3D visualization). It is possible to add user defined renderers for displaying scenario-specific streams of entities, and to add special views for showing the current states of *situation machines*.

The tool, specifically the inner module called *SitMod*, provides both pulling and pushing APIs to services. The pulling API of a *situation model* is done by direct calls to Java methods, and the pushing API by an eventing and listeners mechanism.

Figure 4 shows the SITCOM graphical environment with both 2D and 3D visualizations and one camera view corresponding to a real situation in IBM smart room. Three services are running in this particular configuration: the Occupancy service tracking participants of the meeting (see the output of the Attendance SM in the Situation Summary panel), the Video service allowing on-line displaying of camera streams, and the Connector service for determining the interruptibility of a particular person.

4. Application Design Use-cases

4.1. CHIL Connector

We will illustrate the use of our framework on a CHIL[1] Connector scenario proposed and exploited in [11]. The connector service is responsible for detecting acceptable interruptions (phone call, SMS, targeted audio, etc.) of a particular person in the smart room. During the meeting, for example, a member of the audience might be interrupted by a message during the presentation, whereas the service blocks any calls for the meeting presenter.

In contrast to classic application development, the developer of a context aware application must be concerned with things beyond the plain application logic. This is because

[1]CHIL (Computers in the Human Interaction Loop) – an integrated project (IP 506909) under the EC's Sixth Framework Programme finished in October 2007.

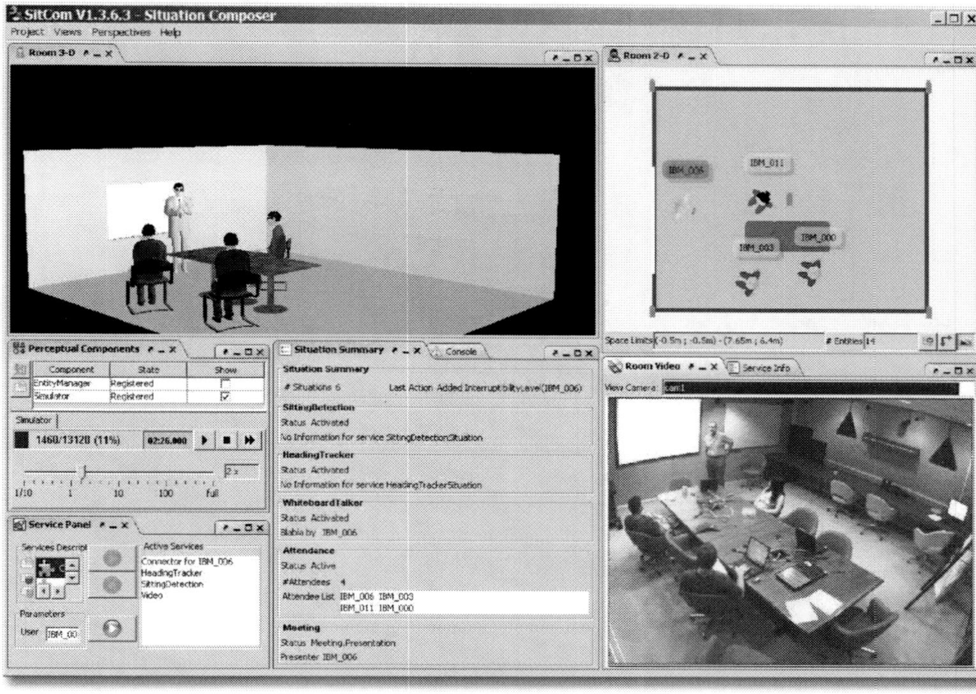

Figure 4. Screenshot of SITCOM running services on meeting data recorded in the IBM smart room.

she must also define what is the contextual information to be exploited by her application, which obviously drives the requirements on what perceptual components are needed to extract such information from the available sensors (see Figure 5). These tasks are supported by the SitCom tool as described further.

4.1.1. Perceptual Components

The smart room is equipped with multiple cameras and microphones on the *sensor* level. The audio and video data are streamed into the following *perceptual components*, as depicted in Figure 5:

- **Body Tracker** is a video-based tracker providing 3D coordinates of the head centroid for each person in the room;
- **Facial Features Tracker** is a video-based face visibility detector, providing nose visibility for each participant from each camera;
- **Automatic Speech Recognition** is providing speech transcription for each participant in the room.

SITCOM tool presents these perceptual components in its GUI, and provides the status info about each component. For example, the Body Tracker and Face Tracker are already

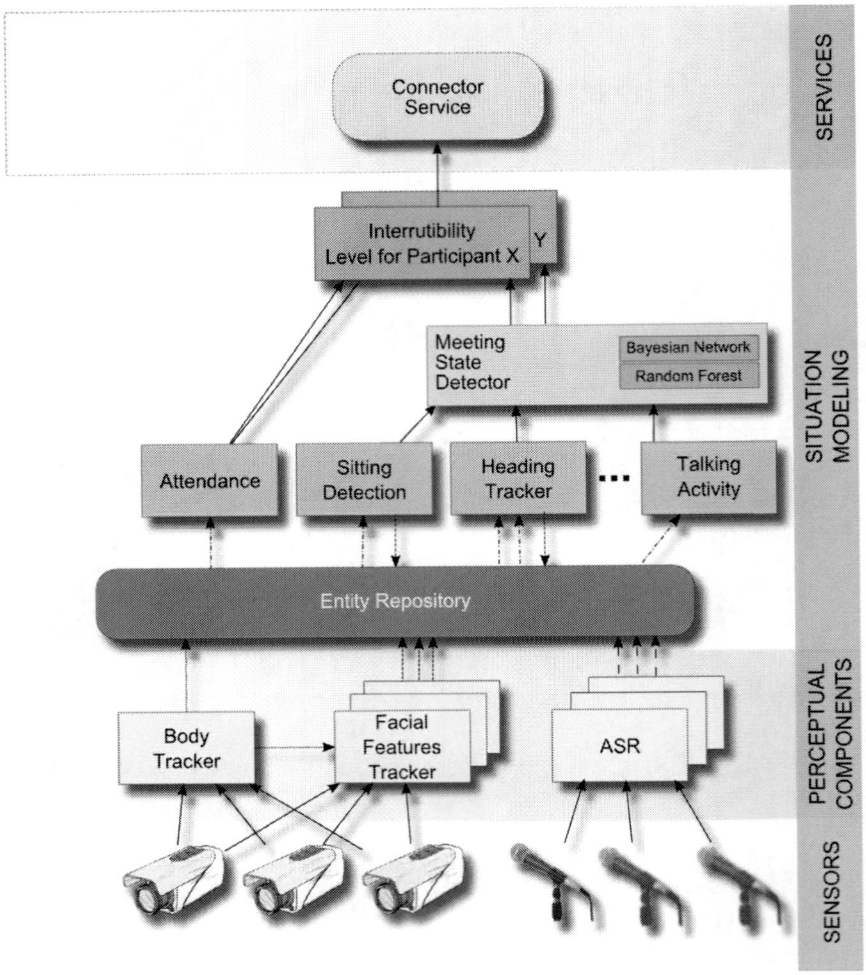

Figure 5. Information flow diagram for meeting scenario.

running and serving input to other applications, while the ASR needs to be launched. Alternatively, SITCOM can simulate the data outputs of these components based on a prerecorded script, if these components are not physically available (e.g., when the developer is working on laptop at home). Later, the simulated components can be replaced by their real counterparts when practical and affordable.

Figure 4 shows the rendering of the aggregated information in a $2D$ fashion for people locations, headings, their speech activity, as well as sitting postures. The $3D$ is useful when annotating information, as it renders a more natural view which helps better assess the meeting status. Note also the view indicating the output of the sensors (in this case the cameras), which is useful for perceptual component developers to assess the quality of their output.

4.1.2. Situation Modeling

To deal with the definition of the relevant contextual information for an application, the Connector developer can use the SITCOM's *situation machines* metaphor introduced earlier. The author can use from an existing set of SMs, such as those below, or create his own by building over the existing ones.

- **Attendance** tracks the number of participants in the room;
- **Motion Detection** reports how many participants have moved over a certain distance threshold in a given time period;
- **Heading Tracker** infers the head orientation for each person from her position and face visibility for all cameras;
- **Attention Direction** tracks the number of participants looking towards the same spot, using their heading information;
- **Sitting Detection** infers whether a particular person is sitting or not from the z (height) coordinate;
- **Crowd Detection** searches for groups of at least 3 people whose relative distance does not exceed a threshold;
- **Talking Activity** tracks the duration of speech activity and number of speaker changes in a given period;
- **Meeting State Detector** infers the current state of the meeting in the room;
- **Interruptibility Level** selects the level of *interruptibility* of a particular participant according to the current meeting state and the current speaker.

The actual implementation of a particular *situation machine* can be part of the Java runtime or it can be a standalone module using a remote API. The use of a statistical classification module for meeting state recognition has been investigated in [12].

Again, through its GUI, we can see all the *situation machine*s and their state, for any particular time in the simulation data. In Figure 4, we can see the middle lower panel showing the currently active set of situation machines, along with their state information. Later, we could use the same status information to follow live what our system is doing, all while recording additional real data.

4.2. Netcarity Smart Home Services

Another use case based on the approach advocated in this paper is a Netcarity system [2] designed to help elderly people to improve their well-being, independence, safety and health at home.

[2]Netcarity is an ongoing integrated project supported by the EC under the Sixth Framework Programme (IST-2006-045508).

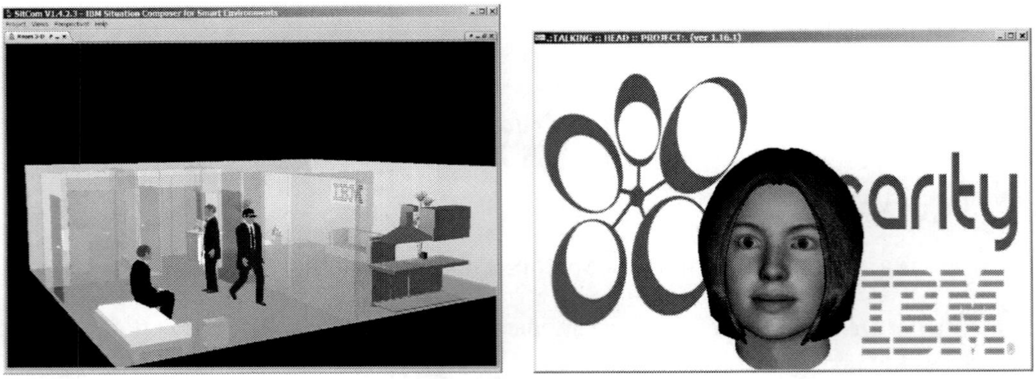

Figure 6. SITCOM's 3D visualization of smart home and Talking Head interface.

The idea of this project is to equip homes of Netcarity clients by a variable set of perceptual technologies and connect them to the central point providing various types of services. The most important activity of the central service point, called *Netcarity Server* (NCS), is to "keep an eye" on the client's health and safety exploiting different perceptual technologies, i.e., an acoustic event detector, body tracker, etc. In addition, Netcarity can provide services to improve activities of daily living of the elderly person. Examples of such additional services are chatting or videoconferencing with family members or friends, an on-line health care service, ordering of food and other goods, etc.

This project is currently in its first stage when the technologies are still under development, but service providers are already eager to try prototypes and discuss the deployment of feature functions with possible clients. This is why we have proposed to use the *Reference Architecture Model*. To support that we have created a SITCOM virtual smart home and several scenarios simulating essential situations in inhabitants' life.

The following sensors, actuators and perceptual components are being simulated in the virtual home environment:

- **Motion Detectors** are installed in each room, monitoring movement in the space.
- **Fire Detector** sensor detecting fire in the flat.
- **Gas Detector** sensor detecting leaking gas in the flat.
- **Remote Door Lock** sensor/actuator capable of reporting open/closed status of the door and locking/unlocking the door's lock.
- **Scene Acoustic Analyzer** analyzing noise and sounds in the scene, such as: someone is walking, speaking, laughing, a door is opening/closing.
- **Body Tracker** is a video-based tracker providing 2D coordinates for each person in the room.
- **Talking Head** actuator giving audio/visual feedback to the inhabitant. It serves as a communication channel from the Netcarity server and as a humanoid interface to other services; see Figure 6.

Currently the following context-aware services are being investigated and developed:

- **Health-Care Service** is a component detecting if the inhabitant is in a dangerous situation, for example she fell down and is unable to call for help, or there is a danger of gas leakage or fire in her flat. It uses and combines information from different sensors and detectors.
- **Morning Wellness** is a health checking procedure done regularly in the morning by the Netcarity client. The system takes the elderly person through several steps (such as weight and blood pressure measurements) and allows him to initialize an audio or video call with a Netcarity assistant or a medical doctor. A multi-modal dialogue system is available to the client providing speech enabled control interface.
- **Door Access Service** is a service connected to a camera mounted close to the entrance of the Netcarity client's flat or home. It informs the client about a visitor providing both video and audio stream from the camera. The client can decide to open the door by herself or consult the Netcarity operator to decide whether it is safe to let the visitor in.

In the actual implementation the sensors and actuators in the client's home are controlled by a *Residential Gateway* which is connected to the Netcarity Server by a secure and authenticated channel. In the simulated case, the residential gateway is replaced by the SITCOM run-time connected to the Netcarity Server. There is no difference in behavior or data types between simulated and real-live smart home from the Netcarity Server side. Multiple instances of SITCOM run-times can simulate a whole network of smart homes and their residential gateways. This setup is currently used for tuning Netcarity Server's rule-based system, fall-detector and acoustic scene analysis modules, and for the verification of proposed communication schemes. Beside the visual feedback during the development, the 3D visualization (Figure 6) obviously serves also as a good demonstration platform.

5. Conclusion

We have introduced an architecture model for multi-modal systems which allows to integrate complex context-acquisition components, situation modeling for data fusion, and the application logic into a highly distributed and heterogeneous interoperable system, preserving an effective separation of the effort of different development groups.

We have presented a unified approach for designing, implementing and testing multi-modal perceptual systems based on the *Reference Architecture Model*.

We have introduced a development environment and tools that have been exercised by many developers over the course of several years. The tooling framework is currently used by more than ten sites across Europe and it has been part of several technology demonstrations.

To support the development withing the *Reference Architecture Model*, we have implemented SITCOM framework – an extensible Java toolkit to help in all phases of the development of the non-trivial life-cycle of context-aware services. We equipped SITCOM with a set of functionalities that we found beneficial in development of perceptual applications:

- it simulates the environment and the perceptual input (manually created or recorded scenarios)
- it provides 2D and 3D visualizations of scenes, situations, and scenarios
- it works as a middleware between user services and the layer of perception (situation modeling)
- it supports the context model builders with a framework for data fusion and filtering
- it serves as an IDE for repetitive testing of context-aware services and applications by replaying recorded or simulated scenarios
- it provides portability between virtual and real devices
- it serves as a tool for an annotation of recorded data

The SitCom environment was successfully applied during the whole development cycle of the CHIL platform and it is now helping to bootstrap the Netcarity system. In many ways, having an integrated tool specialized for context-aware applications has been helpful to identify the necessary pieces for the application, like the needed room information, specific sensor information, or the definition of the participants roles.

Acknowledgment

We would like to acknowledge support of this work by the European Commission under IST FP6 integrated projects CHIL (contract number 506909) and Netcarity (contract number IST-2006-045508).

References

[1] Vince Stanford. Pervasive computing goes to work: Interfacing to the enterprise. *IEEE Pervasive Computing*, **1**(3):6–12, 2002.

[2] H. Nishikawa, S. Yamamoto, M. Tamai, K. Nishigaki, T. Kitani, N. Shibata, K. Yasumoto, and M. Ito. UbiREAL: Realistic smartspace simulator for systematic testing. In *Proc. of the 8th Int'l Conf. on Ubiquitous Computing (UbiComp2006)* , 2006. LNCS4206, pp. 459–476.

[3] Manuel Roman, Christopher K. Hess, Renato Cerqueira, Anand Ranganathan, Roy H. Campbell, and Klara Nahrstedt. Gaia: A middleware infrastructure to enable active spaces. *IEEE Pervasive Computing*, pages 74–83, Oct-Dec 2002.

[4] A. Dey, D. Salber, and G. Abowd. A conceptual framework and a toolkit for supporting the rapid prototyping of context-aware applications. *Human-Computer Interaction (HCI) Journal*, **16**:97–166, 2001.

[5] CHIL Consortium. The CHIL reference model architecture for multi-modal perceptual systems, 2007. http://chil.server.de/servlet/is/6503/.

[6] Vincent Stanford, Martial Michel, and Olivier Galibert. Network transfer of control data: An application of the NIST smart data flow. *Journal of Systemics, Cybernetics and Informatics*, **5**(1), 2005.

[7] Gábor Szeder. The ChilFlow data transfer system, 2008. Software. Web-page: http://www.ipd.uka.de/CHIL/ChilFlow.php.

[8] Nikolaos Dimakis, John Soldatos, Lazaros Polymenakos, Manfred Schenk, Uwe Pfirrmann, and Axel Burkle. Perceptive middleware and intelligent agents enhancing service autonomy in smart spaces. In *IAT '06: Proceedings of the IEEE/WIC/ACM International Conference on Intelligent Agent Technology (IAT 2006 Main Conference Proceedings) (IAT'06)*, pages 276–283, Washington, DC, USA, 2006. IEEE Computer Society.

[9] Pascal Fleury, Jan Cuřín, and Jan Kleindienst. SITCOM - development platform for multimodal perceptual services. In *Proceedings of 3nd International Conference on Industrial Applications of Holonic and Multi-Agent Systems* , Regensburg, Germany, September 2007. V. Marik, V. Vyatkin, A.W. Colombo (Eds.): HoloMAS 2007, *LNAI* **4659**, pp. 104–113, 2007. Springer-Verlag.

[10] J. L. Crowley, J. Coutaz, G. Rey, and P. Reignier. Perceptual components for context aware computing. In *Proceedings of UBICOMP 2002, International Conference on Ubiquitous Computing*, Goteborg, Sweden, September 2002.

[11] M. Danninger, E. Robles, L. Takayama, Q. Wang, T. Kluge, C. Nass, and R. Stiefelhagen. The connector service – predicting availability in mobile contexts. In *Proc. of the 3rd Joint Workshop on Multimodal Interaction and Related Machine Learning Algorithms (MLMI)*, Washington DC, US, 2006.

[12] Jan Cuřín, Pascal Fleury, Jan Kleindienst, and Robert Kessl. Meeting state recognition from visual and aural labels. In *Proceedings of 4th Joint Workshop on Multimodal Interaction and Related Machine Learning Algorithms* , Brno, Czech Republic, June 2007. A. Popescu-Belis, S. Renals, H. Bourlard (Eds.): MLMI 2007, LNCS, vol. 4892, pp. 24–35, 2008. Springer-Verlag.

In: VLSI and Computer Architecture
Editor: Kenzo Watanable

ISBN: 978-1-60692-075-6

Chapter 6

MOSFET's Programmable Conductance: The Way of VLSI Implementation for Emerging Applications from Biologically Plausible Neuromorphic Devices to Mobile Communications*

I.S. Han

Institute for Information Technology Convergence,
Korea Advanced Institute of Science and Technology, Daejeon, South Korea

Abstract

This paper describes a mixed-signal VLSI design utilizing the controlled conductance properties of MOSFETs, for tunable filters, low power amplifiers, and analog-mixed signal processing. The programmable linear conductance is realized by the pair of MOSFETs in the triode region, and it produces the voltage-controlled tunable function or the analog computing element. It demonstrates the application areas of neural synaptic functions, electronic neurons with biologically plausible characteristics, and tunable analogue complex filters for RF communication. The programmable conductance by MOSFETs is analyzed by the experimentation of prepared MOSFETs and the SPICE simulation.

The amplifier based on the proposed MOSFET conductance is employed to enforce the tunable filters for RF applications. It has the tuning range of +/- 50% and the bandwidth over 1 GHz is exhibited. For applications to analog computing, the design of an analog multiplier and biologically plausible neuromorphic devices are illustrated by utilizing MOSFET's programmable conductance.

---

* A version of this chapter was also published in *MOSFETs: Properties, Preparations and Performance,* edited by Noah T. Andre and Lucas M. Simon, published by Nova Science Publishers, Inc. It was submitted for appropriate modifications in an effort to encourage wider dissemination of research.

I. Introduction

The voltage-controlled conductance of MOS transistors has been widely used in the analog signal processing or VLSI circuits of programmable amplifiers, tunable filters, neural computing and others. The MOSFET's conductance in the triode region can be exploited for transconductors as in [1-4], though it is necessary to eliminate the nonlinear characteristics. Various methods were reported to improve the linearity of transconductance, as the advantages of low voltage and low power operation are exploited in the areas of analog complex filter, switched capacitor filter or continuous time filter [5-8]. Many cascode or source degeneration topologies were proposed to improve its linearity of triode-region transconductor , and its application to the tunable filter was introduced [1-9].

Recently the new improvement by a pair of MOSFETs in triode region was presented as a way of implementing the controlled linear conductance for programmable amplifier and tunable filters, while the early design based on a pair of MOS transistors in the triode region was developed for continuous time filters [10, 11]. It can be used to implement the new four quadrant analog multiplication function for modulators, and basic signal processing elements. The voltage controlled linear transconductance of MOSFETs was also adopted to produce the synaptic or neuron function of various applications from robots to regenerative medicine, which was inspired by the biological plausibility and low power supply [12-16]. The neuron circuit based on the MOSFET controlled conductance demonstrated the asynchronous spike generation with a refractory period, the biological behavior of integration-and-firing, and the synaptic computation [17-20].

In this paper, the proposed design principle of MOSFET transconductance is introduced and demonstrated its performance for various applications of a programmable amplifier, tunable complex filters of wireless system, and the neuromorphic device for bio-inspired implementation.

II. MOSFET for Analog VLSI Signal Processing

The expanding demand of modern information technology drives the rapid growth of semiconductor industry and VLSI products of MOS transistors. The analog signal processing has been an essential element for the success of many VLSI products, particularly for the telecommunication and wireless technology. For analogue VLSI signal processing, it went back to the VLSI filter for voice CODEC in the early days of digital communication. For the advantage of digital processing or advanced VLSI computing, the front-end analog filter is required to select only the wanted input signal before the A/D conversion. The integrator circuit of Fig. 1 is a general building block of analog filter, where the additional fabrication process is required for tuning the resistors' value of VLSI. It is commonly resolved by the laser trimming for the exact resistance value, as the standard VLSI process has the inherent constraints.

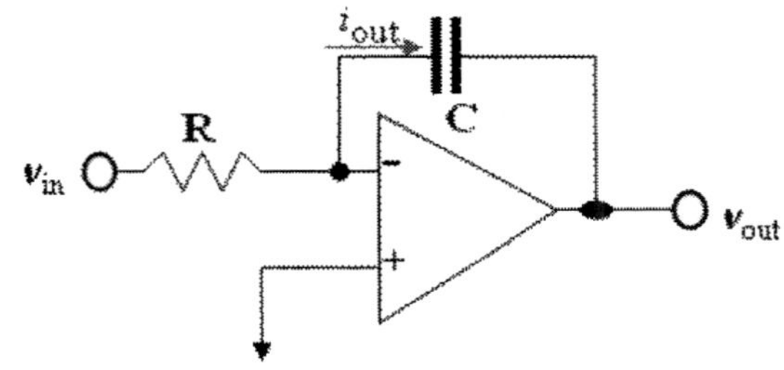

Figure 1. The basic building block of analog filer – an integrator circuit.

The switched capacitor (SC) technique was introduced to improve the problem of precise resistance value of an integrator, based on the SC integrator of Fig. 2.

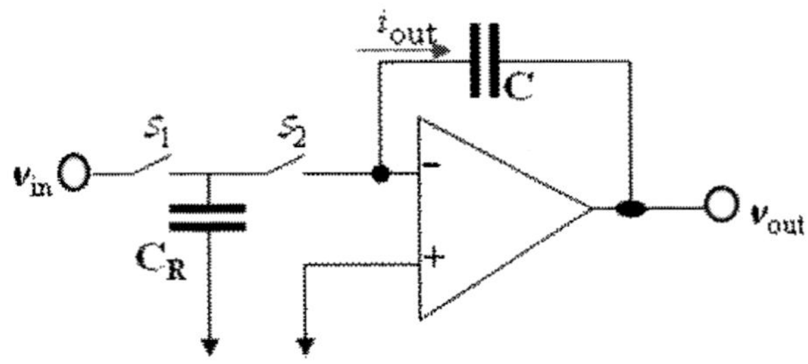

Figure 2. The switched capacitor integrator.

The integrator circuit of Fig. 1 is based on the integration of current i_{out} at capacitor C, which is represented as v_{out}. For the SC integrator of Fig. 2, the current i_{out} is equivalent to the total amount of the electric charge (Q) per second and can be represented as in the following equation (1).

$$i_{out} = f \bullet (v_{in} \bullet C_R) \tag{1}$$

where f is the switching frequency of s_1 and s_2. Switches of s_1 and s_2 are controlled by the non-overlapping clock and the clocking frequency f is assumed to be higher than the Nyquist sampling rate of the input signal v_{in}. It is based on the property that the charge of ($v_{in} \bullet C_R$) contributes to Q at each clock cycle. The switch is implemented by a single MOS transistor or a CMOS transistor pair, and the SC circuit has been widely employed for analog or analog-mixed VLSI products.

However, the SC based VLSI design becomes limited for certain applications, such as digital mobile communication of the high frequency and low power requirements. There are two constraints for the proper SC integrator of Fig. 2, which are the high clocking frequency and a CMOS operational amplifier. For an example, the filter of 1MHz bandwidth is necessary for Wifi applications. It means that the sampling clock becomes much higher compared 4 KHz bandwidth of voice CODEC. In addition to the higher clock frequency, the CMOS operational amplifier is likely prohibitive for such high frequency and low power requirements. Instead of utilizing SC type of analog sampled data type design, the integrator based on transconductance amplifier (Fig. 3) became widely used.

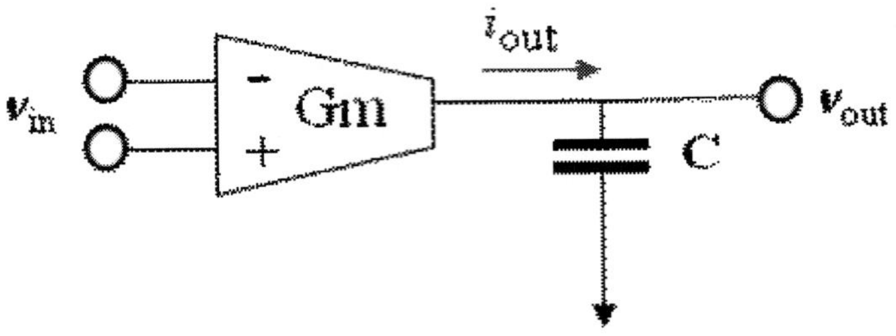

Figure 3. The integrator by a transconductance amplifier.

The advantage of transconductance amplifier is its relatively large bandwidth and the low power consumption. One of key aspects in transconductance amplifier design is the utilization of MOSFET's channel conductance, and there are a few ways of exploiting the channel conductance. The conductance in the triode region has been adopted for realizing the tunable resistance, and the equivalent circuit of Fig. 4 shows a MOSFET's channel characteristics in the triode region.

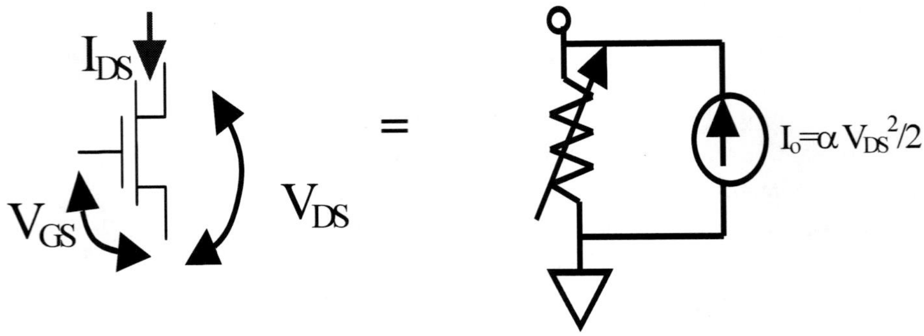

Figure 4. The equivalent circuit of a MOSFET in the triode region.

The I-V characteristics of MOSFET in the triode region can be modeled as in the equation (2).

$$I_{DS} = \alpha\,[\,(V_{GS}-V_T)V_{DS}-V_{DS}^2/2],\ \alpha=\mu C_{OX}W/L \qquad (2)$$

where $V_{GS}-V_T > V_{DS}$. The second order nonlinearity well illustrated in the I-V curves of Fig. 5. The geometry of MOSFET is the [gate width (W)]/[gate length(L)]=1.

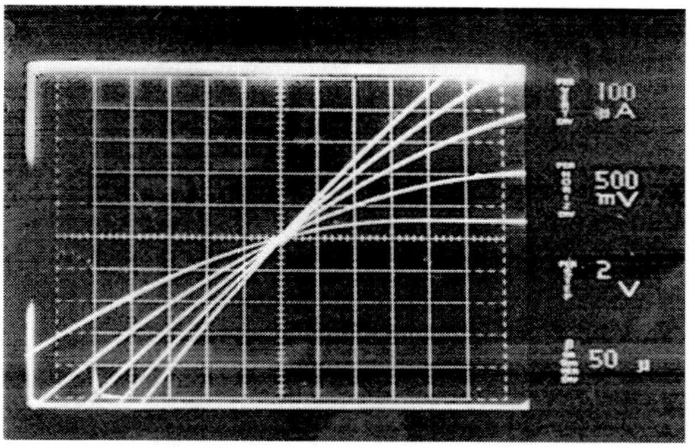

Figure 5. I-V characteristics of a MOSFET in the triode region, W/L=1.

The I-V characteristics of Fig. 5 shows the channel conductance controlled by the gate-source voltage, and it illustrates the feasibility of implementing the programmable conductance by MOSFETs in triode region. The parallel connection of two MOSFETs in Fig. 6 was proposed to improve the linearity of equivalent conductance in Fig. 5, which was motivated to challenge the limited characteristics of SC filters [10].

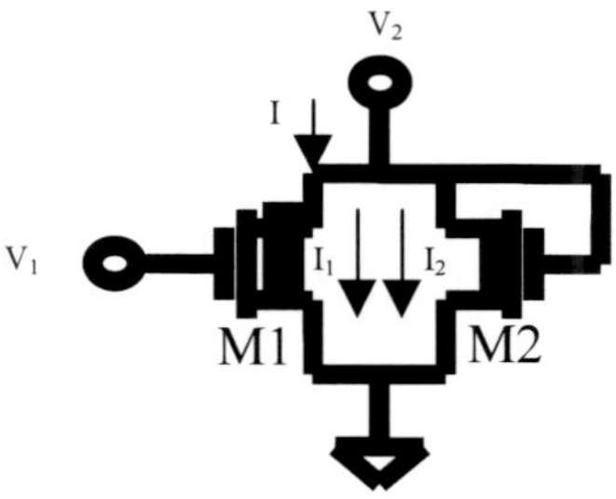

Figure 6. The voltage-controlled linear conductance by two MOSFETs in the triode region.

The operation principle of the circuit in Fig. 6 is to produce the quadratic term of V_2^2 by M2, and compensate the inherent negative quadratic terms of V_2^2 by M1 or M2. The compensation of the nonlinear component in (2) is proved by the equations of (3a), (3b) and (4). The equation (3a) represents the operation of M1, while the equation (3b) for M2. By adding I_1 and I_2, the equivalent conductance can be derived by equation (4) of total current I.

$$I_1 = \alpha[\,(V_1\text{-}V_T)V_2\text{-}V_2^2/2] \quad (3a)$$

$$I_2 = \alpha[\,(V_2\text{-}V_T)V_2\text{-}V_2^2/2] \quad (3b)$$

$$I = \alpha\,(V_1\text{-}2V_T)\,V_2 \quad (4)$$

It is necessary to satisfy the condition for the triode region operation, e.g. the threshold voltage $V_T < 0$. The experimental result is illustrated in Fig. 7, where two MOS transistors are identical to the transistor used for the experimentation in Fig. 5. For the negative input, the swapping of drain and source does not affect the equation (4) representing I, as the change related to the gate-source bias voltage is disappeared after the summation process of I_1 and I_2. The comparison of Fig. 5 and Fig. 7 demonstrated the susbstitute to the resistors or SC-type design by the linear channel conductance of MOSFET in thertriode region. The simple parallel connection of two transistors satisfied the requirements of voice CODEC, though the slight degradation in Fig. 7 was caused by the threshold voltage modulation.

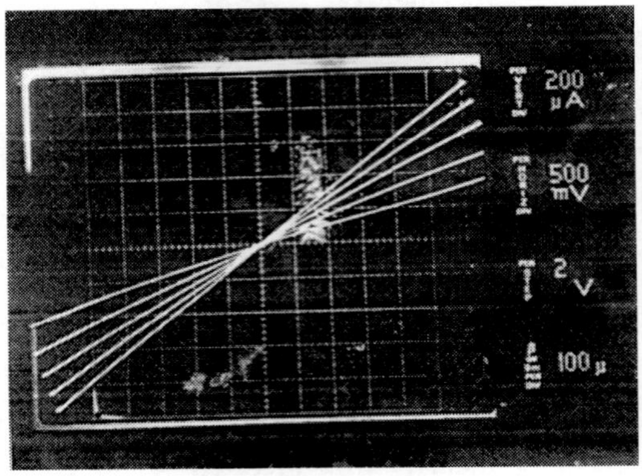

Figure 7. The voltage-controlled linear conductance by two MOSFETs.

There exists the small nonlinearity, as threshold voltage V_T in equation (4) is modulated by the input signal V_2. The change of V_2 drives the change in the source-substrate reverse-biased voltage, as the source voltage is equal to V_2 for the negative input. There is another but minor cause to the nonlinearity, which is due to the modulation of channel carrier mobility. The programmable conductance by the circuit of Fig. 6 demonstrated the improvement of issues by SC circuits, because of the tunable characteristics without relying on the discrete sampled data operation. As in the SC circuits, operational amplifies are necessary for the circuit of Fig. 6, which is still an issue for the modern mobile communication or special purpose applications like large scale analog-mixed VLSI integration.

III. Programmable Linear Conductance by MOSFET in the Triode Region – Principles and Applications

The requirement of general purpose operational amplifiers imposes the constraint on some applications, because of its power consumption, limited frequency bandwidth or etc. An example can be VLSI neural networks, which requires the large scale of multiplication and summation. Instead, the nature of biological neural networks drives the appropriate neuromorphic computation based on the large scale multiplication of neural states and synaptic connections with the excitatory-inhibitory weight value. It is a straightforward implementation to utilize the programmable conductance like I-V characteristics in Fig. 7, as it is similar to the neural networks model. With the large scale requirements like neural networks, it is realistic to implement the programmable linear conductance without using an operational amplifier.

The circuit of Fig. 8(a) was proposed for the programmable conductance without an operational amplifier. The operation principle is based on M1 in the triode region and the current mirror of M3-M5 [11]. The nonlinear term in equation (2) is eliminated by applying the positive and negative voltage of the same amplitude to the drain and the source respectively.

Assuming the drain voltage v and the source voltage $-v$, the drain-source current I of M1 is

$$
\begin{aligned}
I &= \alpha[\,(V_{in}+v-V_T)(2v)-(2v)^2/2] \\
&= \alpha(V_{in}-V_T)(2v)
\end{aligned}
\tag{5}
$$

The output current I_{out} is the mirrored current I by M3 and M5, which is linearly proportional to either V_{in} or $2v$. M2 and M3 are the same ones, which introduce the same voltage drop across them. With the presence of M2 and M3, the same amplitude voltage with the different polarity is applied to the drain and the source of M1 respectively. M4 acts as the offset tuning or reference generator for the bipolar current output.

The circuit of Fig. 8(a) shows the linear characteristics of voltage-controlled transconductance, without an operational amplifier. It suggested a feasible design solution for the large scale VLSI neural networks, with a switch attached to the output. The switch can multiply the synaptic weight (V_{in}) and the spike neural input, and every synapse cell of Fig. 8a can be connected in parallel without the restriction [21, 22]. The bipolar supply voltage is one of key element for the programmable linear conductance of Fig. 8(a), compared to the single voltage supply. For the analog VLSI, the differential inputs are generally preferred to the single-ended signal.

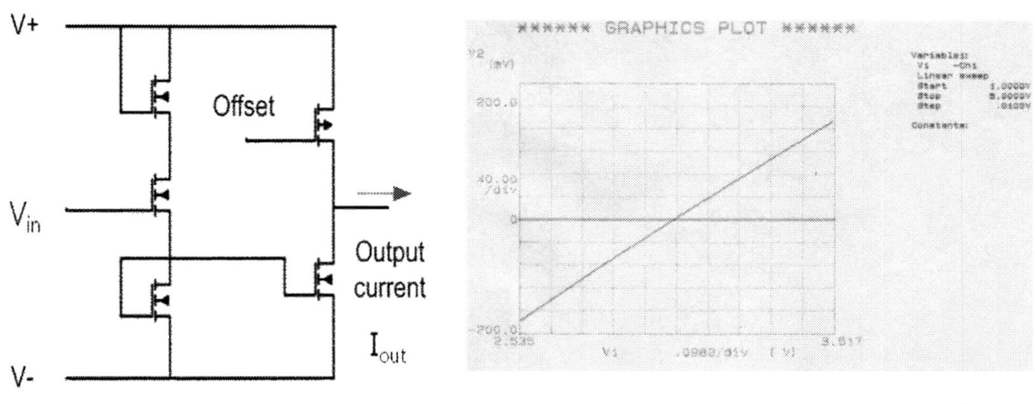

(a) (b)

Figure 8. (a) The linear transconductance circuit based on a MOSFET in the triode region (b) the measured I-V transfer characteristic.

Hence, the transconductor circuit of Fig. 9 was recently designed to accommodate the differential inputs, followed by two previous design methodologies of Fig. 6 and Fig. 8a [11]. The circuit of Fig. 9 is based on the parallel connection of two MOSFETs M1 and M2. Both M1 and M2 operate in the triode region, which are similar to MOSFETs M1 and M2 in Fig. 6. Instead of an operational amplifier, the current mirror configuration was adopted as the circuit of Fig. 8a. The role of M3 is to implement the current mirror, while it is also indispensable for the triode region operation of M2. M2 operates in the triode region with the ordinary positive threshold voltage, as the drain voltage of M2 is kept for the triode-region operation condition by V_{BIAS} of M3.

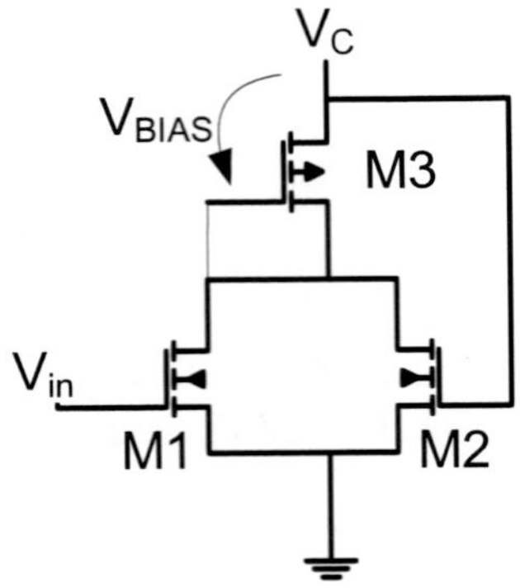

Figure 9. The voltage-controlled linear transconductance circuit based on a pair of MOS transistors in the triode region.

The drain current I_D of M1 and M2 may be expressed as

$$I_{D1} = \alpha\,[(V_{GS} - V_{TH})V_{DS} - V_{DS}^2/2] \qquad (6a)$$

$$I_{D2} = \alpha\,[(V_C - V_{TH})V_{DS} - V_{DS}^2/2] \qquad (6b)$$

where $V_{GS} = V_{inDC} \pm \Delta V_{in}$ is the gate-source voltage of transistor M1, $V_{DS} = V_C - V_{BIAS}$ is the drain-source voltage of transistors M1and M2, V_C is the programming voltage of transconductance, and V_{inDC} is the DC offset of input voltage.

V_{BIAS} of the diode connected M3 operates as the DC voltage, neglecting the minor non-ideal characteristics. Hence, V_{DS} is $(V_C - V_{BIAS})$ and the total current I_D is determined by

$$\begin{aligned} I_D &= I_{D1} + I_{D2} \\ &= \alpha\,(V_{GS} - V_{BIAS} - 2V_{TH})(V_C - V_{BIAS}) \end{aligned} \qquad (7)$$

The analysis of equations (6) and (7) explains that the 2nd order distortion can be removed by a matched pair of M1 and M2 in the triode region. The mobility degradation or variation of bias voltage can cause the slight nonlinearity of transconductance in equation (7). The term of V_{BIAS} in (7) is the gate-source voltage of M3, while the drain current of M3 is I_D. The deviation of V_{BIAS} from the DC voltage is undesirable but exists due to the mobility reduction effect or the nonlinear characteristics of M3 in diode connection. The circuit of Fig. 10 compensates such variation by introducing the differential mode operation, as the variation of V_{BIAS} is almost common between both basic transconductance circuits of Fig. 9. In principle, the circuit of Fig. 10 is based on the parallel operation of basic transconductor circuits in Fig. 9 with the linear and programmable transconducatnce.

Repeated two circuits in Fig. 10 operate in the same condition, other than the small differential input signals applied to the gate of M1 and M1A in Fig. 10. Therefore the gate-source voltage of M3 and M3A is almost same, though the amplitude of small differential signal can cause a slight difference. The transistor pairs of (M3, M4), (M3A, M4A), and (M5A, M5) are connected as the current mirror circuit. M4 copies the drain current I_{D3} of M3 and M5 copies the drain current of M5A, which is equal to the drain current I_{D3A} of M3A. Based on equation (7), I_{D3} and I_{D3A} can be given by

$$I_{D3}=Gm(V_{GS1}-V_{BIAS}-2V_{TH}) \quad (8a)$$

$$I_{D3A}=Gm(V_{GS1A}-V_{BIAS}-2V_{TH}) \quad (8b)$$

Where $V_{GS1}=V_{inDC}+\Delta V_{in}/2$, $V_{GS1A}=V_{inDC}-\Delta V_{in}/2$, and $Gm=K(V_C-V_{BIAS})$.
Therefore, the output current I_{OUT} is

$$I_{OUT}=Gm\,\Delta V_{in} \quad (9)$$

where $I_{OUT}=I_{D3}-I_{D3A}$.

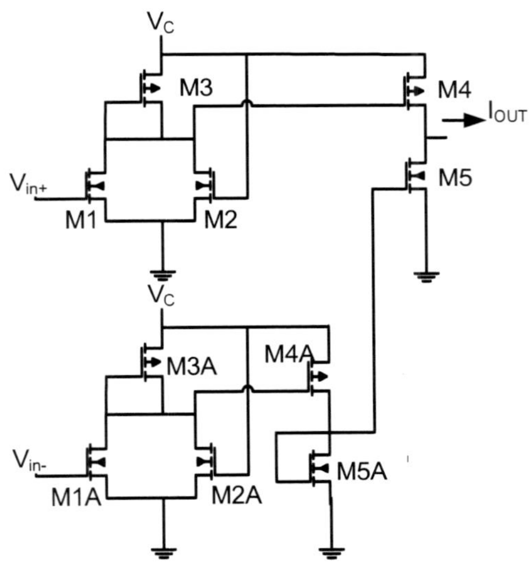

Figure 10. The transconductance amplifier based on the voltage-controlled transconductance circuit of Fig. 9.

The I-V relation of equation (9) is also based on the triode-region operation of transistors M1, M2, M1A, and M2A. By comparing the equations (7) and (9), the variation by the term of V_{BIAS} is less influential in equation (9). For transistors M1 and M1A, the triode-region condition of ($V_{GS}-V_{TH}\geq V_{DS}$) determines the input signal dynamic range. The V_{BIAS}, the gate-source voltage of M3 or M3A maintains the operation of M2 or M2A in the triode region, while M3 or M3A operates in the saturation region with the tuning voltage V_C. Hence, the circuit dynamic range is decided approximately by

$$V_{in} > V_C - V_{BIAS}$$

where V_{BIAS} is larger than V_{TH} of M3 and M3A, V_C is larger than V_{TH} of M2 and M2A. The transconductance amplifier of Fig. 10 was designed for the high frequency application and the simulated result illustrates the operation over the Giga Hertz range in Fig. 11. There are several types of transconductance amplifier with different characteristics and target applications, as there are various requirements with constraints of the linearity, the operation frequency, the tunable characteristics, power consumption, noise characteristics, and so on. The transconductance amplifier circuit of Fig. 10 has the advantage of the high frequency operation with the improved linearity, in addition to the controlled conductance by simple and flexible configuration. The gain bandwidth is observed up to 2 GHz in Fig. 11(a), while the signal dynamic characteristics are shown in the Fig. 11(b).

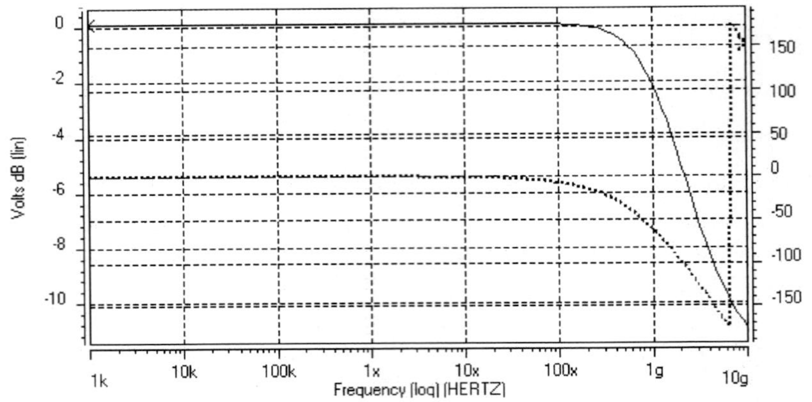

(a)

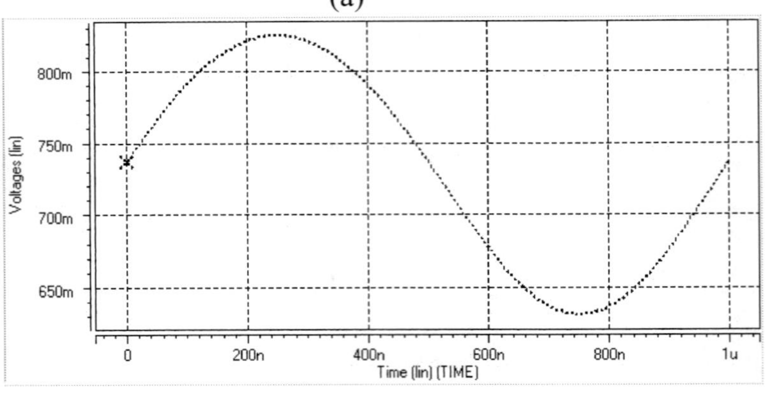

(b)

Figure 11. (a) The frequency characteristics of the transconductance amplifier in Fig. 10, (b) the output waveform dynamics.

The equation (9) implies the analog multiplication, as the term of ($V_C\Delta V_{in}$) in equation (9) represents the mathematical multiplication. Though the digital system or signal processing is dominant nowadays, the analog multiplication is expected to bridge the gap between the high precision digital computing and the large scale special computing [23-25]. The neural networks or convolution decoder is an example of such applications. The offset components of V_C and V_{BIAS} for analog operation can be removed by a pair of same transconductance amplifiers. The differential operation for V_C can be implemented by two transconducatance amplifiers in Fig. 12. It removes the common terms of V_{BIAS} and the offset voltage of V_C from the final output current.

Based on equation (9), the output currents of upper and lower transconductor circuit in Fig. 12 are described as

$$I_{OUT1} = k(v_{iny} - V_{BIAS})\ \Delta v_{inx} \tag{10a}$$

$$I_{OUT2} = k(v_{iny-} - V_{BIAS})\ \Delta v_{inx} \tag{10b}$$

where I_{OUT1} represents the output from left hand side transconductors and I_{OUT2} represents the output of right hand side transconductors. As $\Delta v_{iny} = v_{iny} - v_{iny-}$, the output current I_{OUT} of the circuit of Fig. 12 is determined by

$$I_{OUT} = I_{OUT1} - I_{OUT2} = k(\Delta v_{inx}\Delta v_{iny}) \tag{11}$$

From equation (11), the circuit of Fig. 12 works as the four quadrant analog multiplier with the dynamic range of ($\pm\Delta v_{inx}/2.\ \pm\Delta v_{iny}/2$).

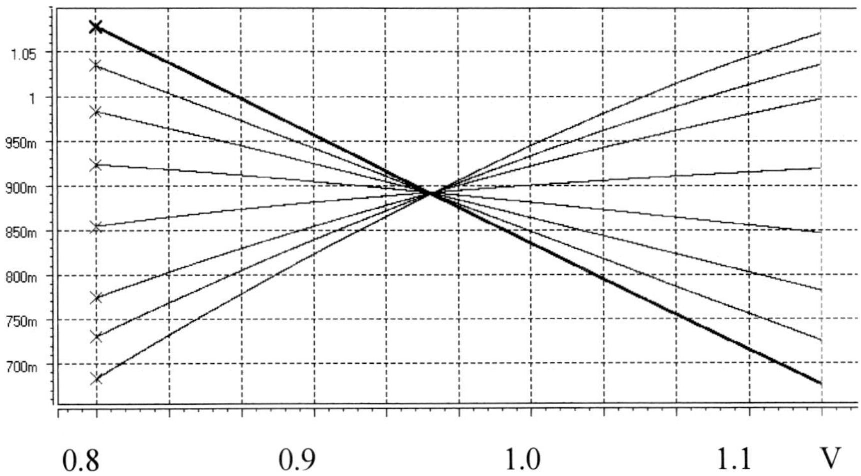

Figure 12. Analog multiplier by two transconductance amplifiers of Fig. 10.

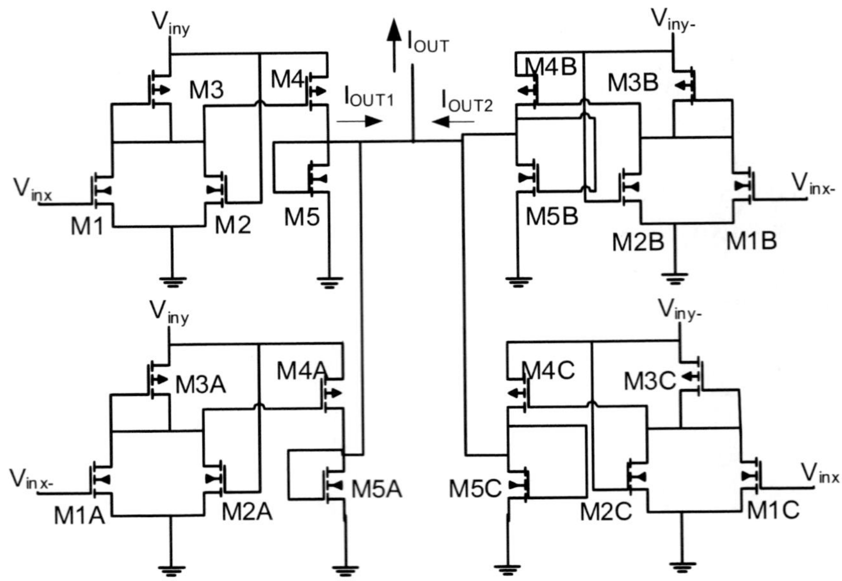

Figure 13. Linearity of analog multiplier of Fig. 12 : x-axis v_{inx} (0.8V-1.14V), y-axis multiplier output (converted to voltage via a load resistor), v_{iny} (0.9V-1.5V).

The simulated result of a proposed analog multiplier is exhibited in Fig. 13, with $\pm\Delta v_{inx}/2 = \pm 0.17$V and $\Delta v_{iny}/2 = \pm 0.3$V. The output current is measured as the voltage via a resistor connected to the output, with the MOSFETs of $W_1/L_1 = W_2/L_2 = W_{1A}/L_{1A} = W_{2A}/L_{2A} = W_{1B}/L_{1B} = W_{2B}/L_{2B} = W_{1C}/L_{1C} = W_{2C}/L_{2C} = 0.9/7.2$, $W_3/L_3 = W_4/L_4 = W_{3A}/L_{3A} = W_{4A}/L_{4A} = W_{3B}/L_{3B} = W_{4B}/L_{4B} = W_{3C}/L_{3C} = W_{4C}/L_{4C} = 3.6/0.36$, $W_5/L_5 = W_{5A}/L_{5A} = W_{5B}/L_{5B} = W_{5C}/L_{5C} = 1.8/0.36$. The overall power consumption of an analog multiplier is 250μW.

The transconductance amplifier of Fig. 14 has the output buffer circuits added to the circuit of Fig. 10, for the complete transconductance amplifier and the balanced structure of filter applications. The transistors M6-M9 in Fig. 14 are added for the output circuit to be independent of the tuning voltage V_C, and the output circuit can be flexible to the application requirements such as the robust current level or dynamic range. The balanced output current can be available from the other transconductor element, by adding the equivalent output circuit. The transistor pairs of (M5, M6) and (M7, M8) are introduced to copy the current in M3 or M4 to the output with the supply voltage of V_{DD}. The current in M3A or M4A is mirrored to the transistor M9, which contributes to the differential output current I_{OUT} as in Fig. 10. The transistor sizes used are $W_1/L_1 = W_2/L_2 = W_{1A}/L_{1A} = W_{2A}/L_{2A} = 0.125$, $W_3/L_3 = W_4/L_4 = W_{3A}/L_{3A} = W_{4A}/L_{4A} = 10$, $W_5/L_5 = W_{5A}/L_{5A} = 5$. The input V_{in} is 1.25V ± 0.2V and the tuning voltage V_C is 1.2V± 0.2V.

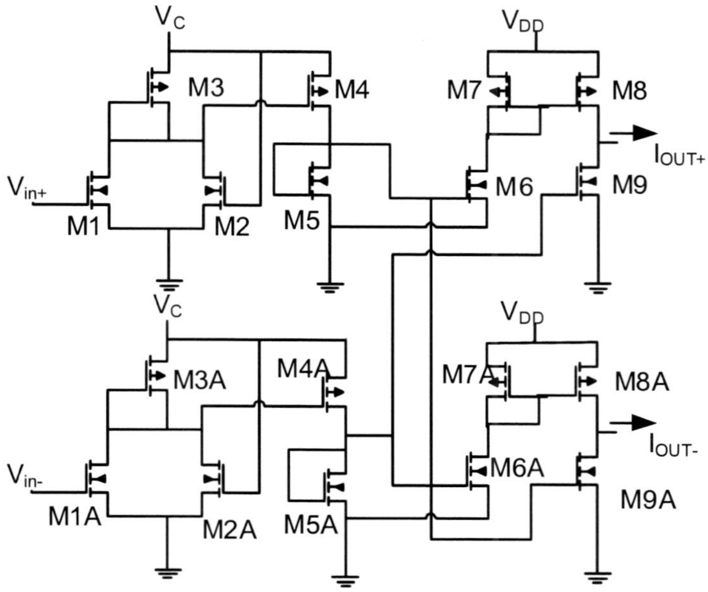

Figure 14. The transconducatnace amplifier with balanced outputs.

The linearity from SPICE simulation is summarized in Table I [11]. The results shown in Table I correspond to the worst case ±1% mismatch of gate geometry (W and L) for all transistors of the circuit in Fig. 14. The simulation is based on V_{DD}=1.8V, V_C=1.0 V, and V_{inDC}=1.25V. From the results in Table I, the total harmonic distortion is less than -65 dB with the input of 0.6 V_{PP}. The input with larger DC value or common-mode can increase the dynamic range. The input common-mode of DC offset of the circuit in Fig. 14 can be varied from 1V to the maximum signal level in the circuit like 1.5V. The simulation result shows the 100μW of power consumption, the tunable transconductance range of ± 50% with the control voltage V_C of 1 V ± 0.2V.

Table I.

V_{in} (V_{PP}) at10 MHz	0.2 V	0.3 V	0.4 V	0.5 V	0.6 V
Total Harmonic Distortion	-70dB	-72dB	-69dB	-68dB	-65dB

The tansconducatnace amplifier in Fig. 14 was evaluated for the GmC filter of 5th order low pass filter. The 1 MHz bandwidth with the Leap-Frog structure was simulated as in Fig. 15.

The linearity of the MOSFET's conductance yielded the total harmonic distortion of the filter as -68 dB with 0.5V input, and it is proved be an efficient way of implementing the programmable conductance [11].

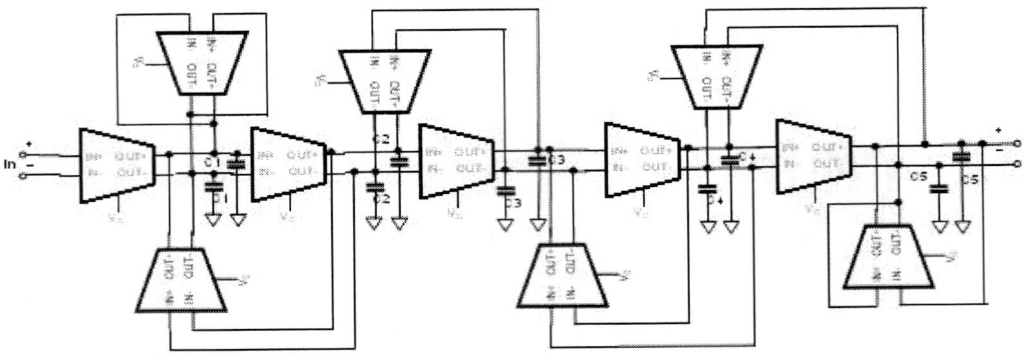

(a)

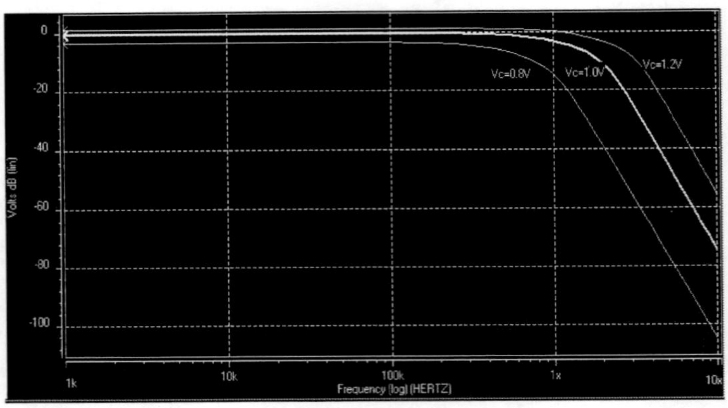

(b)

Figure 15. (a) Fifth-order Gaussian low-pass filter, C1=0.9 pF, C2=1.9 pF, C3=2.6 pF, C4=3.1 pF C5=7.6 pF, using balanced transconductance amplifiers in Fig.5, (b) Cut-off frequency vs. control voltage V_C; 0.5 MHz (0.8V), 1 MHz (1.0V), 1.6 MHz (1.2V).

The conductance by MOSFET in the triode region has other advantages, and the one is the flexible programming for the analog computing or complex analog-mixed signal processing. For an example, an analogue multiplier of Fig. 12 was implemented by the principle based on simple equations (10) and (11). An analog polyphase filter is designed to evaluate the effectiveness of utilizing the programmable conductance, as the tunable analogue filters have been emerged as key issues for advanced wireless applications of Mobile TV, digital RF, and Cognitive Radio.

For implementing an analog polyphase filter, the feedback between I and Q signals requires the higher transconductance coefficient to shift the centre frequency from dc to the range of MHz. Two transconducatnce amplifiers (G2) of Fig. 16(a) correspond to the feedback paths of I and Q signals, where v_{in} represents the I channel signal and jv_{in} of 90 degrees phase shifted v_{in} represents the Q channel signal. The control voltage v_{Ct} is the transconducatnce tuning voltage of G2 to tune the centre frequency of polyphase bandpass filter. The output buffer of transconductance amplifier of Fig. 14 improves the dynamic characteristic by maintaining the consistent condition, even with the dynamical change of V_C. To meet the requirement of a different transconductance value for I and Q channel feedback, a simple way of current amplification is employed at the first stage by multiplying the transistors of M4 and M4A of G2 amplifiers in Fig. 16(a). The G2 amplifier has the 6 times larger gate width of M4 and M4A than G1 amplifiers. The MOSFETs of G1 amplifier are $W_1/L_1 = W_2/L_2 = W_{1A}/L_{1A} = W_{2A}/L_{2A} = 1/8$, $W_3/L_3 = W_4/L_4 = W_{3A}/L_{3A} = W_{4A}/L_{4A} = 10$, $W_5/L_5 = W_6/L_6 = W_9/L_9 = W_{5A}/L_{5A} = W_{6A}/L_{6A} = W_{9A}/L_{9A} = 5$, $W_7/L_7 = W_8/L_8 = W_{7A}/L_{7A} = W_{8A}/L_{8A} = 10$, with the simulated analogue polyphase filter characteristics of Fig. 16(b). The capacitor C is 3.5 pF for a low pass prototype filter of 1 MHz bandwidth.

The overall linear characteristic was evaluated by SPICE simulation, where the total harmonic distortion was less than 0.1 % with the input signal of 0.6 V_{PP}. The supply voltage V_{DD} of CMOS tunable analogue polyphase filter is 1.8V with the power consumption of 100 μW and the control voltage is in the range of 1V ± 0.1V. The simulated characteristics in Fig. 16(b) illustrated that the centre frequency of the pass band can be controlled linearly by the change of control voltage v_{Ct} of G2 transconducatnace amplifiers. The control voltage V_C of transconductance amplifiers G1 is maintained to 1.0V throughout the simulation.

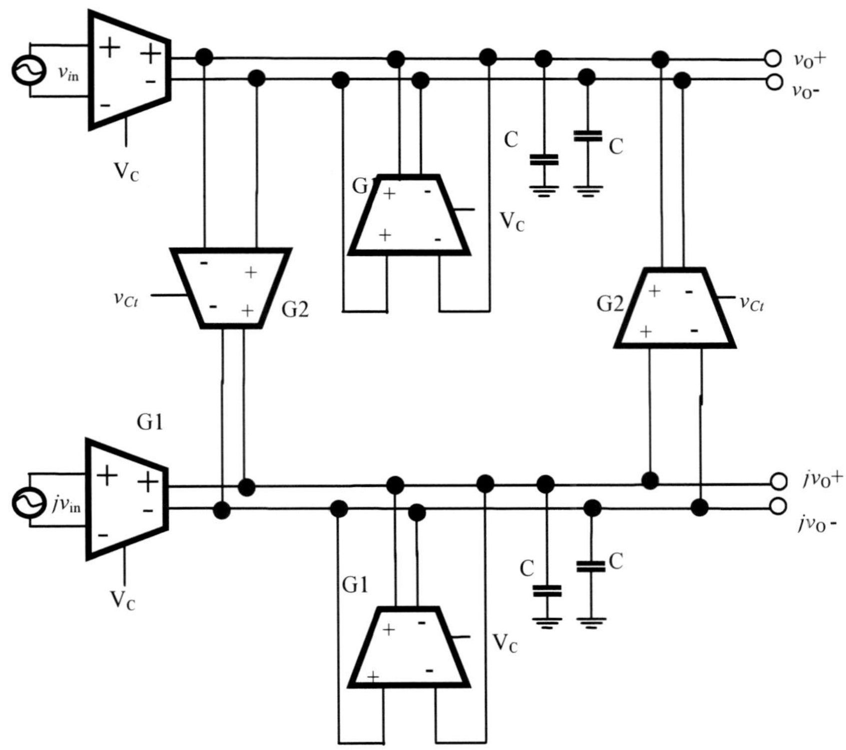

(a)

Figure 16. Continued on next page.

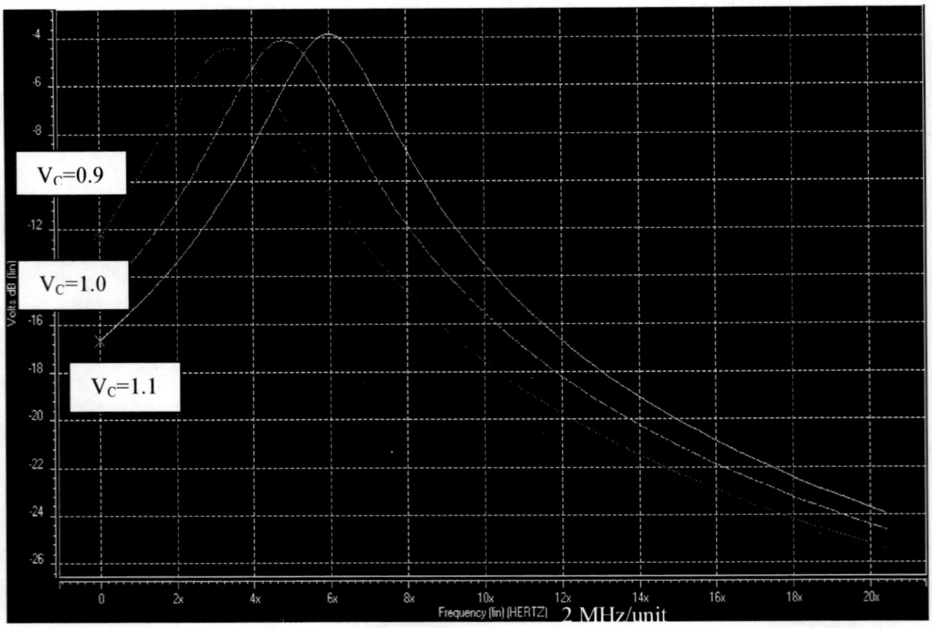

(b)

Figure 16. (a) An analog complex filter using transconducatnce amplifiers of Fig. 14, (b) simulated tuning of analog complex filter: the control voltage v_{Ct} of G2 is in the range of 1V ± 0.1V, the control voltage V_C of G1 is 1.0V.

The new programmable conductance by MOSFETs can be utilized for implementing the biologically inspired applications of neuromorphic signal processing. For neuromorphic applications, the transconductor of Fig. 10 has the advantage in implementing synaptic connections with excitatory and inhibitory weights, based on biologically plausible spike-based neural networks [26]. By adding a MOSFET at the output, the neural computing for synaptic function can be realized by an additional MOSFET for a switch, as the output current simulates the post synaptic state. There is another advantage of implementing the biologically plausible neuron like Hodgkin-Huxley(H-H) formalism as in Fig. 17, which has been well-known for the electrical model based on the controlled conductance for its physiological dynamics [27-31].

The circuit of Fig. 17 works on the dynamics of H-H formalism, with the simplified but simulating the membrane conductance model. The H-H formalism can include two digits of ion numbers depending on the purposes, and the objective of Fig. 17 is to implement the fully asynchronous spike neuron with the refractory period like the biological nature. The circuit operation is illustrated in Fig. 18. In Fig. 18, the membrane potential (a) of neuron (axon) produced the control voltage of the membrane conductance (b). Once the membrane potential reached the threshold of firing the spikes (d), neural input spikes (c) to the neuron won't contribute to increase the membrane potential for a certain period.

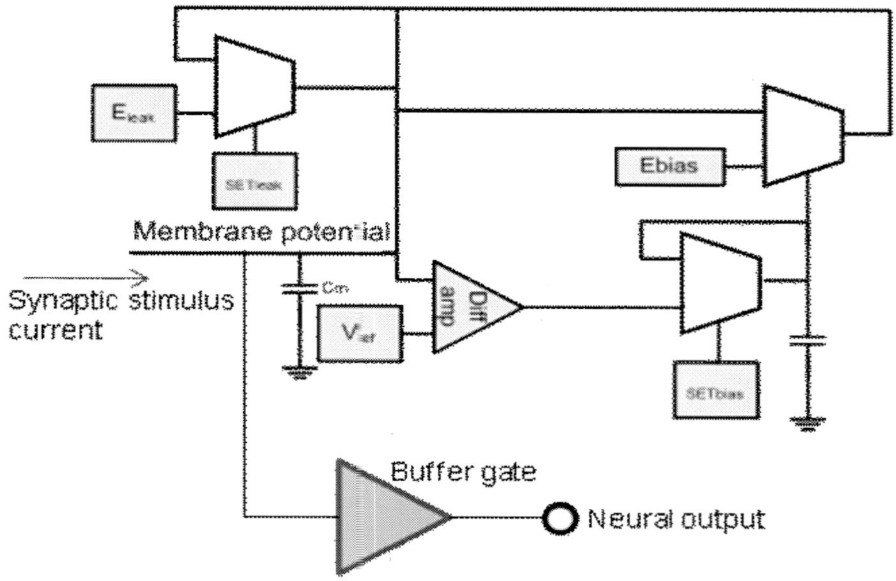

Figure 17. The neuron circuit of Hodgkin-Huxley formalism based the transconductance amplifier.

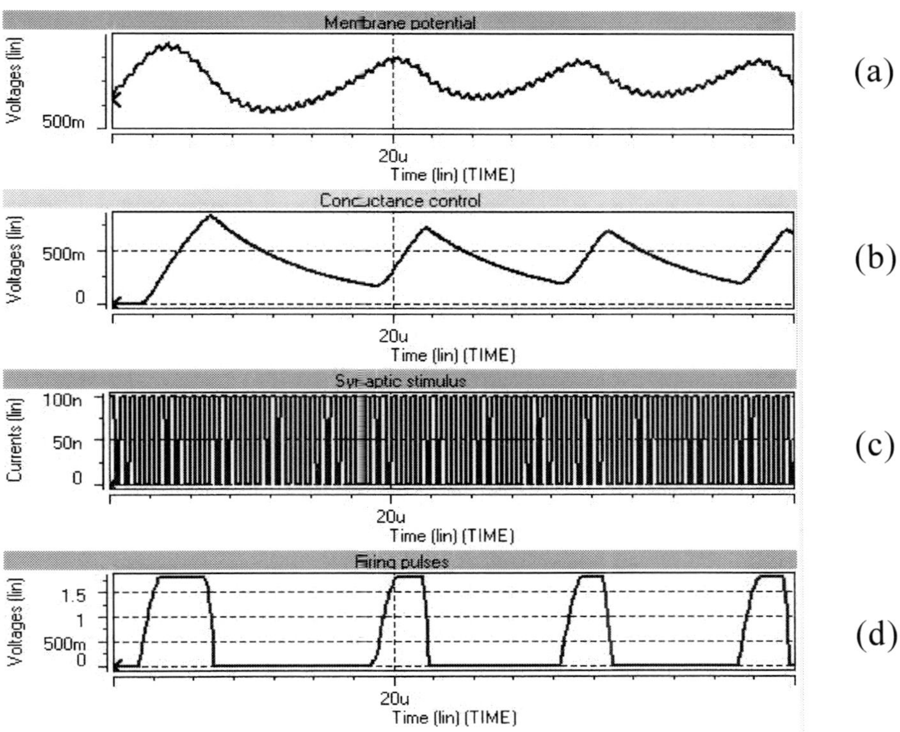

Figure 18. Biologically inspired asynchronous neural spikes with a refractory period, based on Hodgkin-Huxley formalism by MOSFETs' conductance; (a) membrane potential (b)conductance control voltage (c)simulated post-synapse neural input (d)neural spike output with the refractory period.

The transconductor of Fig. 10 was fabricated in CMOS technology, with MOSFETs of $W_1/L_1 = W_2/L_2 = W_{1a}/L_{1a} = W_{2a}/L_{2a} = 0.06$, $W_3/L_3 = W_4/L_4 = W_{3a}/L_{3a} = W_{4a}/L_{4a} = 20$, $W_5/L_5 = W_{5a}/L_{5a} = 10$. The modulated sinusoidal waveform in Fig. 19 shows the tuning and linear characteristics of tunable linear transconductor, where a sinusoidal signal of V_{in} and a sawtooth signal of V_C were applied as input signals. The sawtooth wavefrom in Fig. 19 represents V_C, while a resistor was added to convert the output current to voltage at the output.

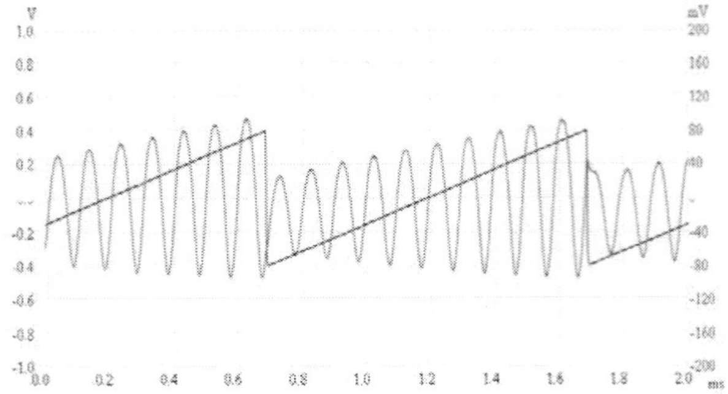

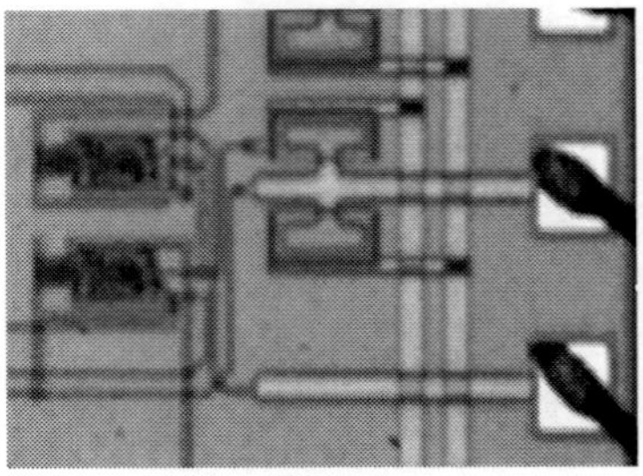

Figure 19. The experimental result and CMOS circuit of voltage-controlled conductance in Fig. 10.

IV. Conclusion

The conductance of MOSFET is proposed as a methodology of realizing analog-mixed VLSI for analog signal processing, analog front-end for innovative mobile information technology, and the biologically inspired systems. In this paper, the ways of utilizing the MOSFET in the triode region are introduced and analyzed for implementing the linear and programmable conductance from the nonlinear physical property of MOS transistors. The recent design concept is based on the pair structure of two transistors in the triode region, and the controlled conductance is realized with the flexibility of adapting to complicated applications.

The special function like analog multiplier can be implemented by repeating the basic circuit.

Example applications are investigated for amplifiers, filters, analog multiplier, and neuromorphic computing. The accuracy or the flexibility is demonstrated by the performance of filters and special analog signal processing, with the tunable range of 100% and the linearity equivalent to - 65dB in total harmonic distortion. The application to the biologically plausible signal processing is demonstrated by implementing the H-H formalism in analog-mixed MOSFET circuit, while the H-H formalism is too complex computing to be efficient by the digital VLSI. The spiking neuron output with the refractory period is simulated by MOSFETs' programmable conductance.

The linear and programmable conductance by MOSFETs is proposed as a new way of the VLSI implementation for analog-mixed applications, mobile communications to biologically inspired information processing. The principle is flexible to extend or modify depending on application specific requirements, as it is based on the simple configuration of two MOS transistors in the triode region.

References

[1] D. V. Morozov and A. S. Korotkov, A realization of Low-Distortion CMOS Transconductance Amplifier, IEEE Trans. Circuits Syst. I, 2001, vol. 48, pp. 1138-1141

[2] M. Kachare et al, A Compact Tunable CMOS Transconductor with High Linearity, *IEEE Trans. Circuits Syst. II*, 2005, vol 52, pp. 82-84

[3] S. Korotkov, D. V. Morozov and R. Unbehauen, Low-voltage continuous-time filter based on a CMOS transconductor with Enhanced Linearity, *International Journal of Electronics and Communications*, 2002, vol 56, pp.416-420

[4] E. A. M. Klumperink and B. Natua, Systematic Comparison of HF CMOS Transconducators, *IEEE Trans. Circuits Syst. II,* 2003, vol 50, pp. 728-741

[5] Q. Huang, A MOSFET-Only Continuous-Time Bandpass Filter, *IEEE J. Solid-State Circuits*, 1997, Vol. 32, pp. 147-158

[6] L. Corban and P. E. Allen, Low-voltage CMOS transconductance cell based on parallel operation of triode and saturation transconductors," *Electron lett.*, 1994, Vol. 14, pp. 1124-1126

[7] S. H. Yang et al,"A Novel CMOS Operational Transconducatance Amplifier Based on a Mobility Compensation Technique," *IEEE Trans. Circuits Syst. II,* 2005, vol 52, pp. 37-42

[8] U. Yodprasit and C. C. Enz, "A 1.5 V 75-dB dynamic range third-order Gm-C filter integrated in a 0.18-um Standard digital CMOS process," *IEEE J. Solid-State Circuits,* 2003, Vol. 7, pp. 1189-1197

[9] M. Banu and Y. Tsividis, "Fully Integrated Active RC Filters in MOS Technology," *IEEE J. Solid-State Circuits, 1983*, Vol. 6, pp. 644-651

[10] S. Han and S. B. Park, "Voltage-controlled linear resistor by two MOS transistors and its application to active RC filter MOS integration," *Proc. IEEE*, 1984, Vol. 11, pp. 1655-1657

[11] S. Han, A novel tunable transconductance amplifier based on voltage-controlled resistance by MOS Transistors, IEEE trans. Circuits Syst. II, 2006, Vol. 53, pp. 662-666

[12] Y. Horio, K. Aihara, and O. Yamamoto, Neuron–Synapse IC Chip-Set for Large-Scale Chaotic Neural Networks, *IEEE Trans on Neural Networks*, 2003, vol. 14, pp1393-1404

[13] M. Milev and M. Hristov, Analog Implementation of ANN With Inherent Quadratic Nonlinearity of the Synapses, *IEEE Trans on Neural Networks,* 2003, vol. 14, pp1187-1200

[14] G. Bugmann, Biologically plausible neural computation, *Biosystems*, 1997, Vol. 40, pp. 11-19

[15] J. Taylor, Paying attention to consciousness, *Progress in Neurobiology*, 2003, vol. 71, pp.305-335

[16] J. Taylor, Private communication, 2004

[17] Christodoulou, G. Bugmann, and T. Clarkson, A spiking neuron model: applications and learning, *Neural networks*, 15, 2002, pp. 891-908

[18] Bartolozzi and G. Indiveri, A neuromorphic selective attention architecture with dynamic synapses and integrate-and fire neurons, *Proceedings of BICS 2004*, Sterling, 2004, BIS2.2

[19] S. Han, Mixed-signal neuron-synapse implementation for large-scale neural networks, *Neurocomputing,* 2006, Vol .69, pp. 1860-1867
[20] S. Han, A pulse-based neural hardware implementation based on the controlled conductance by MOSFET circuit, *Proc. IJCNN*, 2006, pp5100-5106
[21] S. Han and R. Webb, Neural Network Switch Controller with Analogue-Digital Mixed Neural Network VLSI, *Proc. of EANN'97*, 1997, pp. 299-302
[22] S. Han, Neural network VLSI implementation and its applications, *Korea Telecom Journal*, 1997, Vol.2, pp. 12-20
[23] L. Coban, and P. E. Allen, Low-voltage, four quadrant analogue CMOS multiplier, *Electron. Lett.,* 1994, Vol. 30, pp. 1044-1045
[24] Ramirez-Angulo et al, Low-voltage CMOS analogue four quardrant multiplier based on flipped voltage followers, *Electron. Lett.*, 2003, Vol. 39, pp. 1171-1172
[25] Maundy and M. Maini, A comparison of three multipliers based on V_{gs}^2 technique for low voltage applications, *IEEE Trans. Circuit & Systems-I,* 2003, Vol. 50, pp. 937-940
[26] S. Han, Mixed-signal neuron-synapse implementation for large scale neural network, *Proceedings of BICS,* Sterling, 2004, NC4.2
[27] Linares-Barranco et al, A CMOS implementation of FizHugh-Nagumo neuron model, *IEEE JSSC*, 1991, Vol. 26, pp956-965
[28] M. Hausser, The Hodgkin-Huxley theory of action potential, *Nature neuroscience* supplement, 2000, vol. 3, pp. 1165
[29] M. Izhikevich, Which model to use for cortical spiking neurons?, *IEEE trans. Neural Networks*, 2004, vol. 15, pp1063-1070
[30] T. Asai et al, A subthreshold MOS neuron circuit based on the Volterra system, *IEEE trans Neural Networks*, 2003, Vol. 14, pp.1308-1312
[31] M. Simoni et al, A multiconductance silicon neuron with biologically matched dynamics, *IEEE Trans. Biomedical Eng*, 2004, Vol. 51, pp. 342-354

In: VLSI and Computer Architecture
Editors: Kenzo Watanabe

ISBN 978-1-60692-075-6

Chapter 7

VISION-BASED PATH PLANNING WITH ONBOARD VLSI ARRAY PROCESSORS*

***N. Sudha*[†] *and A.R. Mohan*[‡]**
Centre for High Performance Embedded Systems
Nanyang Technological University, Singapore

Abstract

This chapter gives a hardware-efficient algorithm and a VLSI architecture for finding a path for a mobile robot on an environment image. The algorithm constructs a distance map to identify the collision-free region for a given robot and then finds a path on the region. The path obtained from a start to a goal is the shortest path in terms of the number of steps. The time-critical part of the algorithm is mapped on to a two-dimensional cellular array VLSI architecture that consists of a locally interconnected array of identical processing elements. Due to this local interconnection and regular structure, the architecture can be operated at a high speed and is easily scalable. The design has been implemented on the XCV8000 device of Xilinx. The maximum frequency of operation obtained is 246 MHz. This leads to computing a collision-free path on images of size 100×100 in less than 41 μs. The hardware is capable of processing images at a video rate for real-time navigation in a dynamic environment.

Keywords: mobile robot, navigation, Distance map, cellular architecture, VLSI, FGPA

1. Introduction

Navigation is a fundamental task of a mobile robot by which it guides itself through the environment on the basis of sensory information. The potential of computational vision for robotic control is enormous and the vision-based navigation has been aggressively studied

*A version of this chapter was also published in *New Research on Mobile Robots*, edited by Ernest V. Gaines and Lawrence W. Peskov, published by Nova Science Publishers, Inc. It was submitted for appropriate modifications in an effort to encourage wider dissemination of research.

[†]E-mail address: sudha@ntu.edu.sg
[‡]E-mail address: moha0091@ntu.edu.sg

in the last decade [1]. The work has been progressed on two separate fronts: (1) vision-based navigation of indoor robots [2, 3, 4, 5] where the overall knowledge of the environment is available a priori and (2) vision-based navigation of outdoor robots [6, 7, 8] where the partial knowledge of environment is only available.

Many exiting navigation algorithms were designed for implementation in software. In the case of dynamic environment, high-speed navigation is necessary to avoid collision of robots with moving objects. In such cases, its computational requirement exceeds the computing power of present-day general-purpose processor implementing the software algorithm for navigation. To overcome this problem, it is necessary to build specialized hardware for processing at high-speed.

A new vision-based hardware-directed algorithm for tracing a path for a translating and a rotating robot is proposed in this paper. The image of the entire workspace of the robot is grabbed by a camera mounted overhead. The image is then segmented for obstacles and free space and the Euclidean Distance Transform (EDT) of the segmented image is used for planning a path from start to goal pixel. In our earlier work [9], a cellular architecture was designed for EDT. Similar type of architecture is designed for the computationally intensive part of the navigation algorithm. The main advantage of such architecture is that it can be operated at high-speed due to its local interconnections.

1.1. Related Work

While custom hardware implementation for navigation is concerned, work has been progressed on two approaches. One approach is based on visibility graph and the other is based on Voronoi diagram.

In [10], a parallel hardware-directed algorithm has been proposed for the construction of a complete visibility graph. An extension to this work has been attempted in [11]. A hardware-efficient scheme has been proposed for the computation of shortest path from a start and a goal on such visibility graph. However, the visibility graph approach is not image-based and it assumes the approximation of obstacles and robots to enclosing polygons. The appoximation requires adequate preprocessing. It is also necessary to transform the polygonal robot to a point and then to expand the polygonal obstacles with respect to the polygonal robot before constructing the visibility graph. Since the visibility graph methods for navigation [10, 11] are not image-based, they are computationally less intensive. The edges and vertices of the expanded polygonal obstacles have been taken as input for the visibility graph construction by hardware in [10]. The design for an environment with obstacles containing roughly 60 vertices has been implemented on XCV3200E device of Xilinx. The maximum frequency of operation is close to 60 MHz. The complete path planning involves the computation of shortest path on the visibility graph. The work in [11] proposes a custom architecture for the computation of the shortest path based on linear programming. The design for a graph with 58 nodes and 82 edges has been fitted on one XC2V6000 device of Xilinx. The actual time taken for the computation of the visibility graph and the shortest path has not been reported. The visibility graph approach needs modelling the environment before computing the graph.

Another useful geometrical structure for navigation is the Voronoi diagram. Some work on the design of array-type architectures for the construction of Voronoi diagram on an environment image are available in literature [12, 13]. In [12], a cellular architecture has been proposed for constructing a d_4 metric-based Voronoi diagram. The authors have applied it to navigation for a diamond shaped robot on a synthesised image containing simple obstacles. The diagram is constructed considering also the interactions between features belonging to the same obstacle, and therefore, has extra branches. For a simple image, the Voronoi diagram constructed is shown in Figure 1 (a). But the Voronoi diagram is quite complex when an image of real obstacles is considered and the navigation on such a diagram is difficult. Figure 1 (b) shows an example. The cellular architecture for the 32×32 image size has been mapped on to an ASIC chip of dimension 2.35 mm $\times$ 2.29 mm = 5.38 mm^2 which can be operated with a maximum frequency of 60 MHz. However, the actual time taken for the construction of Voronoi diagram has not been reported.

(a)

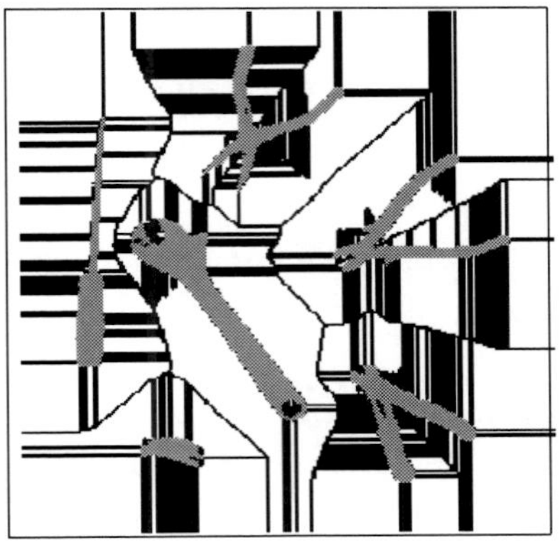

(b)

Figure 1. Voronoi diagrams constructed by the method in [12].

An algorithm for the construction of the exact Voronoi diagram based on Euclidean distance metric and its ASIC implementation are presented in our previous work[13]. The algorithm takes as input a binary image consisting of object and background pixels. Each object in the image is assumed to be a connected component, and the objects are disjoint. In order to construct the Voronoi diagram, each object is dilated iteratively, and the connectivity between the neighboring pixels belonging to the same dilated object is established at each iteration. The iterative dilation process is stopped when the entire image is occupied by the dilated objects. The algorithm has linear time complexity and is suited to a two-dimensional cellular architecture due to the use of local neighborhood calculations on reduced bit-width data. The Voronoi diagram constructed by the algorithm for a binary image is shown in Figure 2. The construction considers the entire obstacle and not its features and hence there are no extra branches. The cellular architecture is in turn complex. Only 16 cells of the architecture have been mapped on to an ASIC chip of dimension 3.16 mm $\times$ 3.16 mm with the maximum frequency of operation 50 MHz. However, navigation has not been attempted using the diagram.

Recently, we presented a solution for navigation on a binary image of environment in [14]. The method constructs the path from the start to the goal pixels on the Euclidean

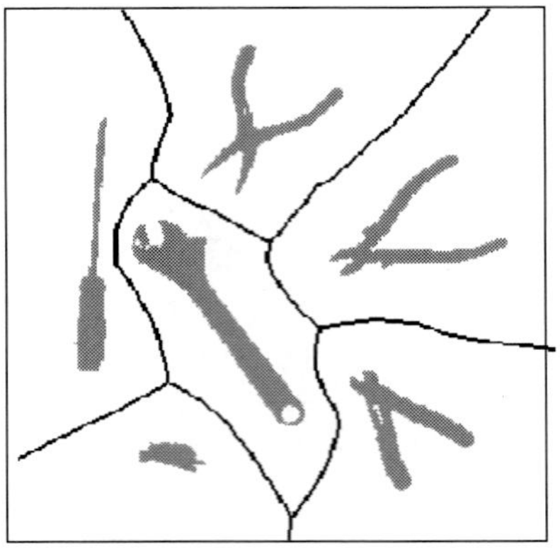
(a)

Figure 2. Voronoi diagram constructed by the method in [13].

distance transform of the image. The transform is used to find the collision-free path. The path obtained from the start to the goal is the shortest in terms of the number of steps. The hardware mapping of the solution was also described. However, the hardware is not time efficient and it is not easily scalable.

In this article, an array architecture based hardware solution is proposed for complete navigation on an environmental image. Due to the local interconnections between array of processors, it can be operated at a high speed and is easily scalable. The method described in the next section for planning a path is similar to [14] but includes the construction of the distance map as well. A novel cellular architecture is then proposed for the complete navigation. Results of FPGA implementation show that the architecture is quite suitable for real-time navigation in a dynamic environment.

2. Robot Navigation using an Environment Image

2.1. Methodology

The method first constructs the distance map for the binary image of an environment to determine the collision-free region on the image. The shortest path is then constructed on the collision-free region.

2.1.1. Construction of Distance Map

For a given binary image of obstacles, the Euclidean Distance Transform (EDT) of the image is first computed as in [9]. The EDT converts the binary image to a multivalued image in which each pixel p is assigned the Euclidean distance between p and the nearest obstacle pixel. In [9], a parallel algorithm which is amenable to VLSI implementation is presented. The salient feature of the algorithm is that the computation of EDT involves only integer arithmetic operations within a small neighborhood of each pixel and hence it is suitable for mapping onto a high-speed array architecture. The algorithm computes distance vector $(\Delta x, \Delta y)$ of pixels where Δx and Δy are the number of rows and columns by which a pixel is displaced from its nearest object pixel. The Euclidean distance is given by $\sqrt{\Delta x^2 + \Delta y^2}$. $(\Delta x, \Delta y)$ of object pixels are initialized to (0,0) and those of free-space pixels are computed iteratively starting from the pixels nearby object and moving towards the far away pixels. At any iteration k, $(\Delta x(p), \Delta y(p))$ of those pixels p whose nearest integer approximation to Euclidean distance $d(p)$ equals k are computed. That is, $d(p)$ lies within $(k-0.5, k+0.5]$. $d(p)$ is not an integer and hence we shall consider $d^2(p)$. $d^2(p)$ lies within $(k^2-k, k^2+k]$ since $d^2(p)$ is an integer. However, $d^2(p)$ is quite large in magnitude and it requires a large storage space in hardware. A new integer quantity $\delta(p)$ which is much smaller than $d^2(p)$ is defined as $(k^2+k) - d^2(p)$. δ is used for the computation of $(\Delta x, \Delta y)$. It is derived as follows.

$d^2(p)$ is derived first and then it is substituted in the definition for $\delta(p)$. $d^2(p)$ can be derived using the already computed $(\Delta x, \Delta y)$ of eight neighbours p_i, $i = 1$ to 8, surrounding p. It is given by $\min[\Delta x_i^2 + \Delta y_i^2]$ where $\Delta x_i = \Delta x(p_i)$ if p_i is in the same row of p. Otherwise, $\Delta x_i = \Delta x(p_i) + 1$. The increment by 1 is due to p being displaced from p_i by one row. Similarly, Δy_i is given in terms of $\Delta y(p_i)$. $d^2(p)$ can be rewritten as $\min[d^2(p_i) + \Delta X_i + \Delta Y_i]$ where $\Delta X_i = 0$ if p_i is in the same row of p. Otherwise, $\Delta X_i = 2\Delta x(p_i) + 1$. $\delta(p)$ is now derived as follows.

$$\begin{aligned}\delta(p) &= k^2 + k - d^2(p) = \max[k^2 + k - d^2(p_i) - \Delta X_i - \Delta Y_i] \\ &= \max[\delta(p_i) - \Delta X_i - \Delta Y_i] = \max[\delta_i] \qquad (1)\end{aligned}$$

$\delta(p) \geq 0$ means $d(p)$ lies below k+0.5.

The iterative computation of $(\Delta x, \Delta y)$ proceeds as follows. δ of the background pixels are initialized to 0. At each iteration k, δ of foreground pixels whose $(\Delta x, \Delta y)$ are not yet known are computed using already computed δ, Δx and Δy of neighbours. If $\delta(p) \geq 0$, then $(\Delta x(p), \Delta y(p))$ corresponds to $(\Delta x_i, \Delta y_i)$ where subscript i pertains to the pixel p_i that gives $\delta(p)$. That is, p_i that satisfies Equation (1). p can be assigned an integer distance d_i which is equal to k. This integer distance is sufficient to determine the collision-free region for the given robot.

Once $(\Delta x, \Delta y)$ of a pixel is known, its δ should be updated for at least two successive iterations as it depends on k. The updating allows use of δ for the computation of $(\Delta x, \Delta y)$ of neighbours. $\delta_k(p)$ at iteration k is derived from $\delta_{k-1}(p)$ at iteration $k-1$ as follows.

$$\begin{aligned}\delta_k(p) &= k^2 + k - d^2(p) = 2k + [(k-1)^2 + (k-1) - d^2(p)] \\ &= 2k + \delta_{k-1}(p) \qquad (2)\end{aligned}$$

The iterative computation of $(\Delta x, \Delta y)$ and δ values of pixels of an image with two background pixels is illustrated in Figure 3. The pixels whose values have been computed at each iteration k are shown in the figure.

To keep track of pixels whose $(\Delta x, \Delta y)$ have been computed, a flag *done* is assigned to each pixel, whose value is set to 1 when the transform values of pixels are computed at any iteration. The $d_i(p)$ given by the iteration number k when *done* is set forms the distance map.

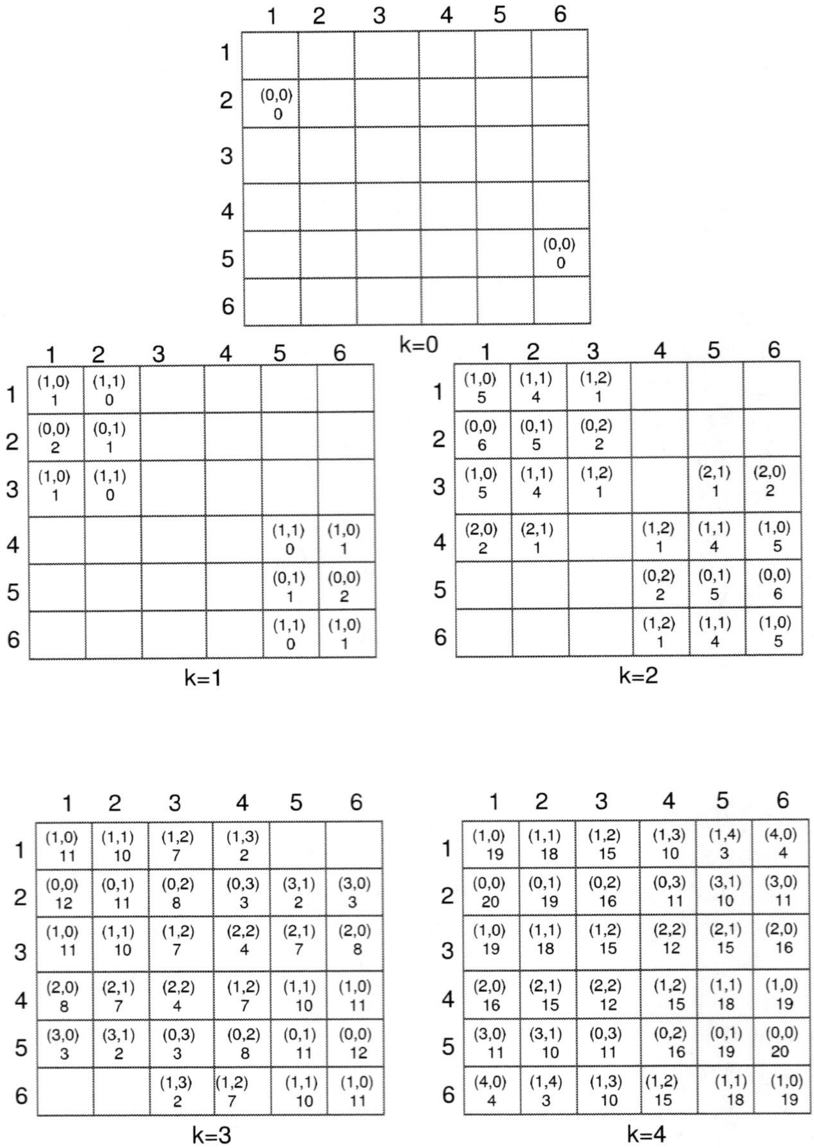

Figure 3. Iterative computation of $(\Delta x, \Delta y)$ and δ of pixels.

2.1.2. Navigation using the Distance Map

The distance map is used to find first the collision-free region for the robot. Let d_{far} be the distance between the center of rotation and the farthest point of the robot from the center. The collision-free region consists of those pixels whose distance values (d_i) are greater than d_{far}. Consider the collision-free region as a graph. Each pixel is represented as a node and it is connected to the eight neighboring pixels surrounding it by edges. The construction of the shortest collision-free path involves the construction of the breadth first search (*bfs*) tree on the collision-free graph with its root being the goal pixel's node. The construction of *bfs* tree is proceeded until the start pixel's node is encountered. The construction involves storing the parent of each pixel. The path is traced by following the parent nodes from the start pixel's node until the root node is reached. The path constructed by the method on the image is a shortest path in terms of the number of pixels. The robot is moved from one pixel to the next in the path by aligning its center of rotation to every pixel.

The method described above is illustrated in Figure 4. Consider a 10×10 image shown in Figure 4 (a). The integer approximation of EDT values are shown in Figure 4 (b). Let d_{far} equal two pixel units. The corresponding collision region whose pixels having integer distance values less than or equal to 2 is as shown in Figure 4 (c) and the collision-free graph is displayed in Figure 4 (d). Consider the pixel at (2,6) as start pixel and the pixel at (9,1) as goal pixel. The *bfs* tree constructed with goal pixel's node as root is shown in Figure 4 (e). The collision-free shortest path on the image is shown in Figure 4 (f) by arrows.

A parallel algorithm is derived for the proposed method.

2.2. Algorithm

Given the binary image of objects, start pixel, goal pixel and d_{far} of a robot, the algorithm iteratively constructs the distance map and then the *bfs* tree. A path for the robot from the start pixel to the goal pixel is derived from the *bfs* tree. We set forth the notations for describing the algorithm.

Notations

$d(p)$	EDT value of a pixel p
$(\Delta x(p), \Delta y(p))$	Distance vector of p
k	iteration number
$\delta(p)$	$k^2 + k - d^2(p)$
$d_i(p)$	integer approximation of $d(p)$
$done(p)$	flag set to 1 if $(\Delta x, \Delta y)$ of p is computed
d_{far}	distance in pixel units between the center of rotation and the farthest point in the robot from the center
$cfr(p)$	flag set to 1 if p is in collision-free region
p_s	start pixel
$visited(p)$	set to 1 when p is visited while constructing *bfs* tree
$parent(p)$	pointer to the parent of p in the *bfs* tree

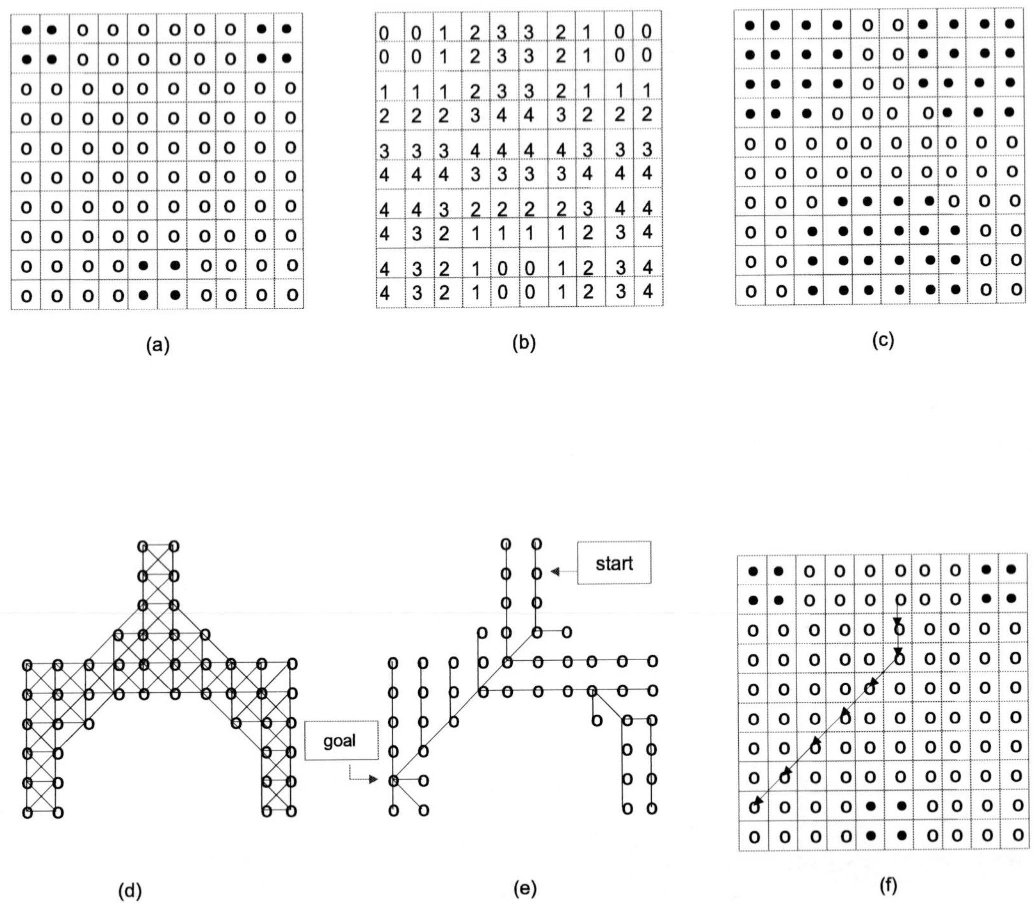

Figure 4. Illustration of the proposed method. (a) Example image. (b) Integer EDT values. (c) Collision region for $d_{far} = 2$ pixel units. (d) Collision-free graph. (e) Constructed *bfs* tree. (f) Collision-free path on the image.

The parent of a pixel in the *bfs* tree is one of the eight neighbors surrounding the pixel in the 3×3 neighborhood N. The pointer value corresponding to each neighbor is assigned as follows.

4	3	2
5	0	1
6	7	8

The pseudocode of algorithm is as follows.

ALGORITHM:

Inputs: EDT of an $n \times n$ image, d_{far}

Outputs: Sequence of moves for the robot from start pixel to goal pixel

Step 1: Initialization

$\Delta x = \Delta y = \delta = 0$ & $done = 1$ for object pixels

$$visited(p) = \begin{cases} 1 & p \text{ is goal pixel} \\ 0 & otherwise \end{cases}$$

$parent(p) = 0$ for all p

Step 2: Find EDT

repeat for k=1,2,...
  for all pixels p **do in parallel**
    Compute Δx_i, Δy_i and δ_i, i = 1 to 8
    δ_m = max $\{\delta_i$ with $done(p_i)\}$.
    if $(done(p) = 0 \wedge \delta_m \geq 0)$
      $\Delta x(p) = \Delta x_m$
      $\Delta y(p) = \Delta y_m$
      $\delta(p) = \delta_m$
      $done(p) = 1$
    else if $done(p) = 1$
      $\delta(p) = \delta(p) + 2k$
    end if
  end for
until $done = 1$ for all pixels

Step 3: Find collision-free region

for every pixel p **do in parallel**
  if $(d_i(p) > d_{far})$ $cfr(p) = 1$

else $cfr(p) = 0$
end for

Step 4: Construction of *bfs* tree

repeat
for every pixel p **do in parallel**
if $(visited(p)=0) \wedge (\exists p_n \in N[visited(p) = 1])$
$visited(p) = 1$
$parent(p)$ is pointed to p_n
end if
end for
until $(visited = 1$ for all p where $cfr = 1)$

Step 5: Find sequence of moves
$p = p_s$
while $(parent(p) \neq 0)$
print $parent(p)$
p = parent pixel of p
end while

The sequence of moves are given by the pointers. The robot can take horizontal, vertical or diagonal step based on the pointer value. It should be noted that once the *bfs* tree is constructed in Step 3, the shortest path from any pixel to the specified goal pixel can be found out.

2.3. Simulation Studies

The algorithm has been coded in C and simulated on a PC with Intel Pentium 4 processor, 2 GB RAM and running Windows XP operating system.

The performance of the algorithm has been tested on different images. Figure 5 shows a binary image of size 130×90 and the resulting collision-free paths for different start and goal pixels. Let the d_{far} be 5. The coordinates of start and goal pixels chosen in Figures 5 (a) and (b) are (50,80) and (100,30), and (5,6) and (100,51) respectively. The number of moves for the robot in these examples are 58 and 109 respectively.

3. Proposed Cellular Array VLSI Architecture

In the algorithm described in Section 2.2., the construction of EDT (Step 2) and *bfs* tree (Step 4) are time consuming since they are iterative procedures and at each iteration, every pixel has to be processed. In this section, a VLSI architecture namely cellular architecture, is designed for the operations from Step 2 to Step 4 in the algorithm. For an $n \times n$ image, the architecture consists of an $n \times n$ array of cells in which each cell is connected to the eight neighboring cells surrounding it. A 4×4 cellular array is shown in figure 6. A cell is a sequential logic consisting of storage elements Δx, Δy, δ, d_i and *done* for the computation

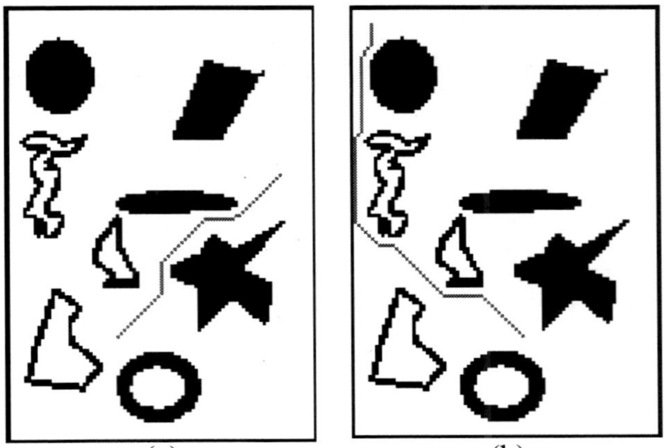

(a) (b)

Figure 5. Simulation results of the algorithm on a test image. The collision-free paths for d_{far}= 5 are shown in red. (a) Start and goal pixels are (50,80) and (100,30) respectively. (b) Start and goal pixels are (5,6) and (100,51) respectively.

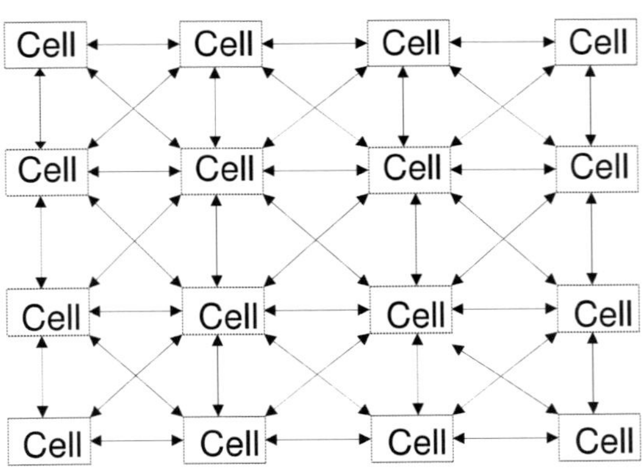

Figure 6. A 5×5 cellular architecture.

of EDT and cfr, $visited$ and $parent$ for constructing a collision-free path using EDT. The sizes of the storage elements depend on the size of the image. The computation of these values is implemented by using data processing elements such as adders, subtracters and comparators, and simple logic gates. The iteration number k is generated by an external counter and it is given to all the cells. The d_{far} value is also fed to all cells. The data paths between cells carry the values Δx, Δy, δ, $done$ and $visited$. The cells are updated synchronously with an external clock. The speed of operation of the cellular architecture depends primarily on the delay due to a cell. The delay due to interconnection between cells is negligible because the cells are locally connected. We have designed a cell in order to estimate the delay in terms of the maximum clock frequency.

3.1. Design of a Cell

The different modules of a cell are shown in Figure 7. The ADD-SUB module computes δ_i,

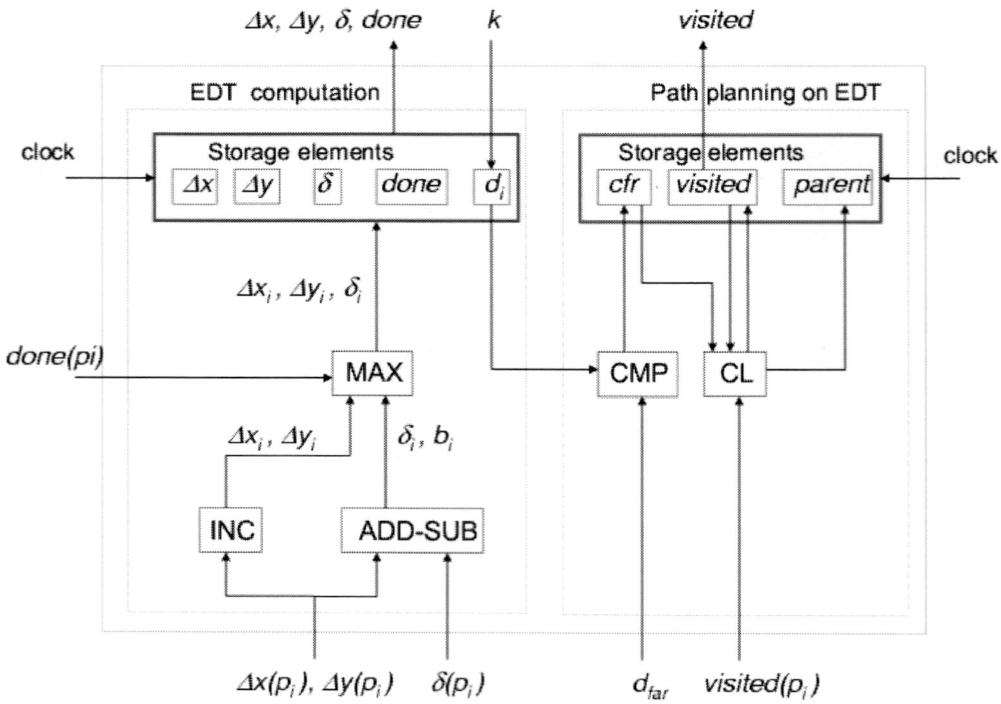

Figure 7. The block diagram of a cell. CL denotes the combinational logic.

$i = 1$ to 8. The computation involves an addition and a subtraction for $i = 2,4,6$ & 8 and only a subtraction for $i = 1,3,5$ & 7. Therefore, eight subtracters and four adders are required to realize this computation. The INC module computes Δx_i and Δy_i. The computation requires twelve incrementers, 6 for implementing $\Delta x_i = |\Delta x(p_i)| + 1$, i=1,2,4,5,6 & 8, and another six for implementing $\Delta y_i = |\Delta y(p_i)| + 1$, i=2,3,4,6,7 & 8.

Once δ_i, Δx_i and Δy_i for i=1 to 8 have been computed, these values are given to a MAX module along with $done(p_i)$ and borrow bits b_i from the subtracters of the ADD-SUB module. The MAX module has seven CMP-MUX (comparator-multiplexer) modules arranged in three levels as shown in Figure 8 to compute $\max[\delta_i]$, i = 1 to 8. In the figure, $\max[\delta_i]$ is denoted by δ_m. Further, the MAX module allows the values of Δx_i and Δy_i, corresponding to the i that yields δ_m. These values are denoted by Δx_m and Δy_m.

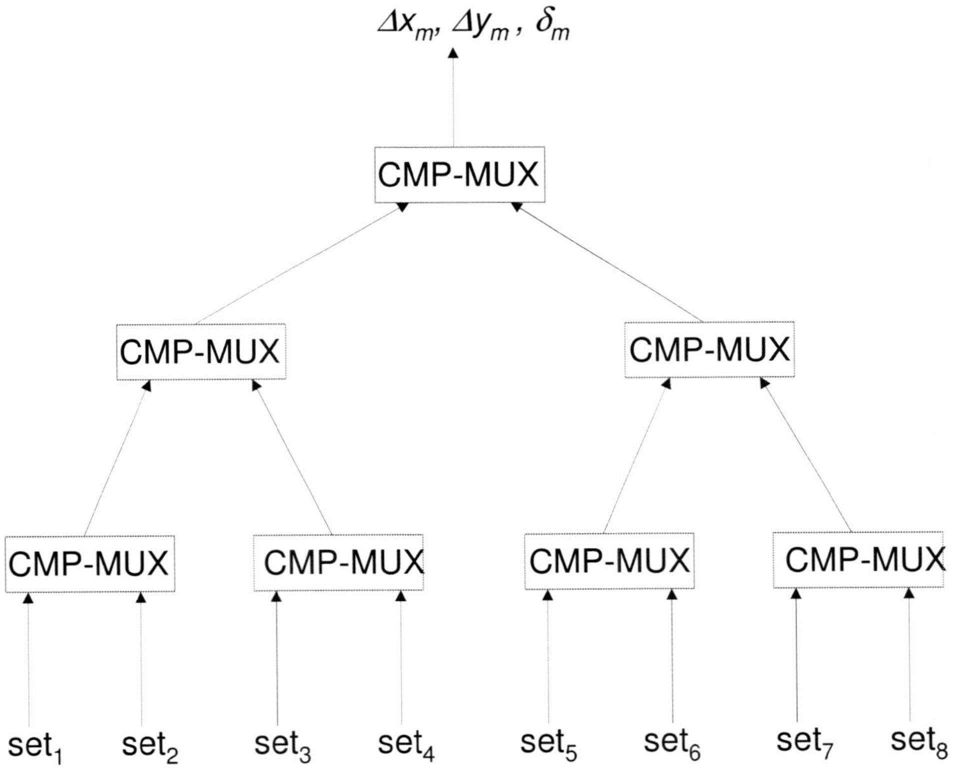

Figure 8. MAX module. set$_i$ indicates the set of inputs $\delta_i, \Delta x_i, \Delta y_i, done(p_i)$ and b_i.

The CMP-MUX module takes two sets of inputs, set_j = $\{\delta_j$, Δx_j, Δy_j, $done(p_j)$, $b_j\}$ and set_k = $\{\delta_k$, Δx_k, Δy_k, $done(p_k)$, $b_k\}$. It has a comparator to compare $\delta_j > \delta_k$ and a multiplexer that outputs either set_j or set_k depending on the value of sel input of multiplexer. set_j is output when sel=0 while set_k is output when sel=1. The sel is generated using the output cmp of comparator and the values of $done$ and borrow bits. The comparator is designed for the simple case of comparing two unsigned binary numbers. $done(p_j)$=1 means the value of δ_j is valid. b_j=1 means δ_j is negative and cmp=1 means $\delta_j > \delta_k$. The value of sel for different cases is tabulated in Table 1.

The δ_m, output by the MAX module, is added to $2k$ where k is the iteration number generated by an external counter. The output $\delta(p)$ of the adder is given as input to the

Table 1. Value of sel for different cases of inputs to CMP-MUX module. "-" indicates "don't care".

$done(p_j)$	$done(p_k)$	b_j	b_k	cmp	sel	comments
0	0	-	-	-	-	Both δ_j & δ_k are invalid
1	0	-	-	-	0	δ_j is only valid
0	1	-	-	-	1	δ_k is only valid
1	1	0	0	1	0	Both are valid and positive. $\delta_j > \delta_k$
1	1	0	0	0	1	Both are valid and positive. $\delta_k > \delta_j$
1	1	1	1	1	0	Both are valid and negative. $\delta_j > \delta_k$
1	1	1	1	0	1	Both are valid and negative. $\delta_k > \delta_j$
1	1	0	1	-	0	δ_j is positive. δ_k is negative
1	1	1	0	-	1	δ_j is negative. δ_k is positive

register δ. The outputs, Δx_m and Δy_m, of the MAX module are given as inputs to the registers Δx and Δy. The $done$ flip-flop is input with logic 1. In the design, the registers and flip-flop are loaded with the available inputs during the rising edge of the clock. The clock is activated only when the following conditions are satisfied.

1. $done$ of cell is not set.

2. At least a value of $done$ of neighbours is logic 1.

3. $\delta(p)$ is positive. In two's complement representation of $\delta(p)$, this means that the MSB is 0.

In the case of register δ, the clock is activated when the conditions 2 and 3 are satisfied. Let $cndn$ be the signal whose value is 1 if the above three conditions are satisfied. $cndn$ can be generated using a simple AND-OR logic circuit.

After the EDT computation, the storage element d_i of each cell receives the integer Euclidean distance value. A comparator compares d_i with the input d_{far} and sets cfr accordingly. The cfr values of all cells are computed in a single clock cycle. The $visited$ and $parent$ values are received by cells in different clock cycles. These values are set based on the $visited$ of neighboring cells. A simple combinational logic derives these values.

3.2. Operation of Hardware

The storage elements are first initialized. The $done$ of cells corresponding to obstacle pixels are initialized to 1 and the $visited$ of goal pixel is set to 1. All the remaining storage elements are initialized to 0. The d_{far} is applied and the architecture is allowed to run in synchronous with the clock. The computation of EDT is carried out during the initial few clock pulses. The number of clock pulses (N_{EDT}) needed to compute EDT is set to d_{max}, the maximum possible distance value, which is $\sqrt{2}n$ in the case of an $n \times n$ image. The cfr computation takes a single clock cycle. The number of clock pulses required for the construction of *bfs* tree (N_{bfs}) for finding a path on it is n^2. This is because in the worst case, the path obtained is a zig-zag one. Hence the total number of clock pulses is given by $N_{pp} = N_{bfs} + N_{EDT} + 1 = n^2 + \sqrt{2}n + 1$.

3.3. FPGA Implementation

The design of the cellular architecture was coded in Verilog HDL [15] and its functional behavior was tested in the Xilinx simulator. The design was tested with different images and the values stored in the storage elements after running the design were checked for correctness with the ones obtained in C implementation. A snapshot of a simulation result is shown in Figure 9. Figure 9 (a) shows a test image grid in which the pixels belonging to objects are marked by 'X'. The start and the goal pixels are marked by 'S' and 'G' respectively. Assuming the d_{far} to be 0, Figure 9 displays the $visited$ values of the pixels near the goal pixel during the construction of the bfs tree. In the figure, $rows[i]cells[j]$ indicate the pixel at row $i + 1$ and column $j + 1$. It can be noticed that the pixels corresponding to the cells are visited at different clock cycles. The $visited$ of the goal pixel has been initialized to logic 1. The $visited$ of pixels marked by 'X' in the chosen window is logic 0 throughout. The $visited$ of the two pixels lying below the goal pixels are set to 1 in the first clock trigger and the $visited$ of the three pixels at the bottom of the window are set in the next clock trigger. After the functional testing, implementation of the design was carried out.

As a single cell decides the frequency of operation of the entire architecture, the design for a single cell has been implemented first to confirm the suitability of the architecture for dynamic situation. The design has been mapped onto one of the target devices of Xilinx. The appropriate device has been chosen taking into consideration the number of logic blocks and pins in the device for implementing the cellular architecture of reasonably large size. The specifications of the target device are as follows - family: Virtex 2; device: XCV8000; package: FF1152. The clock frequency obtained through implementation of a cell is around 246 MHz.

Table 2. Results of implementation of cellular architecture of size 100×100.

Resource	Available	Used	Usage
Slice	46592	13824	29%
Slice Flip Flop	93184	18432	19%
4 input LUT	93184	27648	29%
Bonded IOB	824	2	0%
GCLK	16	1	6%

Once the clock frequency has been estimated, the overall cellular architecture of an $n \times n$ image has been designed by replicating n^2 cells in two dimension and interconnecting them. The other components of the architecture are a clock and a control logic having a counter for generating the iteration number. The functional behavior of the design of the architecture has been tested in ModelSim by giving images of size $n \times n$ as input, allowing the design to run for N_{pp} clock pulses and then verifying the outputs of the registers of the cells. The

X	X	X	X	X	X	X	X
X	X	G	X	X	X		X
X	X			X	X		X
X				X		S	X
X							X
X						X	X
X			X	X	X	X	X
X	X	X	X	X	X	X	X

(a)

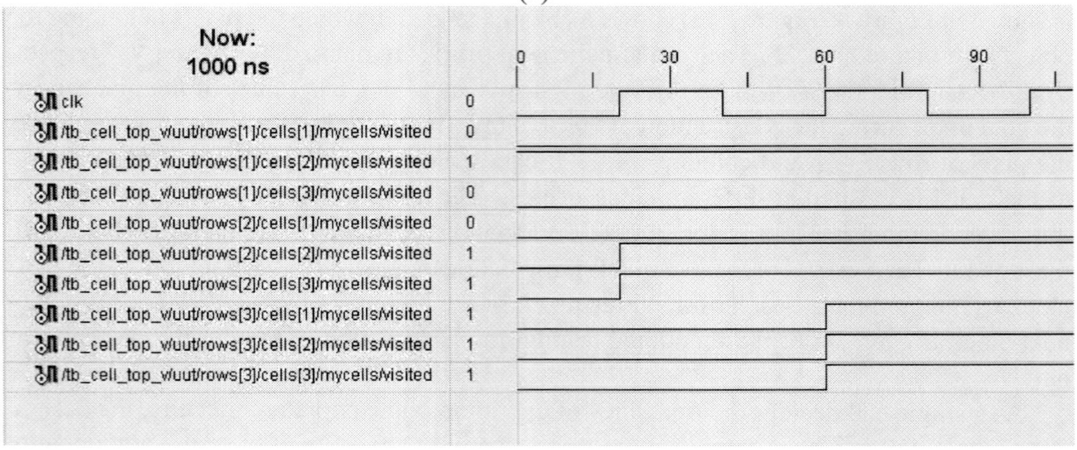

(b)

Figure 9. Simulation results of the cellular architecture. (a) Sample image grid. (b) Snapshot of Simulation displaying the $visited$ values of pixels in the 3×3 window shown in (a)

cellular architecture was then implemented on the same Xilinx device. The implementation results of an array of size 100×100 are given in Table 2. It shows the different hardware resources in the device and their usage.

Since a cellular architecture consists of locally connected identical cells, the delay due to interconnections is negligible. Hence, the time (T_{pp}) taken to compute the path on the image of size $n \times n$ can be estimated from the clock rate (f_c) obtained from the implementation of a cell. The period (T_c) of a clock pulse is $1/f_c$ seconds. Hence, $T_{pp} = N_{pp} * T_c$. The number of images (N_I) processed per second on the cellular architecture is greater than or equal to $\lceil 1/T_{pp} \rceil$.

Some results and comments for an image size 100×100 are as follows. d_{max} for this case is 141.42. The frequency of operation of cell is 246 MHz. T_{pp} is around 41 μs and hence N_I is approximately 24,254. N_I is much greater than the video rate, which is about 30 frames per second and hence the navigation with a cellular architecture is quite suitable for dynamic environment.

4. Conclusions

An algorithm for tracing the shortest path of a translating and a rotating robot on the binary image of an environment is given in this paper. The algorithm first constructs the distance map of the image to obtain a collision-free region and then constructs the bfs tree of pixels in the collision-free region which facilitates in finding the shortest path from a start pixel to the goal pixel. The algorithm has been mapped on to a two dimensional cellular architecture consisting of $n \times n$ array of identical cells. Due to the local interconnections between cells, the architecture can run at a high speed and is easily scalable. The architecture has been implemented on an FPGA device. The device can handle more than 20,000 image frames of size 100×100 in a second. Such a processing speed is useful for real-time navigation or for updating the path in a dynamic environment with obstacles moving even randomly.

References

[1] G. Desouza and A. Kak, "Vision for mobile robot navigation: a survey," *IEEE Transactions on Pattern Analysis and Machine Intelligence*, vol. 24, no. 2, pp. 237–267, 2002.

[2] A. Kosaka and A. Kak, "Fast vision-guided mobile robot navigation using model-based reasoning and prediction of uncertainities," *Computer Vision, Graphics and Image Processing - Image Understanding*, vol. 56, no. 3, pp. 271–329, 1992.

[3] M. Meng and A. Kak, "Mobile robot navigation using neural networks and nonmetrical environment models," *IEEE Control Systems Magazine*, pp. 30–39, 1993.

[4] D. Kim and R. Nevatia, "Representation and computation of the spatial environment for indoor navigation," *Proceedings of International Conference on Computer Vision and Pattern Recognition*, pp. 475–482, 1994.

[5] L. Kavraki, P. Svestka, J. Latombe, and M. Overmars, "Probabilistic roadmaps for path planning in high-dimensional configuration spaces," *IEEE Transactions on Robotics and Automation*, vol. 12, pp. 566–580, 1996.

[6] V. Graefe, "Vision for intelligent road vehicles," *Proceedings of IEEE Symposium on Intelligent Vehicles*, pp. 135–140, 1993.

[7] D. Pomerleau and T. Jochem, "Rapidly adapting machine vision for automated vehicle steering," *IEEE Expert, Intelligent Systems and Their Applications*, vol. 11, April 1996.

[8] E. Dickmanns, "Computer vision and highway automation," *Vehicle System Dynamics*, vol. 31, no. 5, pp. 325–343, 1999.

[9] N. Sudha, S. Nandi, and K. Sridharan, "Cellular architecture for Euclidean distance transformation," *IEE Proceedings - Computers and Digital Techniques*, vol. 147, pp. 335–342, September 2000.

[10] K. Sridharan and T. Priya, "The design of a hardware accelerator for real-time complete visibility graph construction and efficient FPGA implementation," *IEEE Transactions on Industrial Electronics*, vol. 52, pp. 1185–1187, August 2005.

[11] T. Priya, P. R. Kumar, and K. Sridharan, "A hardware-efficient scheme and FPGA realization for computation of single pair shortest path for a mobile automaton," *Microprocessors and Microsystems* , vol. 30, pp. 413–424, 2006.

[12] P. Tzionas, A. Thanailakis, and P. Tsalides, "Collision-free path planning for a diamond-shaped robot using two-dimensional cellular automata," *IEEE Transactions on Robotics and Automation*, vol. 13, pp. 237–250, 1997.

[13] N. Sudha and K. Sridharan, "A high-speed VLSI design and ASIC implementation for constructing Euclidean distance-based discrete voronoi diagram," *IEEE Transactions on Robotics and Automation*, vol. 20, no. 2, pp. 352–358, 2004.

[14] N. Sudha and A. Mohan, "Design of a hardware accelerator for path planning on the euclidean distance transform," *Journal of Systems Architecture*, vol. doi:10.1016/j.sysarc.2007.06.003.

[15] S. Palnitkar, *Verilog HDL: A Guide to Digital Design and Synthesis*. Prentice Hall, 2003.

In: VLSI and Computer Architecture
Editor: Kenzo Watanable
ISBN: 978-1-60692-075-6

Chapter 8

VLSI ARCHITECTURES FOR AUTONOMOUS ROBOTS – A REVIEW*

K. Sridharan[a,†]***, S.K. Lam***[b,‡] ***and T. Srikanthan***[b,#]

[a]School of Computer Engineering
Nanyang Technological University, Singapore
[b]Centre for High Performance Embedded Systems
School of Computer Engineering, Nanyang Technological University
Singapore

Abstract

Mobile robots operating autonomously are valuable in a number of situations. These could involve environments/tasks that are either difficult to reach by humans or those that are dangerous for humans. The latter may include detection of land mines, maintenance of hazardous material etc. Considerable research has been done on sensing, planning and control of robots with different models and systems for each of them. Sensors ranging from sonar to highly sophisticated ones (such as laser range finders) have been studied. A variety of control strategies have also been investigated.

Recently, there has been an increasing emphasis on saving energy during various tasks performed by a mobile robot. A significant source of energy consumption is the embedded processing aboard the robot. Traditional mobile robots typically include a laptop as the embedded computer besides sensors, motors and microcontroller(s) for low-level controls. One approach to reducing energy consumption is by examining architectural alternatives for the efficient implementation of algorithms for mobile robots. Algorithm-centric VLSI architectures for mobile robots with a view to run-time configurability, system integration and design productivity in terms of area, time and energy-efficiency seem appropriate. This review

---

* A version of this chapter was also published in *Autonomous Robots Research Advances,* edited by Weihua Yang, published by Nova Science Publishers, Inc. It was submitted for appropriate modifications in an effort to encourage wider dissemination of research.

† E-mail address: sridhara@ntu.edu.sg; On leave from EE Dept., IIT Madras, India

‡ E-mail address: assklam@ntu.edu.sg

E-mail address: astsrikan@ntu.edu.sg

paper examines the state of the art in VLSI architectures and algorithms for embedded computing on mobile robots and suggests directions for further research.

1. Introduction

Autonomous mobile robots have a wide range of applications. These include surveillance, cleaning, transportation and customer support in indoor environments and mining, firefighting and agriculture in outdoor environments. Autonomous robots are capable of obtaining high-level descriptions of tasks and executing them without continuous human intervention [1].

Autonomous robots operate with a variety of sensors on them. For example, robots used in hospitals for transportation tasks have typically a camera looking to the ceiling to detect lamps as landmarks for localization. Robots used in cleaning applications may have sonar and other mechanisms to enable navigation. Underwater robots also employ sonar for navigation. Many autonomous robots have a combination of different types of sensors such as sonar, IR sensor and laser scanners.

Operation of autonomous robots involves multiple tasks: 1) sensing the environment (2) path planning and 3) issue of motion commands and control. A high level task planner specifies the destination and any constraints on the course (e.g. time). Figure 1 illustrates a possible mobile robot control hierarchy along with the appropriate information flow. It is worth noting that some stages can be skipped for specific tasks. For example, a point to point navigation query may not require a detailed environment map.

Traditionally, autonomous mobile robots have operated with one or more general-purpose processors on-board along with a set of peripherals. With increasing emphasis on extending battery life and reducing payload on autonomous systems, there has been a surge in interest in architecturally-efficient alternatives. The alternatives include Field Programmable Gate Arrays (FPGAs) and Digital Signal Processors (DSPs). There has also been considerable research in development of analog VLSI solutions and hybrid analog/digital VLSI solutions.

Efficient solutions using various processing styles call for careful design of hardware-based schemes and appropriate architectures. Considerable work has been done on using VLSI in various forms. In some scenarios, VLSI has been seen as a means to obtain real-time solutions while in others, the ability to obtain field upgradability and robust solutions has been the driving force.

This survey examines the state of the art in hardware-directed solutions and VLSI architectures for autonomous robots. The survey is divided into three parts: (a) sensing (b) control and (c) navigation of autonomous robots.

The remainder of this article is organized as follows. Next section presents efficient architectures for design and handling of sensors. Section 3 presents VLSI architectures for control of autonomous robots. Section 4 is devoted to survey of VLSI architectures for navigation of mobile robots. Section 5 summarizes the paper and discusses avenues for further research.

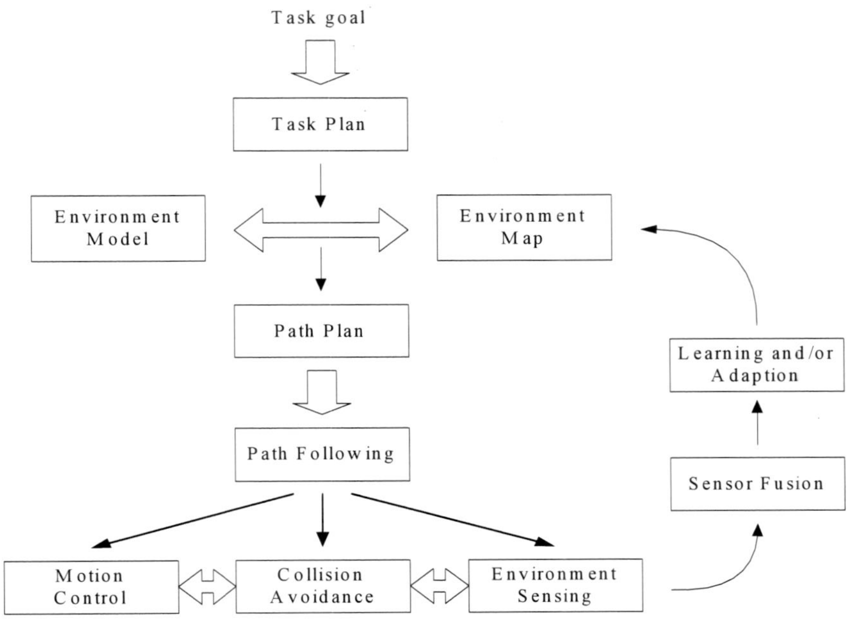

Figure 1. Mobile robot navigation control hierarchy.

2. Architectural Aspects of Sensing in Autonomous Robots

Sensors are central to the operation of an autonomous robot. Sensor design and operation have benefited tremendously from developments in VLSI. In this section, we describe VLSI-based sensor systems for autonomous robots.

An inexpensive and simple sensor appropriate for distance measurement is the ultrasonic range sensor. These sensors have limitations with regard to angle information provided. In other words, bearing angle information using conventional ultrasonic range modules is obtained using the beam angle of one receiver and it is not sufficiently accurate for many applications. While variations have been proposed in terms of number of receiver units, a promising approach is one based on DSPs [2] for real-time sonar echo processing with greater bearing/range accuracy than even laser range finders.

Another piece of work related to ultrasonic sensors is on correlation detection based on an FPGA [3]. The key idea is digital modulation of the signal emitted and the correlation thereof with the echo signal received. The implementation is based on an FPGA and the system is set up readily for multiple transducers. An extension of the work in [3] is presented in [4]. The authors in [4] examine real-time implementation of an efficient Golay correlator as applicable to ultrasonic sensorial systems. The approach in [4] comprises of four transducers and in particular, two emitters and four receivers. Each transducer has an associated set of hardware resources built on a platform based on a Xilinx XC4005 FPGA.

Besides ultrasonic sensors, hardware-directed solutions have been developed for other types of sensor systems. The authors in [5] present an approach for correcting sensor non-uniformities in an infrared system and a hardware architecture for implementing their scheme in real-time. The authors also outline an implementation based on Xilinx XC2V2000 FPGA and point to the advantage of field upgradability for obtaining a robust system.

The development of a reconfigurable and flexible parallel 3D vision system for a mobile robot based on DSPs and FPGAs is described in [6]. The objective here is to take advantage of the capability of XC4005 FPGAs and TMS320C40 for high speed image processing while performing image acquisition using a TMS340C20-based board. Examples of fast reconstruction of 3D scenes are provided in [6].

The authors in [7] analyse the movement of objects in front of an autonomous vehicle using image processing techniques. In particular, they consider the problem of detecting time to impact of an approaching object in road-like environments and develop strategies to minimize the time of calculation. The approach uses a pipelined processing architecture based on programmable devices to calculate the time to impact at a rate of 80 images per second (which they show is adequate for cars moving at speeds close to 100 Km/h). The implementation is based on Altera FLEX8000 family CPLDs.

An active motion vision system designed for use in an autonomous mobile robot is described in [8]. The system has been designed for implementation in a chip-level parallel computer architecture and for synchronous operation with a digital imager at high frame rates. The authors report that tests on synthetic and real images have established robustness to noise and interference.

A recent trend in the domain of umanned air vehicles has been the development of VLSI-based Elementary Motion Detector (EMD) arrays [9]. In particular, the inspiration is from insects that perform complex tasks such as 3-D navigation using optic flow sensors. An FPGA-based implementation of an EMD array is described in [9]. The idea is to take advantage of the ability of an FPGA to receive inputs from several photoreceptors of similar (or different) shapes and sizes located in various parts of the visual field.

While the discussions so far have concentrated on digital VLSI-based solutions, there have also been advances on development of compact, low power sensors in analog and hybrid analog/digital VLSI. One example of this is the work reported in [10]. The authors in [10] discuss the limitations of traditional CCD cameras in terms of how well they compensate for local variations in the illumination of a scene. They describe how biological systems efficiently solve many perceptual tasks. In particular, they discuss how thee image that falls on the retina is not transduced to the visual areas as a set of unrelated pixel readings but is processed locally at each photoreceptor to express neighborhood relationships present in an image. They present details of a mobile robot equipped with a neuromorphic sensors implementing a one-dimensional silicon retina. An analog VLSI chip fabricated using a 2μm CMOS technology is also described.

Another interesting piece of work that also draws its inspiration from biology is presented in [11]. The authors in [11] present a biomimetic sensor that processes an image completely into a control signal to be used directly in the guidance of a robot requiring no microprocessor. The sensor here is an analog VLSI sensor and it transduces a visual image focused on to it, computes small-field visual motion information and passes this information through multiple stages of computation to produce a control signal (in the form of current). The signal responds strongly to small moving objects but only weakly to large-field background motion. The authors report the development of an IC through MOSIS in a 1.5μm CMOS process.

3. VLSI Architectures for Control of Autonomous Robots

In the previous section, we have discussed research on VLSI-based sensors for autonomous robots. Another important component of an autonomous robotic system is control. In this section, we present the state of the art on this subject.

One of the early efforts on development of a real-time control architecture for robotics is described in [12]. The authors in [12] present a restructurable architecture based on a VLSI Robotics Vector Processor (RVP) chip. The RVP structure differs from a systolic array and an asynchronous multiprocessor. The architecture exploits the conditional-free nature of kinematics and dynamics computations (where the sequence of operations at the lowest level is not a function of the input data from the sensors). The architecture is tailored to extract parallelism in low-level matrix/vector operations characteristic of kinematic and dynamics computations required for real-time control. The authors present a floorplan for the RVP and indicate that the entire design would require a single VLSI chip of dimensions 7.9 mm x 9.2 mm using 1.2 μm CMOS technology and operate as a single board multiple-RVP system on a mobile robot.

Several other efforts have followed taking advantage of developments in FPGA and DSP technology as well as advances in analog VLSI. The authors in [13] present the design and implementation of a mobile robot subsumption architecture in which the computing elements of control are based on FPGAs. The advantages of using an FPGA for control are outlined and the design of a behavioural control system using FPGA is described. The authors further justify the platform of choice by the observation that behaviours operate in parallel.

A fuzzy control strategy for autonomous vehicle guidance is presented in [14]. The authors in [14] implement a fuzzy co-processor for their vehicle guidance application using an FPGA. The fuzzy co-processor executes the fuzzy algorithm using a fuzzy rule base memory. The implementation uses the Altera EPF10K50RC240-3 device. The authors indicate that the hardware fuzzy co-processor executes the fuzzy algorithm fast enough to manage correctly the vehicle guidance. Development of a prototype using a radio-controlled miniature car is also described.

Biologically inspired systems have also been developed for control tasks. The authors in [15] present the design of a biologically inspired system for tracking objects in a visual scene. Neural oscillators are used to drive DC motors to provide a compact single-chip system for tracking bright objects. A sensorimotor control vehicle has been developed and described by the authors. The work is believed to be one of the early efforts to fully integrate and apply analog VLSI sensorimotor control to a mobile vehicle.

Besides land vehicles and robots, control of Autonomous Underwater Vehicles (AUVs) has also been studied from the VLSI point of view. The authors in [16] propose an intelligent system for failure detection and control in AUVs where the vehicle could continue exploration in case of minor failures in sensors and control surfaces. The system integrates the adaptability of an artificial neural network and the inferencing ability of a fuzzy rule based expert system on a single VLSI chip. The entire system is designed for use as a low level diagnoser in an overall control system for AUVs.

A DSP-based control system for a mobile robot is described in [17]. The authors present the details of a prototype mobile platform with a TI TMS320F2812 DSP. The platform also has an omni-directional vision system. The DSP controller processes data from various sensors and communicates to a host PC. The DSP replaces a traditional microcontroller.

An extension of the work presented in [14] is described in [18]. The authors in [18] present a strategy to implement human-like driving skills by an autonomous Car-Like Mobile Robot (CLMR). The authors present the notion of Autonomous Fuzzy Behaviour Control (AFBC) and describe the design procedure of human-like AFBC and its implementation on an FPGA. In particular, the authors use Altera EPF602ACT144-3 for the experiment board. The authors also present the setup of the CLMR and experiments of the AFBC on the CLMR. The experiments demonstrate the effectiveness of the proposed AFBC scheme for practical application to real car maneuvers.

Wall-following control by a mobile robot using an FPGA has been investigated in [19]. In particular, the authors describe the design and implementation of a fuzzy logic control algorithm for a robot on a FPGA. The hardware design is based on the Nios processor and a module based on the Altera's APEX20K200E device has been developed. Experimental results on a car-like mobile robot have also been presented to show the effectiveness of the robot to follow a wall in an unknown and changing environment.

Research has also been done on using hybrid analog/digital VLSI for mobile robot control. The authors in [20] consider the problem of control of a legged robot. They present the design of a cellular neural network based chip for locomotion pattern generation. The rhythmic locomotion pattern is generated using an analog approach while the feedback issues are addressed by a digital unit. The details of implementation of the chip and experiments on a legged robot are also described.

4. VLSI Architectures for Navigation of Autonomous Robots

Autonomous robotic systems equipped with one or more sensors perform a variety of tasks including exploration and navigation. Many environments involve robots operating amidst multiple dynamic objects. The literature on VLSI for navigation-related tasks is limited. A summary of research on parallel algorithms and architectures for robotic planning is presented. We also review some of our research in detail.

A review of parallel processing approaches for roadmap, cell decomposition and potential field methods that is well suited for implementation on parallel architectures and multiprocessors can be found in [21].

Neural network based implementations for robot navigation has also been reported. Neural networks have the unique features of self-learning and adaptive capabilities that make them a good candidate for robot navigation. A ring systolic architecture for implementing Artificial Neural Networks for robotic processing has been proposed in [22].

Research based on distributed representation of the *C-space* [23] and cellular automata architectures [24][25] have been conducted for robotic planning. Cellular automata are mathematical models of physical system in which the space and time are discrete. In its simplest form, a cellular automata consists of a line of cells, with each cell initially having a value. At each discrete time step, the value of each cell is updated according to a definite rule. The new value on each of this cell is specified in terms of the values of its neighboring cells [24]. The method used to map an environment for a robot path planning approach using cellular automata is fairly standard for different implementations. Given the binary image representing the map of the robot's configuration space, the map is divided into regular square grids. The map is then transformed into a binary array. Each grid signifying either a free space or an obstacle will be represented as a cell in the cellular automata structure.

A path from the initial and goal configuration can then be obtained by searching these grids in a systematic fashion. The final values in each cell determine the search algorithm to be used. For example, in [24], strengths are defined on each cell and the strength is spread through the whole area according to some pre-defined rules. A path is then found by searching in parallel in the strength space. The authors in [25] present an approach using Voronoi diagrams for path planning. While Voronoi diagrams can help in generating a path for a robot that maintains a safe distance from all obstacles for path planning purposes, their construction is non-trivial and the metric used for the construction influences the quality of the diagram.

Other Voronoi-like structures for path planning have also been studied from a VLSI point of view. The authors in [26] have presented a wavefront-type of propagation method for path planning and a systolic array architecture. The authors also outline details of a chip developed for this purpose.

Another prominent entity in environment map-based methods for robotic navigation is the visibility graph [27]. The high complexity of sequential construction of various types of visibility graphs has motivated development of parallel algorithms. Work on general-purpose architectures is reported in [28]. The algorithms are efficient from an asymptotic viewpoint. In particular, the authors present $O(\log m)$ time algorithms for solving several visibility problems on an m-node polygon in the CREW-PRAM model of computation. Empirical performance details are not available.

We now review our research on VLSI architectures for a type of visibility graph known as the tangent graph. Tangent graph based data structure has been readily used in motion planning for mobile robots and robot manipulators. The complexity of the tangent graph grows exponentially as the robot's configuration space increases in dimension. The ability to construct larger number of tangents at high speed thus becomes crucial to facilitate dynamic motion planning where online avoidance is necessary. We present an efficient scheme for the construction of tangent graphs for an environment consisting of both non-convex and convex obstacles. The technique for tangent graph construction, described in [29][30], is based on a gradient computation approach that encompasses binary search, logarithmic approximation, and half-plane computation modules.

We will limit our discussion to the problem of constructing tangents between a point v to a polygonal obstacle P in a two-dimensional *C-space*. It is worth noting that the proposed method can be easily extended to a *C-space* of higher dimensions using the concepts discussed in [31].

Gradient-Based Tangent Graph Computation

The following definitions and theorem in [32] associated with the tangent graph generation are appropriate for our discussion. The proof of Theorem 1 is presented in [32] and is hence omitted here.

Definition 1: HP^{+}_{sg} and HP^{-}_{sg} represent the positive and negative half-planes that result from dividing a two-dimensional space into two regions along the line *sg*. A third point is necessary to render either half-plane positive or negative.

Definition 2: A vertex *v* of an *r*-sided polygon is said to be the *farthest front vertex* with respect to a point *x* if all vertices v_i for $i = 1, 2, \ldots, r$ are confined entirely either to HP^{+}_{xv} or HP^{-}_{xv}.

Theorem 1: Relative to a point *x* in the plane, the two farthest front vertices of polygon *P* comprise the optimal pair of available via points to circumvent *P*.

The proposed technique for our VLSI-based design is based on the observation that the farthest front vertices for a given point *v* are those for which the line gradient (tangents from *v*) is either the maximum or the minimum. The gradient computation and the maximum/minimum identification are performed for each of the vertices on the polygon *P*. A binary search approach was employed in [30] to reduce the complexity constructing the gradients from a point *v* to a two-dimensional polygon to $O(\log m)$, if *m* is the maximum number of vertices in each polygon. The conventional method of computing the gradient *s* is using equation (1). It, however, incurs a division, which does not lend itself well to hardware implementation.

$$s = \frac{y_a - y_b}{x_a - x_b} \quad (1)$$

Our VLSI-efficient scheme is based on the observation that the actual gradient values of all the vertices are not required. The proposed solution employs the logarithmic function that exhibits a monotonic characteristic with the conventional computation of *s*. As shown in equation (2), logarithm computations relieve us of the costly divider in hardware. Note that although the gradient can assume a positive or negative value, the polarity information is not shown in (2).

$$\log_2 s = \log_2\left(\left|y_a - y_b\right|\right) - \log_2\left(\left|x_a - x_b\right|\right) \quad (2)$$

A novel hardware implementation for logarithmic approximation was presented in [30]. The proposed design employs a LUT (Look-Up Table) and a simple computational unit to compute the logarithmic gradients as described in (2). This approach leads to a fast and cost efficient implementation with low approximation error. In [30], a strategy was discussed to ensure collision-free traversal of the robot by taking into account the approximation error.

Upon identifying the farthest front vertices, each of the nodes in the graph for the environment stores the corresponding gradients to each of the obstacles in logarithm representation ($\log_2 s$). The visibility segments from *v* to the polygon corresponding to the maximum and minimum gradient represent the tangents required for the path search process. The next step involves validation of the computer tangents

The computed tangents must be validated as they may intersect with other obstacles. Consider the case in Figure 2, where the segment connecting node v to a farthest front vertex v^1_{P2} of obstacle P_2 is blocked by obstacle P_1. These obstructed links should be eliminated as they form an invalid tangent, which will entail undesirable collisions. The computed tangents can be validated by looking at the signs using the half-plane equation given by (3) that simply expresses the fact that a point (x, y) lies on a line joining (x_a, y_a) and (x_b, y_b) (if two points lie on the same side of line joining (x_a, y_a) and (x_b, y_b), then we have two inequalities of the same type).

$$(x \times y_a) - (x \times y_b) + (y \times x_b) - (y \times x_a) - (y_a \times x_b) + (x_a \times y_b) = 0 \quad (3)$$

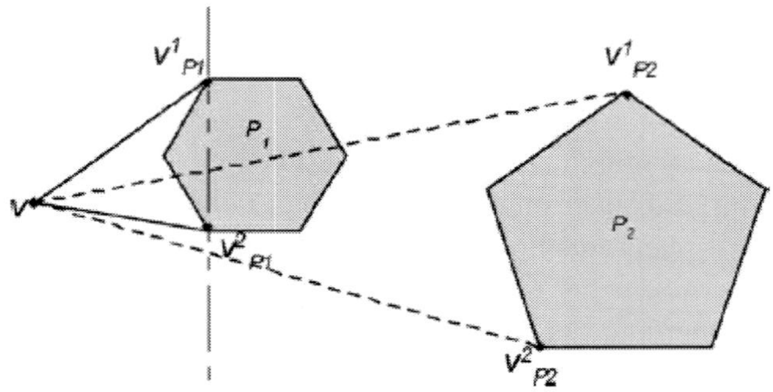

Figure 2. Validating the computed tangents.

Theorem 2: Let us denote by l the segment formed between a node v and its farthest front vertex v_{ffv} that resides on obstacle O. Node v also forms links with farthest front vertices $v_{\max}$ and $v_{\min}$ on obstacle O' (where $O' \neq O$) and the maximum and minimum gradient of these segments (in logarithmic representation) are denoted as $s_{\max}$ and $s_{\min}$, respectively. An obstacle O' is said to obstruct a segment l if both of the following are true:

1) When the gradient of l lies between $s_{\max}$ and $s_{\min}$.
2) When the coordinates of vertices $v_{\max}$ and $v_{\min}$ are substituted into the half-plane equation (3) as (x_a, y_a) and (x_b, y_b) along with the vertices v and v_{ffv} individually, vertices v and v_{ffv} are in different half-plane regions.

Proof:

If the gradient of l does not lie between $s_{\max}$ and $s_{\min}$, then the gradient of l must be greater or less than $s_{\max}$ and $s_{\min}$. This implies that l does not intersect O', and therefore, segment l cannot be obstructed by obstacle O'.

If the gradient of l lies between s_{max} and s_{min}, two cases arise. In the first case, vertex v_{ffv} and v are located in the same half-plane region with respect to the line $v_{max}v_{min}$. Segment l cannot be obstructed by *O'* as segment l does not intersect with $v_{max}v_{min}$. In the second case, vertex v_{ffv} and v are located in different half-plane regions relative to $v_{max}v_{min}$. Segment l intersects with the line $v_{max}v_{min}$, and thus, it is obstructed by *O'*.

In the example above, let us assume that the farthest front vertices for obstacle P_1 have been computed to be v_{P1}^1 and v_{P1}^2. The link formed between v and v_{P2}^1 is obstructed by P_1, since (1) the gradient of vv_{P2}^1 lies between the gradient of vv_{P1}^1 and vv_{P1}^2 and (2) v_{P2}^1 lies on a different half-plane region from v with respect to the line form with end-points v_{P1}^1 and v_{P1}^2. On the other hand, v_{P2}^2 is not obstructed by P_1, since it does not satisfy the first condition of Theorem 2.

The algorithm can be summarized as follows:

Step 1: Perform binary search to find the farthest front vertices for each graph node v for each obstacle. Store the gradients using logarithmic representation.

Step 2: Use the half-plane equation to eliminate obstructed segments for each node v.

The validation of the tangents in Step 2 must be performed for every obstacle in the environment. If the number of *C-obstacles* is c, the second step has a complexity of $O(c)$. The algorithm given exhibits a high degree of parallelism. While Step 2 can only be executed after Step 1, operations within each of these steps can be performed independently. It is noteworthy that the degree of parallelism is limited only by the constraints imposed on hardware resources. With adequate hardware resources, binary search can be done in parallel. Similarly, one can find obstructed links for each node v in parallel.

The proposed technique for gradient-based tangent construction can be easily extended to environments comprising of non-convex-shaped obstacles, by employing a pre-processing step to construct a convex boundary for the non-convex obstacles [30]. Methods of identifying the convex boundary of a non-convex obstacle can vary based on the level of required accuracy and computational complexity of the path traversal process. For example, the pre-processing stage can include methods to identify the convex hull of non-convex obstacles. This will lead to optimal path traversals but require a higher computational complexity. Alternatively, simpler methods can be adopted to identify a rectangular bounding box for the non-convex obstacles when the accuracy of path traversals can be compromised in place of computational efficiency.

We now present the VLSI architecture for the system. Figure 3 illustrates the overall architecture of the system, which comprises of three primary computation units: 1) Binary Search Unit, 2) Half-Plane Computation Unit, and 3) Logarithmic Approximation Unit. The external memory stores the coordinates of all the vertices, the vertices of the respective obstacles, and the maximum/minimum gradients along with the farthest front vertices of each vertex with respect to the obstacles in the *C-space*.

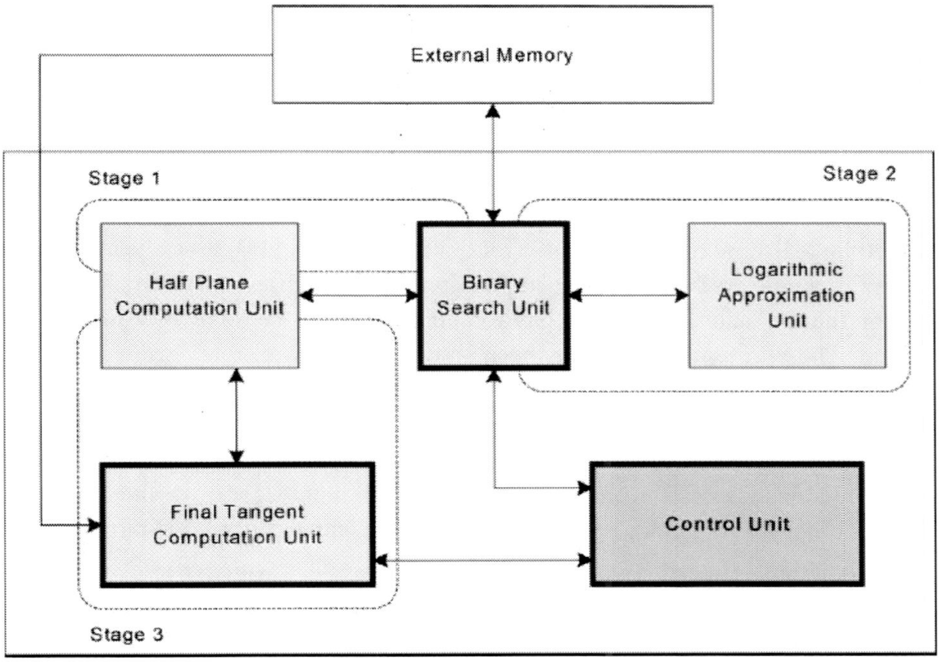

Figure 3. Architecture of the tangent computation system.

The Binary Search Unit is used in two stages of the algorithm. In the first stage, it is used along with the Half-Plane Computation Unit to identify the edge of an obstacle that intersects with a line adjoining a vertex and the first vertex of the obstacle. In the second stage, the Binary Search Unit is used along with the Logarithmic Approximation Unit to determine the maximum and minimum gradients. The Half-Plane Computation Unit is also used along with the Final Tangent Computation Unit (which comprises of a comparator and some registers) to validate the computed tangents. The Logarithmic Approximation Unit is used to determine the logarithmic approximation of the gradient with end-points of vertices (x_a, y_a) and (x_b, y_b) (as explained in equation (2)). The approximated gradient values are then used to determine the maximum/minimum gradient. Implementation details for the Binary Search Unit, Half-Plane Computation Unit, and Logarithmic Approximation Unit can be found in [30].

The modules have been coded in VHDL and synthesized using Synopsys Design Compiler 2001.08-SP1 with Compass Passport CB35OS142 standard cell library in 0.35 µm CMOS process. In comparison to the projected performance of the tangent graph computation approach in [33] on an Intel Pentium IV processor, the proposed VLSI implementation exhibits a speedup of 13. Moreover, the proposed architecture does not require the extensive resource support unlike the Pentium IV processor. This, together with the potential for additional speed-up through parallel computations, makes the hardware implementation appropriate for low-cost applications that demand high performance.

The reliability of the proposed hardware implementation for tangent graph construction is also high as it eliminates any intermediate layers such as an operating system. Unlike general-purpose processor based implementations, the proposed hardware alternative lends well for a highly responsive system with relatively lower clock rates, thereby resulting in a highly reliable and power-efficient solution, which is crucial to robust path planning applications.

We now discuss some of our other research contributions in VLSI for robotic planning. In a dynamic environment, the visibility graph must be recomputed whenever the obstacles move from a position that was previously occupied or when new obstacles appear. A naive solution is to re-create the visibility graph/tangent graph from scratch. However, due to the complexity of the computation, even hardware solutions for the naive approach may not be able to meet the real-time requirements for robot navigation. In [34], it has been demonstrated that efficient dynamic visibility graph computation can be realized by generating the 'virtual rectangle' on the fly such that only the environment needed to facilitate onward traversal can be identified. The proposed method has been shown to lend well for VLSI as it facilitates a high-degree of parallelism at the architecture level.

In [35], a parallel algorithm and FPGA realization have been presented for the construction of a reduced visibility graph. The proposed method, which is based on the approach in [36], relies on the efficient assignment of binary codes to the vertices of the obstacles in order to determine visible segments in the reduced visibility graph. The computational complexity of the parallel algorithm is $O\left(c^2 \log(n/c)\right)$, where c is the number of obstacles and n is the total number of vertices.

The approaches discussed so far aim to speed up the process of environment mapping. As discussed earlier, a path planning process is required in the navigation system to identify the traversal paths. Algorithms that have been commonly employed for path-planning include breadth-first search, depth-first search, Dijkstra's shortest path and A*. The attractiveness of a hardware portable algorithm for path-planning lies not only in the low complexity and low resource implementation, but chiefly in the potential for parallelism. For that matter, efforts to formulate a parallel Dijkstra's algorithm have not been overlooked. In [37], a compilation of several interesting methods to formulate parallel strategies for the Dijkstra's algorithm was discussed.

Besides parallel algorithms and hardware-efficient solutions for robotic planning, there have been studies on taking advantage of other aspects of reconfigurable computing platforms. We provide some background next on reconfiguration and then discuss specific efforts in robotics using reconfigurable hardware.

Advanced reconfigurable architectures provide for run-time hardware adaptability [38][39][40] that promise higher performance and power benefits. Currently, commercially available Xilinx FPGA devices (e.g. Xilinx Virtex II and Virtex II Pro) support dynamic partial reconfiguration that enables part of the device to be reconfigured while the rest of the area remains operational. In addition, Xilinx has also proposed two standard flows for the partial reconfiguration process: 1) Difference-based flow and 2) Module-based flow [41]. The Xilinx Virtex II Pro FPGAs incorporate the ICAP (Internal Configuration Access Port) that provides configuration access to the FPGA logic. This facilitates self-reconfiguration, whereby the reconfiguration control can be implemented within the FPGA device, either through the available soft/hard processor cores or a dedicated hardware controller. Dynamic runtime reconfiguration for robot navigation is a relatively new and promising area of a research due to the possibilities of realizing high performance, low cost and power efficient systems.

In [42], a multi-agent system paradigm is realized on a self-reconfigurable platform, which integrates high performance microprocessor and FPGA logic, to enable the mobile robot to adapt to the dynamic environment on the fly. Each agent comprises of a hardware task that can be reconfigured whenever they are required. The author proposed two reconfiguration models: 1) Virtual agent reconfiguration and 2) Pipeline reconfiguration architecture, to demonstrate the feasibility of the system for mobile robot navigation. A cellular spiking neural network has been implemented on the FPGA for obstacle avoidance in [43]. The hardware implementation consists of a 2D array of cells where each cell is a spiking neuron. The connectivity of the neurons can be dynamically changed to cater to the run-time requirements.

In [44], an approach to incorporate partial reconfiguration in an environment with multiple mobile robots is presented. The authors present a case study involving autonomous fault handling and reconfiguration to show the behavior reconfiguration in a team of two robots (denoted by R_1 and R_2). Both perform wall following initially. Thereafter, when the IR sensor on R_2 fails, it sends an error message to R_1. R_1 on receipt of the error message formulates a leader/follower behaviour and configures R_2 to follower behaviour. The authors present snaps of the actual point where reconfiguration takes place. They also present results of FPGA usage.

An approach and architecture for dynamic reconfiguration on autonomous mini-robots are presented in [45]. The authors in [45] consider two application examples, one dealing with reconfigurable digital controllers and the second dealing with image processing to illustrate their ideas. Two robotic platforms are also described.

5. Summary and Extensions

This review has summarized VLSI architectures for sensing, control and navigation of autonomous robots. Both analog and digital VLSI-based approaches have been discussed. It is clear that advances in VLSI have had a tremendous impact on various aspects of autonomous robotic systems.

There are a number of avenues that are worth investigating. Sensing strategies can be enhanced by investigating hardware mapping of a variety of other techniques developed in the domain of communication and signal processing.

VLSI-efficient sensor-based construction of different geometric structures is another interesting area of research. VLSI for humanoid robots operating in dynamic environments would be also worth investigating.

With regard to planning, the discussion has concentrated on parallel and hardware-efficient solutions for deterministic schemes. One area that has become popular in recent years for planning is probabilistic algorithms. Hardware-efficient mapping of probabilistic solutions would be valuable for study.

Another promising line of research is design of biologically-inspired solutions for robotic planning problems. Further work on exploring applications of dynamic reconfiguration for autonomous robots also seems worthwhile.

References

[1] Jean-Claude Latombe, Robot Motion Planning, Kluwer Academic Publishers, 1991

[2] Andrew Heale and Lindsay Kleeman, "A Real Time DSP Sonar Echo Processor", *Proceedings of the 2000 IEEE/RSJ International Conference on Intelligent Robots and Systems*, 2000, pp. 1261-1266.

[3] J. Urena, M. Mazo, J.J. Garcia, A. Hernandez and E. Bueno, "Correlation Detector Based on FPGA for Ultrasonic Sensors", *Microprocessors and Microsystems,* 1999, pp. 25-33.

[4] Hernandez, J. Urena, D. Hernanz, J.J. Garcia, M. Mazo, J.P. Derutin, J. Serot and S.E. Palazuelos, Real-time Implementation of an Efficient Golay Correlator (EGC) Applied to Ultrasonic Sensorial Systems, *Microprocessors and Microsystems*, 2003, pp. 397-406.

[5] Kumar, S. Sarkar and R.P. Agarwal, "A Novel Algorithm and FPGA Based Adaptable Architecture for Correcting Sensor Non-uniformities in Infrared System", *Microprocessors and Microsystems*, 2007, pp. 402-407.

[6] K.M. Hou, A. Belloum, E. Yao, X.W. Tu, M. Shawky, M. Meribout, J.L. Mayorquim, A. Trihandoyo and B. Jardin, "A Reconfigurable and Flexible Parallel 3D Vision System for a Mobile Robot", *Proceedings of IEEE International Conference on Computer Architectures for Machine Perception*, 1993, pp. 215-221.

[7] F. Pardo, I. Llorens, F. Mico and J.A. Boluda, "Space Variant Vision and Pipelined Architecture for Time to Impact Computation", *Proceedings of IEEE International Workshop on Computer Architectures for Machine Perception*, 2000, pp. 122-126.

[8] M. Mikhalsky and J. Sitte, "Active Motion Vision for Distance Estimation in Autonomous Mobile Robots", *Proceedings of the 2004 IEEE Conference on Robotics, Automation and Mechatronics,* Singapore, 2004, pp. 1060-1065.

[9] F. Aubepart and N. Franceschini, "Bio-inspired Optic Flow Sensors Based on FPGA: Application to Micro-Air Vehicles", *Microprocessors and Microsystems*, 2007 pp. 408-419.

[10] G. Indiveri and P. Verschure, "Autonomous Vehicle Guidance Using Analog VLSI Neuromorphic Sensors", *Proceedings of ICANN'97, Lecture Notes in Computer Science,* Vol. 1327, 1997, pp. 811-816

[11] C.M. Higgins and V. Pant, "A Biomimetic VLSI Sensor for Visual Tracking of Small Moving Targets", *IEEE Transactions on Circuits and Systems-Part I,* Vol. 51, No. 12, December 2004, pp. 2384-2394.

[12] P. Sadayappan, Y.L.C. Ling, K.W. Olson and D.E. Orin, "A Restructurable VLSI Robotics Vector Processor Architecture for Real-Time Control", *IEEE Transactions on Robotics and Automation,* Vol. 3, No. 5, Oct. 1989, pp. 583-599.

[13] A. Kongmunvattana and P. Chongstitvatana, "A FPGA-Based Behavioral Control System for a Mobile Robot", *Proceedings of 1998 IEEE Asia-Pacific Conference on Circuits and Systems*, 1998 pp. 759-762.

[14] J.L. Arroyabe, G. Aranguren, L.A.L. Nozal and J.L. Martin, "Autonomous Vehicle Guidance with Fuzzy Algorithm", Proceedings of 26th *Annual Conference of IEEE Industrial Electronics Society*, 2000, pp. 1503-1508.

[15] D.M. Wilson, E.D. Blom, M.A. Marra III and B.L. Walcott, "Direct Sensorimotor Control for Low-Cost Mobile Tracking Applications", *IEEE Transactions on Industrial Electronics*, Vol. 47, No. 4, August 2000, pp. 939-950.

[16] N. Ranganathan, M.I. Patel and R. Sathyamurthy, "An Intelligent System for Failure Detection and Control in an Autonomous Underwater Vehicle", *IEEE Transactions on Systems, Man, and Cybernetics – Part A*, Vol. 31, No. 6, November 2001, pp. 762-767.

[17] J. Xiao, A. Calle, J. Ye and Z. Zhu, "A Mobile Robot Platform with DSP-based Controller and Omnidirectional Vision System", *Proceedings of the 2004 IEEE International Conference on Robotics and Biomimetics*, 2004, pp. 844-848.

[18] T.H.S. Li, S.J. Chang and Y.X. Chen, "Implementation of Human-Like Driving Skills by Autonomous Fuzzy Behavior Control on an FPGA-Based Car-Like Mobile Robot", *IEEE Transactions on Industrial Electronics*, Vol. 50, No. 5, Oct. 2003, pp. 867-880.

[19] M.S. Masmoudi, I. Song, F. Karray, M. Masmoudi and N. Derbel, "FPGA Implementation of Fuzzy Wall-Following Control", *Proceedings of IEEE 16th International Conference on Microelectronics*, 2004, pp. 133-139.

[20] P. Arena, L. Fortuna, M. Frasca and L. Patane, "A CNN-Based Chip for Robot Locomotion Control", *IEEE Transactions on Circuits and Systems- Part I*, Vol. 52, No. 9, September 2005, pp. 1862-1871

[21] Dominik Henrich, "A Review of Parallel Processing Approaches to Motion Planning", *IEEE International Conference on Robotics and Automation*, April 1996, pp. 3289-3294

[22] S. Kung and J. Hwang, "Neural Network Architectures for Robot Applications", *IEEE Transactions on Robotics and Automation,* Vol. 5, October 1989, pp. 641-657

[23] Chang Shu and Hilary Buxton, "A Parallel Path Planning Algorithm for Mobile Robots", International Conference. on Automation, Robotics and Computer Vision, 1990, pp. 489-493

[24] Chang Shu and Hilary Buxton, "Dynamic Motion Planning Using a Distributed Representation", *Journal of Intelligent and Robotic Systems*, November 1995, pp 241-262

[25] P. Tzionas, Ph. Tsalides and A. Thanailakis, "Collision-Free Path Planning for a Diamond-Shaped Robot Using Two-Dimensional Cellular Automata", *IEEE Transactions on Robotics and Automation*, Vol. 13, No. 2, April 1997

[26] N. Ranganathan, B. Parthasarathy and K. Hughes, "A Parallel Algorithm and Architecture for Robot Path Planning", *Proceedings of IEEE Eighth Parallel Processing Symposium*, 1994, pp. 275-279.

[27] Tomas Lozano-Perez and M. Wesley, "An Algorithm for Planning Collision-Free Paths among Polyhedral Obstacles", Commun. ACM 22, October 1979, pp. 560–570

[28] Michael T. Goodrich, Steven B. Shauck and Sumanta Guha, "Parallel Methods for Visibility and Shortest Path Problems in Simple Polygons", *Proceedings of the Sixth Annual Symposium on Computational Geometry*, 1990, pp. 73-82

[29] S. K. Lam, K. Sridharan and T. Srikanthan, "Hardware Efficient Schemes for Logarithmic Approximation and Binary Search with Application to Visibility Graph Construction", *IEEE Transactions on Industrial Electronics*, Vol. 51, No. 6, December 2004, pp. 1346-1348

[30] S. K. Lam, K. Sridharan and T. Srikanthan, "VLSI-Efficient Schemes for High-Speed Construction of Tangent Graph", *Journal of Robotics and Autonomous Systems*, Vol. 51, No. 4, June 2005, pp. 248-260

[31] A.B. Doyle and D.I. Jones, "A Tangent based Method for Robot Path Planning", *IEEE International Conference on Robotics and Automation*, 1994, pp. 1561–1566.

[32] Jason A. Janet, Ren C. Luo and Michael G. Kay, "T-Vectors Make Autonomous Mobile Robot Motion Planning and Self-Referencing More Efficient", *IEEE International Conference on Intelligent Robots and Systems,* 1994, pp. 587–594

[33] Yun-Hui Liu and Suguru Arimoto, "Computation of the Tangent Graph of Polygonal Obstacles by Moving-Line Processing", *IEEE Transactions on Robotics and Automation*, Vol. 10, December 1994, pp. 823–830

[34] S. K. Lam and T. Srikanthan, "High-Speed Environment Representation Scheme for Dynamic Path Planning", *Journal of Intelligent & Robotic Systems*, Vol. 32, No. 3, November 2001, pp. 307-319

[35] K. Sridharan and T. K. Priya, "A Hardware Accelerator and FPGA Realization for Reduced Visibility Graph Construction Using Efficient Bit Representations", *IEEE Transactions on Industrial Electronics*, Vol. 54, No. 3, June 2007, pp. 1800-1804

[36] Jason A. Janet, Ren C. Luo and Michael G. Kay, "Autonomous Mobile Robot Motion Planning and Geometric Beacon Collection Using Traversability Vectors", *IEEE Transactions on Robotics and Automation,* Vol. 13, No. 1, 1997, pp. 132–140

[37] Vipin Kumar, Ananth Grama, Anshul Gupta and George Karypis, "Introduction to Parallel Computing: Design and Analysis of Algorithms", Benjamin/Cummings Publishing Company, 1994

[38] Reiner Hartenstein, "The Microprocessor is No Longer General Purpose: Why Future Reconfigurable Platforms Will Win", *IEEE International Conference on Innovative Systems in Silicon*, October 1997, pp. 2-12.

[39] Scott Hauck, "The Future of Reconfigurable Systems", Proc. 5th Canadian Workshop on Field Programmable Devices, June 1998

[40] Katherine Compton and Scott Hauck, "Reconfigurable Computing: A Survey of Systems and Software", ACM Computing Surveys, Vol. 34, No. 2, June 2002, pp. 171-210

[41] Xilinx Application Note, "Two Flows for Partial Reconfiguration: Module Based or Difference Based", XAPP290, Version 1.2, September 2004

[42] Yan Meng, "A Dynamic Self Reconfigurable Robot System", *IEEE International Conference on Advanced Intelligent Mechatronics*, July 2005, pp. 1541-1546

[43] Daniel Roggen, Stephane Hofmann, Yann Thoma and Dario Floreano, "Hardware Spiking Neural Network with Run-Time Reconfigurable Connectivity in Autonomous Robot", *IEEE Proceedings of the NASA/DoD Conference on Evolvable Hardware*, 2003

[44] V. Tadigotla, L. Sliger and S. Commuri, "FPGA Implementation of Dynamic Run-Time Behavior Reconfiguration in Robots", *IEEE International Symposium on Intelligent Control*, October 2006, pp. 1220-1225

[45] Carlos Paiz, Teerapat Chinapirom, Ulf Witkowski and Mario Porrmann, "Dynamically Reconfigurable Hardware for Autonomous Mini Robots", *IEEE Conference of Industrial Electronics*, November 2006, pp. 3981-3986

In: VLSI and Computer Architecture
Editor: Kenzo Watanable
ISBN: 978-1-60692-075-6

Chapter 9

DESIGN OF AN ENHANCED MAC ARCHITECTURE FOR MULTI-HOP WIRELESS NETWORKS*

***Ralph Bernasconi*[1]*, Silvia Giordano*[2]*, Alessandro Puiatti*[3]**
DTI, University of Applied Science (SUPSI), Via Cantonale,
Gallera 2 – 6928 Manno, Switzerland
***Raffaele Bruno*[4]*, Marco Conti*[5] *and Enrico Gregori*[6]**
IIT Institute, National Research Council (CNR), Via G. Moruzzi, 1 - 56124 Pisa, Italy

Abstract

This chapter presents a very unique work at the MAC layer that spans from analytical and simulative investigations, to the architectural design and implementation of an enhanced MAC protocol and the consequent experiments in real scenarios. This work was spiked by some seminal papers, which demonstrated the shortcomings of using the 802.11 technology for distributed mobile environments like multi-hop ad hoc networks. Thus, we decided to redesign the MAC architecture and to realize a prototype of a new MAC protocol integrating features and mechanisms more suitable for ad hoc communications. This work was very successful as, on one side, we designed a very flexible architecture that overcomes some of the drawbacks of traditional 802.11 solutions and, on the other side, we concretely implemented and tested an enhanced 802.11 backoff algorithm, showing the alignment of the analytical-simulative work with the implementation-experimental one.

---

* A version of this chapter was also published in *Multi-hop Ad Hoc Networks from Theory to Reality,* edited by Marco Conti, Jon Crowcroft and Andrea Passarella, published by Nova Science Publishers, Inc. It was submitted for appropriate modifications in an effort to encourage wider dissemination of research.

[1] E-mail address: ralph.bernasconi@supsi.ch
[2] E-mail address: silvia.giordano@supsi.ch
[3] E-mail address: alessandro.puiatti@supsi.ch
[4] E-mail address: raffaele.bruno@iit.cnr.it
[5] E-mail address: marco.conti@iit.cnr.it
[6] E-mail address: enrico.gregori@iit.cnr.it

1 Introduction

Instantly deployable, multi-hop wireless networks can be both fully self-organized (ad hoc networks) and connected to the wired backbone via multiple wireless hops (multi-hop wireless LANs). Potential applications of this type of wireless networks include rescue, civilian and military networks, where users self-organize to communicate or to access the Internet; and multi-hop Hot Spots, i.e., wireless networks deployed in public areas, in which traditional WLAN technologies are augmented with ad hoc wireless communications. Today, the de facto standard for building these types of wireless networks is the IEEE 802.11 technology [1]. In fact, the availability of inexpensive and highly interoperable networking solutions based on this technology, and the growing trend of providing built-in 802.11-compliant wireless network cards into mobile computing platforms, make the 802.11 technology the most interesting off-the-shelf enabler for multi-hop ad hoc networks.

However, the adoption in the 802.11 standard of a CSMA/CA-based MAC protocol, as well as the intrinsic impairments of a wireless medium arise several technical challenges which should be addressed to guarantee high-performance communications in ad hoc networks [3,12,13,15,16]. Two main reasons were pointed out in literature to explain the poor performance of the 802.11 MAC protocol in ad-hoc and multi-hop networks:

1. *Multi-hop communications*: The wireless medium has neither absolute nor nearly observable boundaries outside of which the stations are known to be unable to receive frames. When using a carrier-sensing random access protocol to coordinate the channel access, this condition generates complex phenomena, such as hidden station and exposed station problems. In addition, the typical channel impairments of a wireless medium, such as flat or frequency selective fading, multi-path and interference, negatively affect the MAC protocol performance.
2. *Backoff*: The channel access scheme is regulated by the exponential backoff: nodes failing to obtain the channel have to backoff a random time before trying again. It is widely recognized that, depending on the network configuration, the standard IEEE 802.11 protocol can be not fair and operate very far from the theoretical limit of the wireless network. While this unfairness is somehow controlled in the infrastructure scenarios, it can drastically grow in distributed ones. Furthermore, both unfairness and low channel utilization impact upper layer protocols, especially at transport layer if TCP is used. These phenomena have been shown through simulations [7,8], and appear even worst when tested in real experiments [2,5,13];

Several research projects have been funded to investigate possible enhancements of the IEEE 802.11 MAC protocol so as to solve the aforementioned issues. In this chapter, we aim at presenting the activities carried out in the framework of the MobileMAN project, which have led to the architectural design and implementation of an enhanced MAC protocol more suitable for ad hoc environments. In particular, the MobileMAN project is an initiative funded by the European FET FP6 Programme with the primary technical objective of investigating the potentialities of the Mobile Ad hoc NETwork (MANET) paradigm, defining and developing a metropolitan, self-organizing, and totally wireless network. As one of the major aims of the MobileMAN project was to perform experiments in real scenarios, we decided to redesign the MAC architecture and to realize a prototype implementing the new MAC protocol specified for the MobileMAN network.

The building block of the enhanced MAC protocol we implemented in hardware is the Asymptotically Optimal Backoff (AOB) mechanism [9], which dynamically adapts the backoff window size to the current network contention level and guarantees that an IEEE 802.11 WLAN asymptotically achieves its optimal channel utilization. The AOB protocol has been selected as the reference MAC protocol for the MobileMAN network because it relies only on topology-blind estimates of the network status based on the standard physical carrier sensing activity. Hence, it appears as a suitable and robust solution for multi-hop configurations. Several extensions for the AOB protocol have been proposed in the framework of the MobileMAN project such as to make it more efficient and fair in multi-hop ad hoc environments. In this chapter, we do not go into details of the various proposed mechanisms, but we specifically focus on describing the architecture of our enhanced IEEE 802.11 wireless network card and on showing experimental results proving the effectiveness of the implemented solutions [11]. Note that our Medium Access platform has been designed to be a versatile architecture that could be used for implementing and testing: 1) backoff algorithms more adequate to multi-hop operations; 2) dynamic channel switching schemes to exploit channel quality diversity; 3) efficient packet forwarding schemes inside the MAC layer; and 4) cross-layering optimizations through the exploitations of topology information provided by the routing layer. In this chapter, we outline our activity concerning the point 1) above. Specifically, we present our card architecture and we describe how the AOB protocol has been implemented in our MAC platform. Moreover, we describe the implementation of a *credit-based strategy* used to extend the contention control algorithm adopted by the AOB protocol, such as to improve its efficiency, which has been proposed in [10]. The experimental results obtained comparing our enhanced MAC card with traditional IEEE 802.11 wireless cards, show the significant per-station throughput improvement ensured by the enhanced MAC protocol. Furthermore, they open promising directions to investigate additional enhancements, as discussed in Section 5.

The rest of this chapter is organized as follows: in Section 2 we outline our implementation and the main hardware and firmware design choices. Section 3 describes the algorithms that have been implemented in the network card. In Section 4 we present the measurement environment and we report the results of our real experiments, discussing the most relevant points. Section 5 concludes this chapter with some further discussion and detailed description of the ongoing and future work.

2 Hardware Design of the MobileMAN NIC

Generally speaking, a wireless NIC has three main functional blocs: the MAC, the Baseband (BB) and the Radio Frequency (RF). Since the main part of the conceptual work conducted in our activities is concentrated on the MAC protocol, we decided to use off-the-shelf solutions for the BB and RF parts. For these reasons, we acquired a board, called DT20 modem, produced by the Elektrobit, which implements the 802.11 PHY layer with the Prism I chipset produced by the Intersil. Note that at the time we started the card development, this company was the world leader manufacturer of the chipsets for wireless network interface cards.

Concerning the MAC protocol, given that our goal was to develop a new backoff algorithm over the 802.11 standard and not to entirely re-design the standard channel access mechanisms, we tried to find a flexible development platform providing an implementation of the legacy 802.11 standard. Unfortunately, the platforms provided by the major producers of wireless NICs were too expensive or with a very limited set of possible enhancements. Thus, we were forced to implement the 802.11 MAC standard from scratch. In addition, we needed a development platform ensuring a great programming flexibility. For these reasons, we tried to find a development platform that could fulfil the following constraints:

- an easy, well known and tested development environment to speed up as much as possible the implementation of the 802.11 standard,
- the possibility to develop some MAC functionalities directly in hardware to fulfil the timing constraints imposed by the 802.11 standard, [1]
- a processor with high performances for new and future implementations.

In the end, the solution that best fitted our criteria was the Orsys Micro-line C6713Compact DSP board. The hardware overview of the enhanced wireless network interface card, integrating both the DSP board and the DT20 modem is shown in Figure 1.

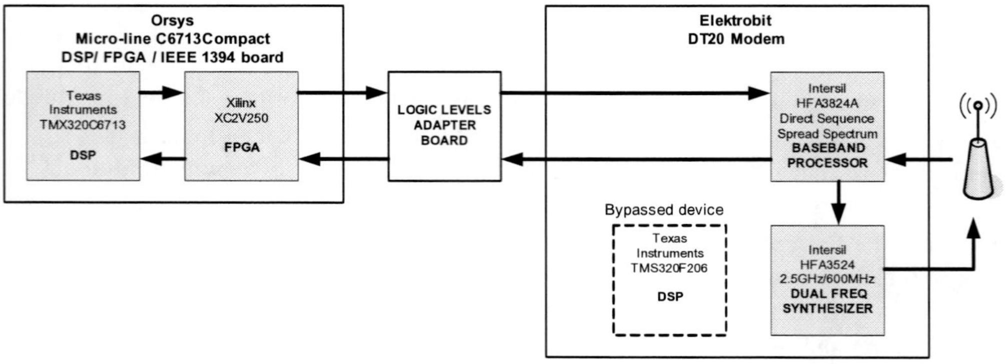

Figure 1. Overview of the enhanced 802.11 Wireless Network Interface (PHY).

The DSP board integrates a Texas Instruments TMX320C6713 DSP and an FPGA device (Xilinx XC2V250), which is very important for the implementation of the protocol functionalities characterized by stringent time constraints. Due to the fact that the DSP board and the DT20 modem board have different logic levels, 3.3V and 5V respectively, a logic levels adapter has been developed to allow the communication between the boards.

2.1 Implementation

The part of the 802.11 MAC protocol implemented in the C6713 DSP has been realized in standard C. On the other hand, the communication layer between the DSP and the modem has been developed on the FPGA device. A more detailed overview of the interfaces between the modules implementing our network interface is illustrated in Figure 2.

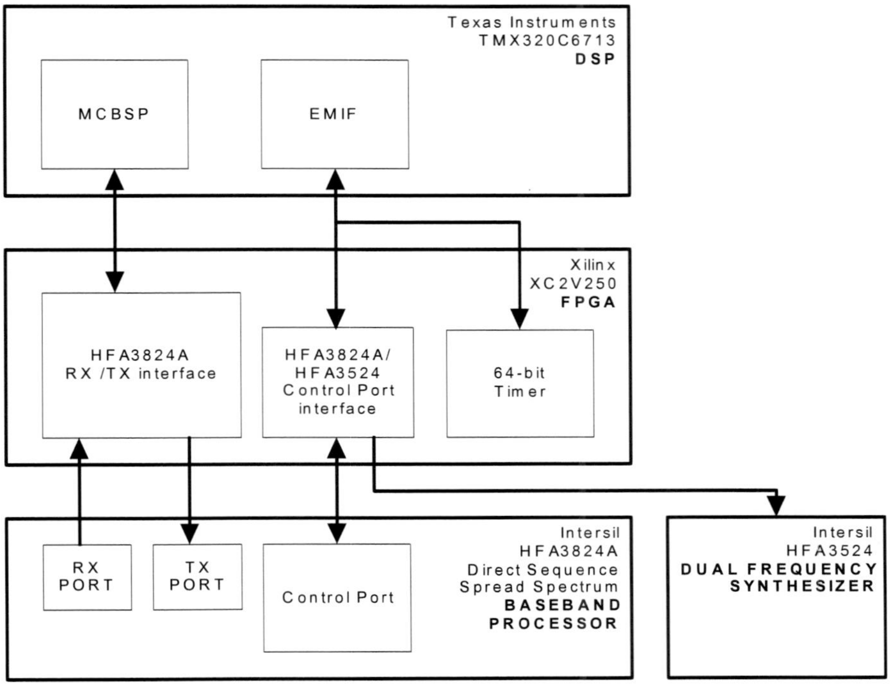

Figure 2. Logic blocks diagram of the MAC implementation. Note that only three functional blocks have been implemented in the FPGA.

The specific interfaces are:

- **HFA3824A RX/TX Interface**: this block operates as glue logic between the MCBSP serial interface available on the DSP and the serial Receive & Transmit ports of the HFA3824A Baseband Processor.
- **HFA3824A/HFA3524 Control port interface**: this block is used as an interface between the DSP and the Control port of the HFA3824A device. Through this interface, the Baseband Processor and the Dual Frequency Synthesizer can be configured.
- **64-BIT TIMER**: this is a 64-bit timer implemented for the MSDU management at the end of 802.11 frame transmission and reception events.

The firmware was realized in such a way to maintain the maximum possible level of abstraction and to minimize the software re-design in case of change of the development platform. Thus, only few software components are specific to the C6713Compact board; among these: timing considerations, available DSP resources, configuration and control related to the specific implementation (i.e. we could not implement a general abstraction at the source code level).

The PHY firmware is subdivided into the following components:

- *MAC firmware*: is the hard real-time software, which allows packets (fragments) to be physically transmitted and received to and from the RF interface. This part implements both the 802.11 legacy standard and the new backoff algorithm in order to allow mixed environment experiments, where enhanced systems co-operate with standard off-the-shelf components.
- *Host interface firmware*. This software component is less stringent in terms of real-time requirements.
- *Packet data structure*. This data structure is the communication channel between the MAC firmware and the Host interface firmware. It is a vital part of the MobileMAN project since it allows the cross-layering functionalities between PHY/MAC and upper layers.

Note that the firmware comes without an operating system, which is not needed for the implementation of the standard 802.11 Frame Exchange Sequence and relative tasks (fragmentation, de-fragmentation, fragment cache control, etc.). This facilitates the portability of the source code. On the actual system (C6713Compact board), the firmware occupies about 125 Kbytes and can reside completely in the DSP internal RAM, at run time.

The system may be used in a lab environment (through the development system and the JTAG interface), as for instance during synthetic traffic tests, but it may also be tested in a real environment by using the high speed IEEE1394 bus, which allows the full speed connection with a host PC. A specific PC application has been also developed to control and test the NIC when it is running as a stand-alone system (i.e., without connecting a channel emulator and using the TI Code Composer as control environment). With this small and simple application, MAC parameters (for example, station MAC address, signal quality thresholds, synthetic packets generation control) are fully accessible and they can be easily changed using a PC connected to the system with a RS-232 cable. Commands to the MAC system can be fully edited and sent with specific parameters as shown in Figure 3 and Figure 4.

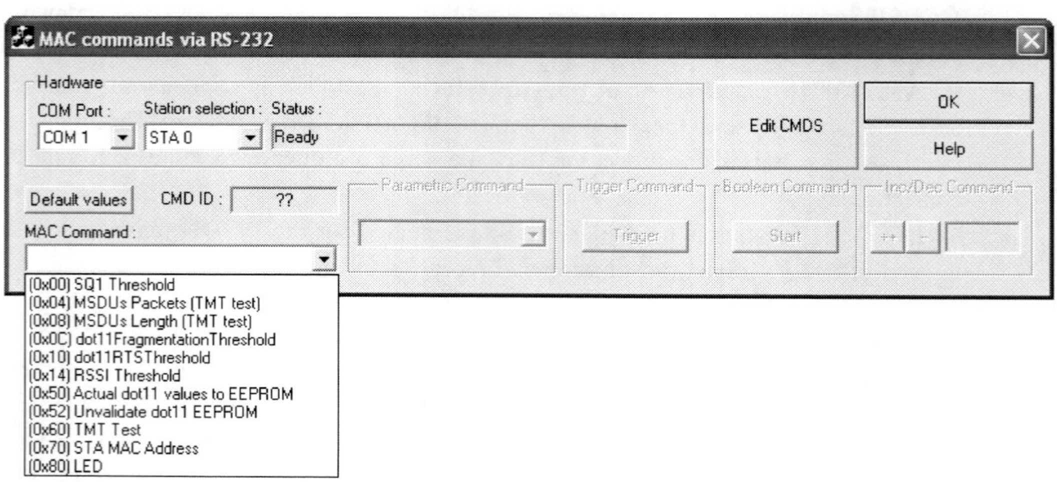

Figure 3. MAC commands via RS-232.

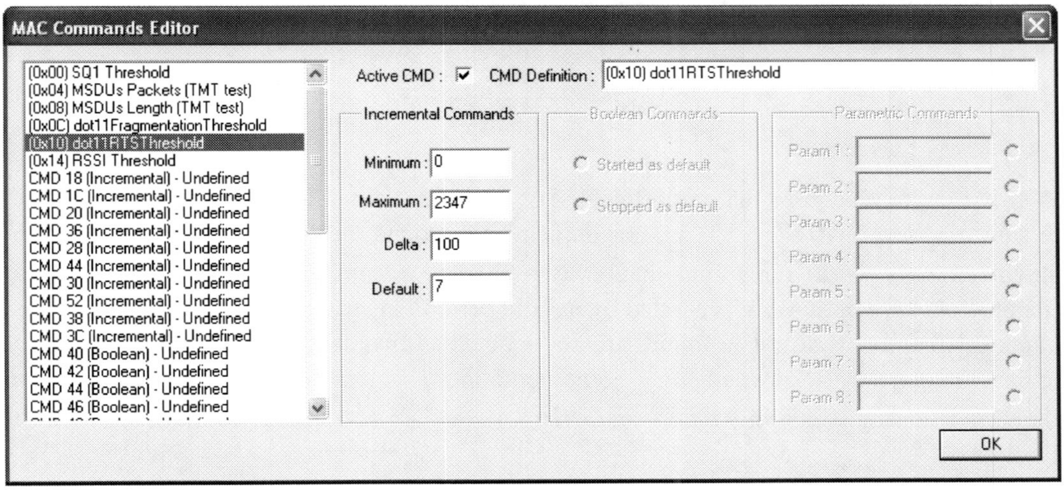

Figure 4. MAC commands editor.

3 MAC Protocol Implementation

In this section, we outline the different modules that have been developed in the MobileMAN NIC card to implement the AOB protocol as defined in [6,9] and the credit-based enhancements as specified in [10].

The AOB protocol and its extensions are based on the notion of slot utilization. For this reason, the first component that has been developed is used for the run-time estimation of this important measure. Specifically, we do not estimate the aggregate slot utilization, as done in [9], but we split it into two contributions: the internal slot utilization (SU_{int}) and the external slot utilization (SU_{ext}), so as to differentiate between the contribution to the channel occupation due to the node's transmissions and to its neighbours' transmissions. This differentiation is motivated by the need to keep our implementation as much flexible as possible to allow future modifications as the one described in [17]. Another variation with respect to the original AOB algorithm is the time interval over which we compute the slot utilization. In fact, the original AOB algorithm computes the slot utilization after each backoff interval, while in our implementation we used a constant observation period T of 100ms. This choice is motivated by the need to avoid frequent slot utilization computations, which could interfere with the time constraints of the atomic MAC operations (e.g., RTS/CTS exchange). During each time window T, each station counts the number n_{tx} of performed transmission attempts, the number n_{rx} of the channel occupations (either correct frames or collided frames) observed by this node, and the number n_{idle} of slots during which the medium has been perceived idle during this interval (including DIFS and EIFS periods). It is straightforward to note that the total number of busy periods during the observation time T is equal to $n_{tx} + n_{rx}$. From these measurements, the two slot utilization values are computed as follows:

$$SU_{\text{int}} = \frac{n_{tx}}{n_{idle} + n_{tx} + n_{rx}} \tag{1.a}$$

$$SU_{\text{ext}} = \frac{n_{rx}}{n_{idle} + n_{tx} + n_{rx}} \tag{1.b}$$

It is easy to recognize that the original SU value as defined in [6] can be computed as the sum of SU_{int} and SU_{int}. Thus, our implementation and the original AOB protocol are equivalent. It is worth remarking that: *i)* two channel occupations should be considered separated only when they are separated by an idle period longer than the $DIFS$ period. This guarantees that the MAC ACK frames are not counted as channel occupations different from the data frames they acknowledge. *ii)* To compute the T_{idle} it is necessary to count also the idle periods during which the $DIFS$ and $EIFS$ timers are active, and not only the backoff idle slots.

Using formulas (1a) and (1b) we compute a single sample of the slot utilization. However, to avoid sharp fluctuations in the slot utilization estimates we should average these single measures. Our design choice is to compute the SU index by applying an exponentially weighted moving average (EWMA) filter to the sequence of samples. Specifically, let assume that the station is observing the channel during the *i*-th observation period. Then, it follows that:

$$\overline{SU}_{\text{int}}^{(i)} = \alpha_1 \cdot \overline{SU}_{\text{int}}^{(i-1)} + (1-\alpha_1) \cdot SU_{\text{int}}^{(i)} \tag{2a}$$

$$\overline{SU}_{\text{ext}}^{(i)} = \alpha_1 \cdot \overline{SU}_{\text{ext}}^{(i-1)} + (1-\alpha_1) \cdot SU_{\text{ext}}^{(i)} \tag{2b}$$

where α_1 is the smoothing factor, $\overline{SU}_{\text{int}}^{(i)}$ ($\overline{SU}_{\text{ext}}^{(i)}$) is the average internal (external) slot utilization estimated at the end of the *i*-th observation period, and $SU_{\text{int}}^{(i)}$ ($SU_{\text{ext}}^{(i)}$) is the internal (external) slot utilization measured during the *i*-th observation period using formula (1a) ((1b)).

Exploiting the $\overline{SU}_{\text{int}}$ and $\overline{SU}_{\text{ext}}$ estimates we can easily compute the probability P_T of executing a transmission attempt granted by the standard backoff process by implementing the classical formula proposed in [9]:

$$P_T = 1 - \left[\min\left(1, \frac{\overline{SU}_{\text{int}} + \overline{SU}_{\text{ext}}}{ACL(q)}\right)\right]^{N\_A} \tag{3}$$

Since the $ACL(q)$ value depends almost only on the average frame size q and it does not depend on the number of stations in the network, as proved in [9], the table of $ACL(q)$ values for different frame sizes can be stored *a priori* inside the node transmitter. Similarly to the slot utilization computed in formulas (1a) and (1b), formula (3) could induce sharp fluctuations in the P_T estimate. For this reason, the average P_T value is also computed using an EWMA function. Specifically, when the j-th backoff interval is terminated (i.e., the backoff counter is zero). Then, it follows:

$$\overline{P_T}^{(j)} = \alpha_2 \cdot \overline{P_T}^{(j-1)} + (1-\alpha_2) \cdot P_T^{(j)} \tag{4}$$

where α_2 is the smoothing factor, $\overline{P_T}^{(j)}$ is the average probability of transmission a station uses to decide whether performing the transmission attempt or not, and $P_T^{(j)}$ is the probability of transmission computed according to formula (3). It is worth noting that it should be $\alpha_2 > \alpha_1$ because the P_T value is updated after each backoff interval, therefore significantly more often that the SU, which is updated only after each observation interval T (for instance, in our tests we used $\alpha_1 = 0.9$ and $\alpha_2 = 0.95$).

To implement the original AOB protocol, it is only necessary to compute the P_T value according to formula (3), and to maintain the $\overline{SU}_{\text{int}}$ and $\overline{SU}_{\text{ext}}$ estimates using formulas (2a) and (2b). Figure 5 depicts the flow diagram illustrating the different components that have been defined to implement the AOB protocol, and the relationships between the logic blocks.

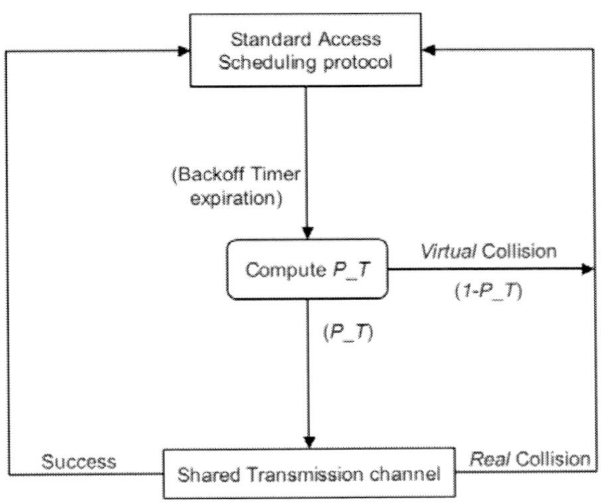

Figure 5. Block diagram of the implemented AOB protocol.

However, to implement the extensions to the AOB protocol designed in the MobileMAN project, we have to develop a module capable of collecting credits. As described in [10,17], each station should earn credits when it releases a transmission opportunity granted by the standard basic access mechanism. These credits, in turn, are spent to perform additional transmission attempts. More precisely, let us assume that the j-th backoff interval is terminated (i.e., the backoff counter is zero), and that the backoff timer was uniformly selected in the range $[0,\ldots,CW(k)-1]$, in which $CW(k) = \min(2^{k-1}, 2^{k_{MAX}}) \cdot CW_{MIN}$. If, according to the probability of transmission P_T, the station releases its transmission opportunity granted by the standard backoff procedure, the new contention window used to reschedule the frame transmission will be $CW(k+1)=\min\left(2^{k},2^{k_{MAX}}\right)\cdot CW_{MIN}$. Thus, after the virtual collision the number of credits CR collected by that station will be:

$$CR = CR_{old} + \min\left(2^{k},2^{k_{\max}}\right), \tag{5}$$

where CR_{old} is the number of credits owned by the station before the virtual collision.

Each station should use the collected credits to perform consecutive transmission attempts separated by $SIFS$ intervals. The analytical and simulative studies conducted in [10,17] have demonstrated that the use of multiple consecutive transmissions regulated by the credits is an effective technique to solve fairness issues arising when the AOB protocol is used in multi-hop networks or heterogeneous WLANs. In addition, using frame bursting is also beneficial to improve the efficiency of the 802.11 MAC protocol and to increase the throughput performances. In the NIC card we have implemented the logic required to support the frame bursting.

As explained in [10], the number of credits needed to perform consecutive transmissions should depend on the average backoff interval. More precisely, each station estimates the average backoff interval that the standard backoff scheme would use in the case that no filtering of the channel access is implemented. To accomplish this estimation, it is useful to remind that the collisions suffered from stations using the AOB protocol can be either virtual collisions, when the station voluntarily defer a transmission attempt, or real collisions, when the station perform the transmission attempt but it does not receive the MAC ACK frame. Let us assume that the total number of transmission opportunities assigned to a station before the successful transmission is K, and that K_{rc} is the number of real collisions occurred. Hence, $K-K_{rc}$ is the number of virtual collisions occurred, i.e., the transmission opportunities released by the station. Denoting with $\overline{CW}^{(j)}_{enh}$, the average contention window estimated after the *j*-th successful transmission, and with $\overline{CW}^{(j)}_{std}$ the average contention window of the equivalent standard MAC protocol estimated after the *j*-th successful transmission, we have that:

$$\overline{CW}^{(j)}_{enh} = \alpha_2 \cdot \overline{CW}^{(j-1)}_{enh} + (1-\alpha_2)\cdot \sum_{k=1}^{K} CW(k) \tag{6.a}$$

$$\overline{CW}^{(j)}_{std} = \alpha_2 \cdot \overline{CW}^{(j-1)}_{std} + (1-\alpha_2)\cdot \sum_{k=1}^{K_{rc}} CW(k) \tag{6.b}$$

The $\overline{CW}^{(j)}_{std}$ value will be used as a threshold to decide if the station has enough credits to perform a transmission attempt. We denote with AOB-CR the standard AOB protocol enhanced with the capabilities of collecting credit and using these credits to regulate the frame bursting. Figure 6 depicts the flow diagram outlining the different components that have been defined to implement the AOB-CR MAC protocol, and the relationships between the blocks.

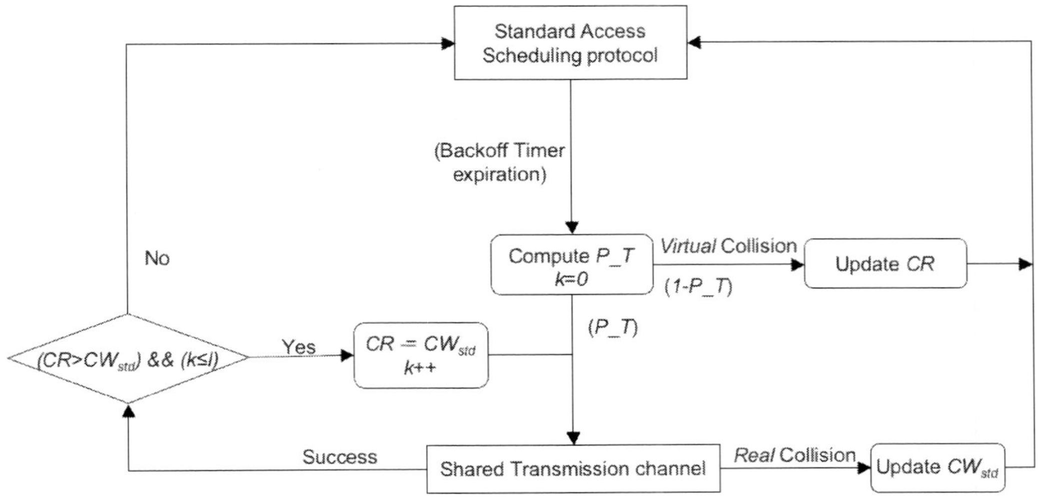

Figure 6. AOB-CR protocol with credit collection and frame bursting.

As shown in Figure 6, when the station performs a successful transmission attempt, it should compare the available credits against the CW_{std} threshold, computed according to formula (6.b). If $CR > CW_{std}$, the station should immediately perform a new transmission attempt separated from the previous one by a $SIFS$ interval. The threshold l limits the maximum burst size. In other words, no more than l consecutive frames can be transmitted before the standard backoff procedure is applied again. It is worth pointing out that transmitting a burst of frames should not affect the computation of the slot utilization. This implies that the entire burst is counted once in the computation of the n_{tx} value. Similarly, all the other stations consider the entire burst as a single channel occupation and they increment the n_{rx} value only once.

4 Experimental Results

In order to validate our enhanced architecture we carried out comparative tests of the performance obtained by the legacy IEEE 802.11 backoff mechanism and the enhanced ones, i.e., the AOB protocol and the AOB-CR protocol. In all the sets of experiments we used our NIC implementation. All the tests were performed in a laboratory environment, considering ad hoc networks in single-hop configurations. Nodes are communicating in ad hoc mode and the traffic was artificially generated. In our scenarios we used a maximum of four stations, due to hardware limitations. However, this is not seen as a problem, because we were already able to demonstrate the performance of our solution and the coherence with previously performed simulations.

As discussed in Section, the average backoff value that maximizes the channel utilization is almost independent of the network configuration (number of competing stations), but depends only on the average packet sizes [8]. Therefore, the $ACL(q)$ value for the frame sizes used in our experiments can be pre-computed and loaded in the MAC firmware. The implementation in the FPGA of the algorithm defining the $ACL(q)$ value in order to compute it at run-time, is an ongoing activity.

We used different scenarios (2, 3 and 4 stations), in order to study at the same time, the performance of our implementation and the correspondence with the previous simulation results. The stations are identically programmed to continuously send 500-byte long MSDUs (MSDU denotes the frame payload). The consecutive MSDU transmissions are separated by at least one backoff interval and we did not use the RTS/CTS handshake, or the fragmentation. The minimum contention window was set to $8 \cdot t_{slot}$ (160 μsec), and all values were computed in stationary conditions. The nodes topology is illustrated in Figure 7. All the experimental results we show in the following were obtained by computing the average over five replications of the same test.

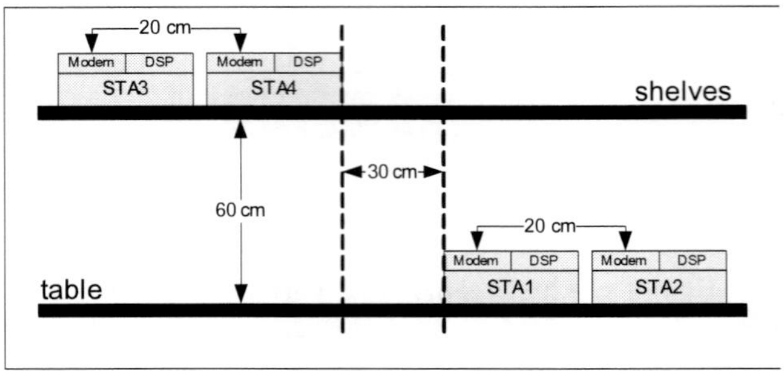

Figure 7. Node topology used in the measurements.

As already demonstrated in [9] and [10] the AOB mechanism introduces a minimum overhead that could negatively affect the performance of the communications between two stations. However, the frame bursting is useful to reduce the protocol overheads because it permits transmitting frames with null backoff. Thus, our first set of experiments was carried out to verify the performance decrease caused by the AOB protocol in a network configuration where two stations are performing a bidirectional communication, as illustrated in Figure 8. In addition, we conducted similar tests to validate if the AOB-CR protocol is effective in improving the MAC protocol efficiency.

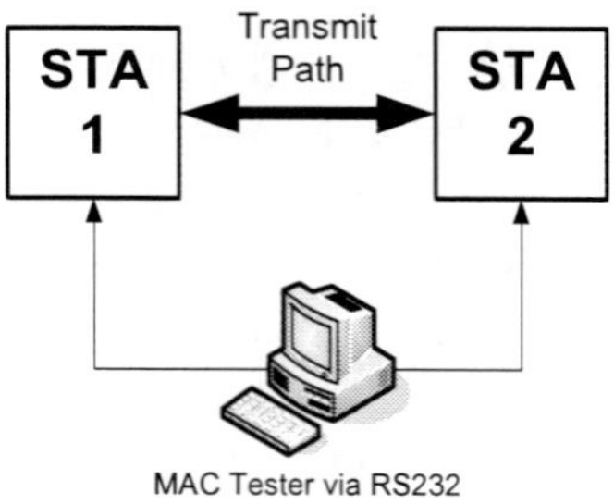

Figure 8. Bidirectional communications with two stations.

The results we obtained in this two-station configuration are reported in Table I. In particular, the throughput at time kT (where T is the sampling period equal to 100 ms) is computed as

$$TP[kT] = DT[kT] - RC[kT] \, , \tag{7}$$

where $DT[kT]$ is the total number of frames sent to a station (either acknowledged or not acknowledged frames), while $RC[kT]$ is the number of real collisions (not acknowledged frames). The average throughput values for each station are evaluated internally by the DSP (thanks to the implemented buffer) after 8 minutes of continuous transmissions. After some computations, the throughput value is sent to a PC through the available RS-232 channel. To validate the stochastic correctness of our result, both the mean and standard deviation of the throughput measurements are reported in the following tables.

From the numerical results listed in Table I, we can observe that the throughput decrease in the case of two competing stations is lower than 3% when using the AOB protocol. However, the AOB-CR mechanism is capable of improving the MAC protocol efficiency, ensuring a 10% improvement in the throughput performance.

Table I. Results for the two-station scenario.

	Average	*Standard Deviation*	*Throughput increase*
Standard 802.11 MAC protocol	1546.19 Kbps	108 bps	-
AOB protocol	1510.62 Kbps	256 bps	-2.3%
AOB-CR protocol	1694.93 Kpbs	91 bps	+9.6%

In the second set of experiments we considered a network configuration with three stations, as depicted in Figure 9.

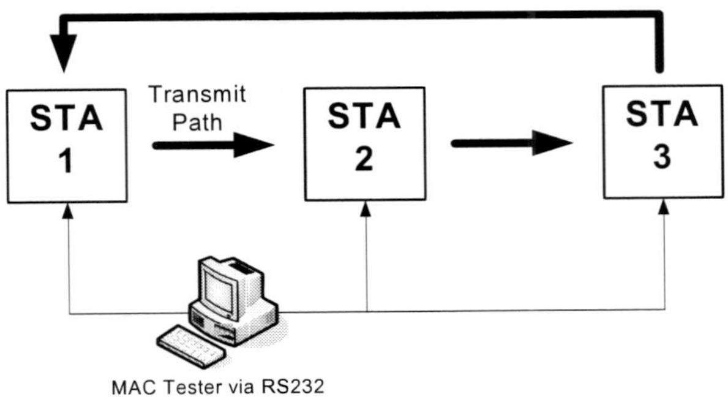

Figure 9. Three-station scenario.

The experimental results we obtained in the three-station configuration are reported in Table II. We can note that with three competing stations, the throughput decrease with the AOB protocol is almost negligible. On the other hand, it is further confirmed that the AOB-CR protocol guarantees a significant improvement with respect to the standard 802.11 MAC protocol.

Table II. Results for the three-station scenario.

	Average	*Standard Deviation*	*Throughput increase*
Standard 802.11 MAC protocol	1521.32 Kbps	208 bps	-
AOB protocol	1517.36 Kbps	974 bps	-0.26%
AOB-CR protocol	1706.34 Kpbs	279 bps	+12.1%

Finally, the last set of experiments was carried out in the four-station scenario depicted in Figure 10, and the experimental results we measured are listed in Table III.

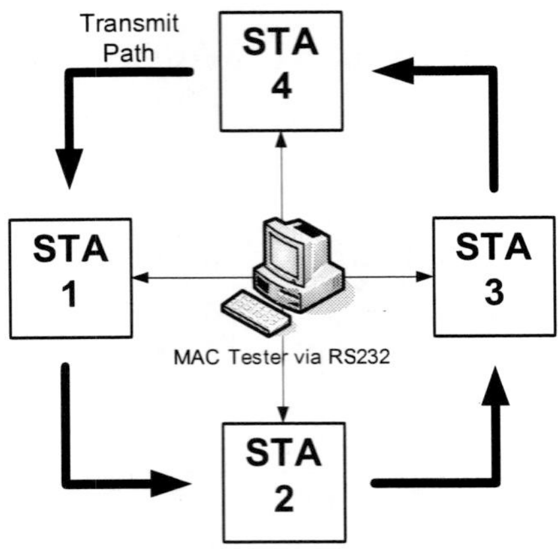

Figure 10. Four-station scenario.

These results confirm the positive trends shown in the previous experiments. In particular, with four competing stations, the AOB protocol provides a higher throughput than the standard MAC protocol. The reason is that the filtering on channel access reduces the collision probability such as that the stations can utilize more efficiently the channel resources. Furthermore, the AOB-CR protocol continues to show better performance that the basic AOB mechanism. In the four-stations scenario the throughput increase provided by the AOB-CR protocol over the standard 802.11 MAC protocol is about 17%.

Table III. Results for the four-stations scenario.

	Average	*Standard Deviation*	*Throughput increase*
Standard 802.11 MAC protocol	1434.31 Kbps	290 bps	-
AOB protocol	1504.0 Kbps	242 bps	+4.85 %
AOB-CR protocol	1681.03 Kpbs	451 bps	+17.5%

The shown results clearly demonstrate that the AOB MAC protocol improves the per-station throughput as the number of stations increases, such as to approximate the maximum channel utilization. On the other hand, the introduction of credit-based frame bursting capabilities permits to further increase the MAC protocol efficiency.

5 Conclusions

Experiments were carried out with the implementation of an enhanced IEEE 802.11 MAC card adopting the optimizations designed in [9] and [10]. The card is still fully compatible with current implementations of the IEEE 802.11 technology because the radio part is compliant to the 802.11 standard. However, the experimental results show that the enhanced mechanism outperforms the standard 802.11 MAC protocol in real scenarios. We have also shown that the advantages of this mechanism go further than the high contention scenarios (e.g., ad hoc networks), for which it was designed, because it is also effective in lessening the negative impact of the external interferences, which traditionally decrease the performances of wireless networks in any environment.

As discussed in the introduction, the wireless network interface is designed for supporting several types of experiments with different enhanced MAC protocols. To this aim, we are exploring several directions, some of them have been outlined in [5] and [17]:

- ***Routing***: In addition to the backoff issues, there is another problem that limits the performance of wireless traffic in WLANs and ad hoc networks: the routing. Indeed, each packet received from the wireless interface must be passed up to the routing layer (in order to discover the next hop), and further down to the same wireless interface for transferring it to the next hop. In the meantime the SIFS time has gone and the station must run again the backoff mechanisms in order to acquire the channel to forwarding the packet. This adds undesirable delay and overhead at both MAC and routing layer. For these reasons we aim at experimenting mechanisms capable of executing routing operations inside the MAC, layer. The solutions we intend to experiment will be based on next-hop address lookup performed in conjunction with a path strategy as, for example, the fixed length labels architecture defined in [4]. Basically, the packet forwarding protocol builds on the IEEE 802.11 DCF MAC protocol by exploiting some additional information delivered in the control packets (RTS/CTS) to allow the forwarding node to determine the next hop node while contending for the channel. Moreover, we intend to add a communication interface between the MAC layer and the routing layer, such as to allow the routing protocol to take its routing decisions exploiting channel information.
- ***Cross-layering***: Besides the routing, research has shown that several mechanisms can profit from the knowledge of some parameters that are typically confined at the MAC layer, such as transport, power management, cooperation, etc. Indeed, several network architecture designers foresee the access to various MAC parameters for a full integration of the mechanisms traditionally working at different layers of the protocol stack. This will be enabled by a cross-layering architecture as the one proposed in the MobileMAN project [14]. In this architecture the shared memory component acts as an exchange area of networking information (parameters, status,

etc.) for all the layers. This allows the MAC layer to distribute “physical” information up to the higher levels, as well as to profit from some higher layer elaborations that are too complex to be performed at MAC. A typical example is the interaction between MAC, routing and transport information for congestion and network utilization purposes. If the transport is aware of the links’ status, it can distinguish between congestion due to physical failures and congestion due to the amount of traffic, acting consequently. Similarly, the routing can decide different routing paths or strategies, and the MAC can modify the distribution of some information as consequence.

References

[1] ANSI/IEEE Std. 802.11: Wireless LAN Medium Access Control (MAC) and Physical Layer (PHY) Specifications, August 1999

[2] G. Anastasi, E. Borgia, M. Conti, and E. Gregori, “Wi-Fi in Ad Hoc Mode: A Measurement Study,” in *Proc. of IEEE PerCom 2004*, Orlando, FL, March 14–17 2004, pp. 145–154.

[3] G. Anastasi, M. Conti, and E. Gregori, “IEEE 802.11 Ad Hoc Networks: Protocols, Performance and Open Issues,” in *Ad Hoc Networking*, New York, NY: IEEE Press and John Wiley&Sons, 2003

[4] Acharya, A. Misra, S. Bansal. A label-switching packet forwarding architecture for multi-hop wireless LANs, in *Proc. of WoWMoM 2002*

[5] R. Bernasconi, R. Bruno, I. Defilippis, S. Giordano, A. Puiatti “Experiments with an enhanced MAC architecture for multi-hop wireless net-works”, *Proc. of REALMAN 2005*, Santorini 2005

[6] L. Bononi, M. Conti, L. Donatiello. Design and Performance Evaluation of a Distributed Contention Control (DCC) Mechanism for IEEE 802.11 Wireless Local Area Networks. *Journal of Parallel and Distributed Computing*, **60**(4):407-430, April 2000.

[7] R. Bruno, M. Conti, E. Gregori. Optimization of Efficiency and Energy Consumption in p-Persistent CSMABased Wireless LANs. *IEEE Trans. Mob. Comp.*, **1**(1):10-31, March 2002

[8] R. Bruno, M. Conti, E. Gregori. Optimal Capacity of p-Persistent CSMA Protocols. *IEEE Commun. Lett.*, **7**(3):139-141, March 2003

[9] L. Bononi, M. Conti, E. Gregori. Run-Time Optimization of IEEE 802.11 Wireless LANs performance. *IEEE Trans. Parallel Distrib. Syst.*, **15**(1):66-80, January 2004.

[10] R. Bruno, M. Conti, and E. Gregori, “Distributed Contention Control in Heterogeneous 802.11b WLANs,” in *Proc. of WONS 2005*, St Moritz, Switzerland, January 19–21 2005, pp. 190–199.

[11] Ralf Bernasconi, Raffaele Bruno, Ivan Defilippis, Silvia Giordano, and Alessandro Puiatti, “Experiments with an enhanced MAC architecture for multi-hop wireless networks”, in Proc. of REALMAN2005, Santorini, Greece, July 14

[12] Chaudet, D. Dhoutaut, and I. Guérin Lassous, “Performance issues with IEEE 802.11 in ad hoc networking,” *IEEE Communication Magazine*, 2005, vol.43, no.7 - July 2005.

[13] Chaudet, D. Dhoutaut, and I. Guérin Lassous, "Experiments of some performance issues with IEEE 802.11b in ad hoc networks," in *Proc. of WONS 2005*, St Moritz, Switzerland, January 19–21 2005, pp. 158–163

[14] M. Conti, S. Giordano, G. Maselli, G. Turi. Cross-Layering in Mobile Ad Hoc Network Design, *IEEE Computer*, **37**(2):48–51, February 2004

[15] S. Xu, T. Saadawi. Does the IEEE 802.11 MAC Protocol Work Well in Multihop Wireless Ad Hoc Networks?, *IEEE Communications Magazine*, **39**(6):130-137, June 2001.

[16] S. Xu and T. Saadawi, "Revealing the problems with 802.11 medium access control protocol in multi-hop wireless ad hoc networks," *Computer Networks*, vol. 38, pp. 531–548, Mar. 2002.

[17] Raffaele Bruno, Claude Chaudet, Marco Conti, Enrico Gregori, "A Novel Fair Medium Access Control for 802.11-based Multi-Hop Ad hoc Networks", in *Proc. of IEEE LANMAN 2005*, Chania, Crete, September 18-21, 2005.

INDEX

M

N

O

P

T

U

V

W

X

Y